Lipids in Cereal Technology

FOOD SCIENCE AND TECHNOLOGY

A SERIES OF MONOGRAPHS

A complete list of the books in this series appears at the end of the volume.

Lipids in Cereal Technology

edited by

P. J. Barnes

RHM Research Ltd
The Lord Rank Research Centre
High Wycombe
Buckinghamshire, UK

1983

ACADEMIC PRESS

A Subsidiary of Harcourt Brace Jovanovich, Publishers
New York London
Paris San Diego San Francisco
São Paulo Sydney Tokyo Toronto

ACADEMIC PRESS INC. (LONDON) LTD.
24/28 Oval Road
London NW1

United States Edition published by
ACADEMIC PRESS INC.
111 Fifth Avenue
New York, New York 10003

British Library Cataloguing in Publication Data

Lipids in cereal technology.
1. Cereal products 2. Lipids
I. Barnes, P. J.
641.3′31 TX557

ISBN 0-12-079020-3

LCCCN 83-71857

Photoset by Preface Ltd., Salisbury, Wilts.
and printed in Great Britain by
Galliard (Printers) Ltd, Great Yarmouth

Contributors

R. A. Anderson, *Northern Regional Research Center, Agricultural Research Service, US Department of Agriculture, Peoria, Illinois, USA*

R. E. Angold, *RHM Research Ltd, The Lord Rank Research Centre, Lincoln Road, High Wycombe, Bucks., HP12 3QR, UK*

D. J. Baisted, *Department of Biochemistry and Biophysics, Oregon State University, Corvallis, Oregon 97331, USA*

P. J. Barnes, *RHM Research Ltd, The Lord Rank Research Centre, Lincoln Road, High Wycombe, Bucks., HP12 3QR, UK*

O. K. Chung, *US Grain Marketing Research Laboratory, Agricultural Research Service, US Department of Agriculture, Manhattan, Kansas 66502, USA*

N. A. Clarke, *Department of Biochemistry and Soil Science, University College of North Wales, Memorial Buildings, Deiniol Road, Bangor, Gwynedd LL57 2UW, UK*

R. Drapron, *Institut National de la Recherche Agronomique, Laboratoire de Technologie Alimentaire, F-91305 Massy, France*

P. J. Frazier, *Dalgety Spillers Ltd, Group Research Laboratory, Station Road, Cambridge, CB1 2JN, UK*

T. Galliard, *RHM Research Ltd, The Lord Rank Research Centre, Lincoln Road, High Wycombe, Bucks., HP12 3QR*

E. G. Hammond, *Department of Food Technology, Iowa State University, Ames, Iowa 50011, USA*

B. O. Juliano, *Chemistry Department, The International Rice Research Institute, Los Baños, Laguna, Philippines*

D. L. Laidman, *Department of Biochemistry and Soil Science, University College of North Wales, Memorial Buildings, Deiniol Road, Bangor, Gwynedd, LL57 2UW, UK*

B. Laignelet, *Institut National de la Recherche Agronomique, Laboratoire de Technologie des Céréales, 9 Place Viala, 34060 Montpellier Cédex, France*

K. Larsson, *Department of Food Technology, University of Lund, Box 740, S-220 07 Lund, Sweden*

F. MacRitchie, *CSIRO, Wheat Research Unit, North Ryde, New South Wales 2113, Australia*

W. R. Morrison, *Department of Bioscience and Biotechnology, University of Strathclyde, Glasgow, G1 1SD, Scotland, UK*

T. L. Mounts, *Northern Regional Research Center, Agricultural Research Service, US Department of Agriculture, Peoria, Illinois, 61604, USA*

J. Nicolas, *Institut National de la Recherche Agronomique, Station de Technologie des Produits Végétaux, Domaine St. Paul B.P. 91, F-84140 Montfavet, France*

Y. Pomeranz, *US Grain Marketing Research Laboratory, Agricultural Research Service, US Department of Agriculture, Manhattan, Kansas 66502, USA*

G. Shearer, *Food Laboratory, Ministry of Agriculture, Fisheries and Food, Colney Lane, Norwich, NR4 7UA, UK*

M. J. Warwick, *Food Laboratory, Ministry of Agriculture, Fisheries and Food, Colney Lane, Norwich, NR4 7UR, UK*

E. J. Weber, *Agricultural Research Service, US Department of Agriculture, S-320 Turner Hall, University of Illinois, 1102 South Goodwin Street, Urbana, Illinois 61801, USA*

M. C. Wilkinson, *Department of Biochemistry and Soil Science, University College of North Wales, Memorial Buildings, Deiniol Road, Bangor, Gwynedd LL57 2UW, UK*

Preface

The grasses, or *Gramineae*, are ecologically and agriculturally the most important family of plants in the world. Cereal grasses and herbage grasses are the main sources of food for human beings and domesticated animals. The cereals are annual grasses whose relatively large grains allowed the development of technologies as diverse as milling, baking, malting and brewing. Starch and proteins are the major components of cereal grains in quantitative terms, and it is not surprising that they have received most attention from cereal scientists. However, we have become increasingly aware that the relatively small quantities of lipids in cereal grains are, nonetheless, important, particularly through their ability to modify the properties of starch and proteins. In the case of corn oil, rice bran oil and wheat germ oil, the cereal lipids themselves represent the final product. In this book an international group of experts bring together their knowledge and experience to provide a comprehensive review of cereal lipids and the role they play in cereal processing and products.

The book begins with the more fundamental aspects of cereal grain lipids and enzymes and then continues with specific cereals, processing and cereal products. Early chapters describe the composition and distribution of lipids in the grain, the biochemical changes that occur when the grain germinates and the biochemistry of the enzymes involved in lipid degradation. In a series of chapters concerned with wheat, the significance of lipids in milling, flour storage, baking and pasta manufacture is discussed. Further chapters are concerned with individual cereals including maize, rice, oats and barley together with corn oil, wheat germ oil and other cereal products. There is no chapter devoted solely to lipids in malting and brewing; there is still a need for more research on this topic and for someone to draw together and critically review all the available information. It is pleasing to know that while this book was being written a major study of lipids and lipid-degrading enzymes in malting began at the UK Brewing Research Foundation. The book is intended for food scientists and technologists, especially those concerned with cereals and edible oils, but should also be of interest to other scientists working with lipids, lipid-degrading enzymes and plant biochemistry.

I am indebted to the twenty-two scientists whose contributions, and more importantly whose research, made this book possible. My thanks are also due to the Graphics Department of the Lord Rank Research Centre for the final drawing of the illustrations for Chapters 2, 3 and 7, to Mrs M. W. Castle for preparation of the manuscripts for my own chapters and to Dr J. Edelman, Director of Research at the Lord Rank Research Centre for encouraging me to take on this task. Finally, I must thank my wife for the support, understanding and enthusiasm that she shows for all my projects.

September, 1983 P. J. Barnes
High Wycombe

Contents

Abbreviations

A number of abbreviations for the names of lipids are used in the text. Each name is given in full when the abbreviation appears for the first time in each chapter. The following is a general guide to the more frequent abbreviations for the convenience of the reader.

FA = fatty acids. FFA = free (unesterified) fatty acids. FAME = fatty acid methyl esters. The standard notation is used for fatty acids, i.e. palmitate (16:0), stearate (18:0), oleate (18:1), linoleate (18:2) and linolenate (18:3).

NL = non-polar lipids, including FFA. GL = glycolipids (any lipid containing sugar). PL = phospholipids (any lipid containing phosphorus). NSL = non-starch lipids. SL = starch lipids.

TG = triglyceride (triacylglycerol). DG = diglyceride (diacylglycerol). MG = monoglyceride (monoacylglycerol).

PC, PE, etc. = phosphatidylcholine, phosphatidylethanolamine, etc.

LPC = lysophosphatidylcholine; APE = *N*-acylphosphatidylethanolamine.

DGDG = diglycosyldiglyceride (diglycosyldiacylglycerol); MGDG = monoglycosyldiglyceride; MGMG = monoglycosylmonoglyceride. AMGDG = 6-O-acyl MGDG.

SE = steryl ester; SG = steryl glycoside; ASG = 6-O-acyl SG.

α-T, β-T, etc. = α-tocopherol, β-tocopherol, etc. α-T-3, β-T-3, etc. = α-tocotrienol, β-tocotrienol, etc. Tocol is used when no distinction is made between saturated (tocopherol) and unsaturated (tocotrienol) side-chains.

1 The Structure of the Cereal Grain

R. E. ANGOLD

The Lord Rank Research Centre, High Wycombe, Bucks, U.K.

I INTRODUCTION

All cereals belong to the family of plants known as the *Gramineae*, the grasses. Cereals are grown in virtually all regions of the world where plants grow and 70% of the world's cultivated acreage is devoted to cereal production. The component of the cereal plant of economic interest is the grain, which is a fruit in the strict botanical sense. It is a dry fruit, and at maturity the fruit wall remains attached to the seed, protecting it. The dryness of the mature cereal fruit makes it particularly valuable to Man since it can be stored for long periods without further processing.

Detailed descriptions of specific tissues and the structure of particular species of cereal will be found in the chapters which follow. This chapter will convey the botanists' understanding of the way in which the cereal grain is formed and relate the structure of the mature grain to the industrial processing that it will experience.

"Lipids in Cereal Technology"
ISBN 0-12-079020-3

The fruit of the cereal is a caryopsis. This contains a single seed. The seed contains the embryo (or germ) and the endosperm, the starchy storage tissue. The seed is covered by the pericarp layers which are derived from the ovary wall. The pericarp does not split to liberate the seed, but dries out at maturity forming a thin, tough protective layer. The caryopsis is the fruit of all the cereals except for one species of millet, the African millet, *Eleusine coracana*, the fruit of which is an utricle. Here, the pericarp does not completely envelop the seed but is confined to the attachment end.

II THE STRUCTURE OF THE CARYOPSIS

The structure of the caryopsis is shown in Figs. 1.1 and 1.2. The example is wheat, but it would serve for any of the cereals with a relatively small change of shape and proportion – except for those cereals with no crease, such as maize (*Zea mays*) which is shown in Fig. 1.3. Rice also has no crease; a diagram of the caryopsis of rice, enveloped in the hull (the interlocked lemma and palea from the floret) can be found in Chapter 15.

Each flower or floret contains an ovary which has a single ovule (Fig. 1.4). At fertilization the two male nuclei from the pollen tube enter the embryo sac, one fusing with the egg cell which will become the embryo, the other fusing with two polar nuclei to form the triple nucleus which will become the endosperm tissue. The aleurone layer is the outermost layer of the endosperm tissue, and like the starchy endosperm, it is triploid, having the three complements of chromosomes, two maternal and one paternal. The bulk of the endosperm is the starchy endosperm which at maturity is packed with starch and occupies the major portion of the caryopsis. It is the principal food reserve which will be utilized by the embryo at germination.

The embryo possesses one cotyledon or seed leaf. This is modified to form the scutellum: the absorptive tissue which lies between the embryo axis and the starchy endosperm. It can be regarded as a "placenta" since it is the route by which the endosperm food reserves are transferred to the embryo axis at germination. (It is not simply a passive organ of transport since it also produces enzymes which assist in the breakdown of the reserves.)

The main axis of the embryo consists of a single shoot and a single root. The shoot is capped by the coleoptile which acts as a protective sheath for the primary leaf and the shoot apex at germination. It is this emergent shoot, enveloped in the coleoptile, that forms the "acrospire" of malted

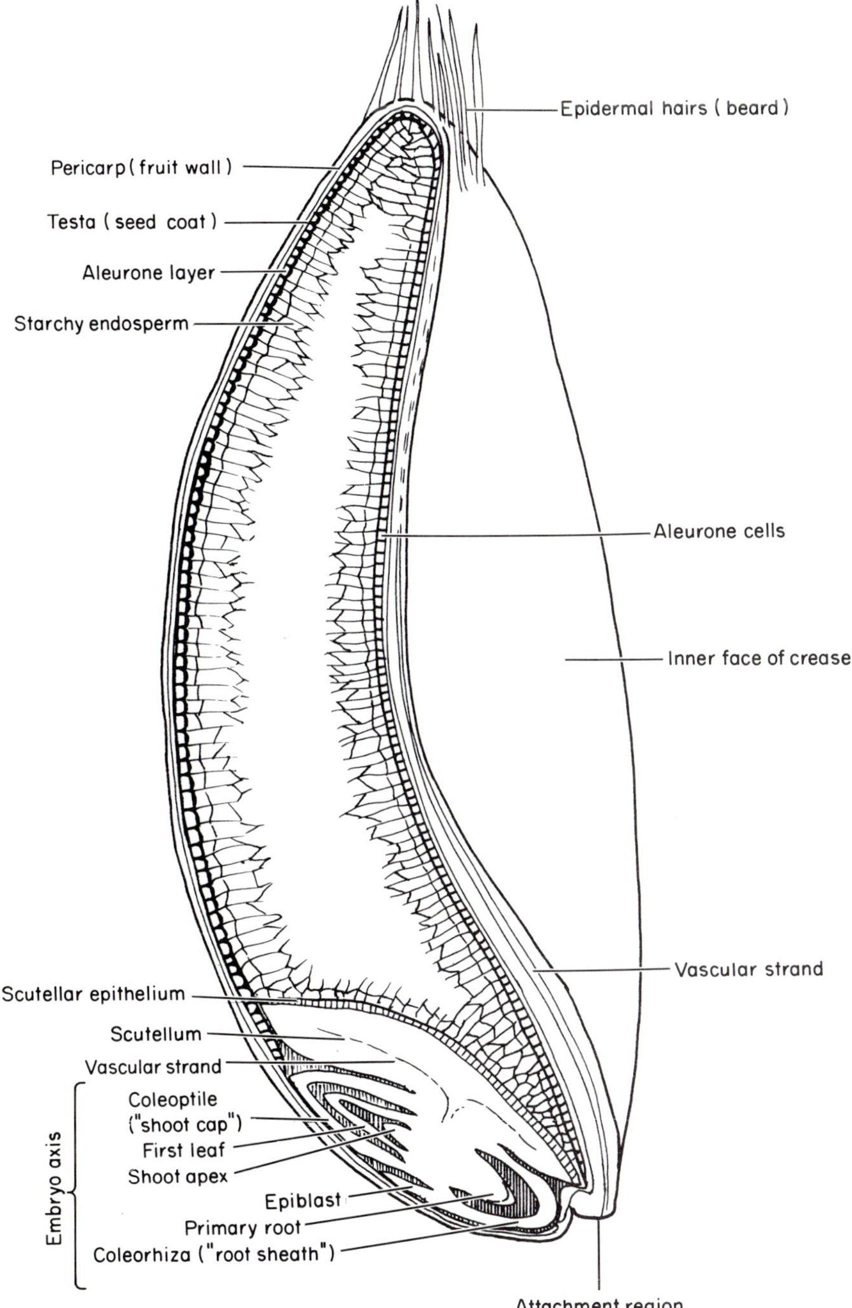

Fig. 1.1 Wheat grain – median longitudinal section.

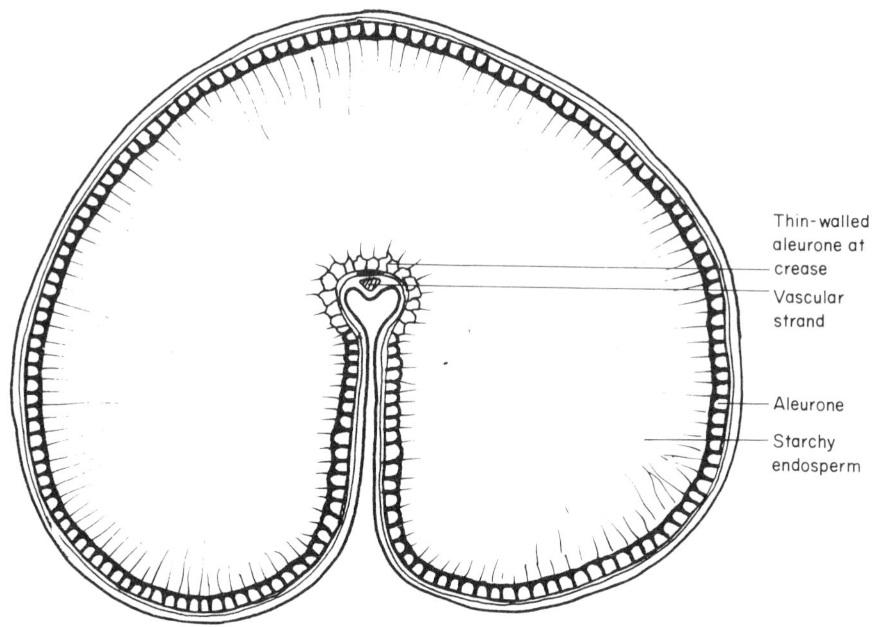

Fig. 1.2 Wheat grain – transverse section.

barley. The root is enveloped in a root sheath or coleorhiza. The coleoptile and coleorhiza are typical of the *Gramineae*.

The embryo and endosperm are enveloped in nucellar tissue, which was the layer of cells surrounding the embryo sac within the ovule. At maturity the nucellus is crushed and only the cell walls remain. The outermost layer of the seed proper is the testa. This is derived from the integuments which surrounded the embryo sac and nucellus. The integuments, at that stage, are two layers of thin walled cells which surround the entire embryo sac except for the micropyle, a small pore at the basal end of the ovary through which the pollen tube enters on its way to fertilize the egg cell. The integuments, like the nucellus, become extended and crushed by the developing and expanding endosperm and embryo; at maturity they are collapsed and only the cuticle remains as a distinct layer. It stains strongly with lipid stains and forms a waxy, water-repellent zone which surrounds the embryo and endosperm except at the position of the micropyle. This testa (seed coat) is the outermost layer of the seed proper, all the tissues that surround it are, strictly, the fruit wall.

As the embryo develops the endosperm tissue expands stretching the nucellus, integuments and pericarp. Free nuclear divisions in the endo-

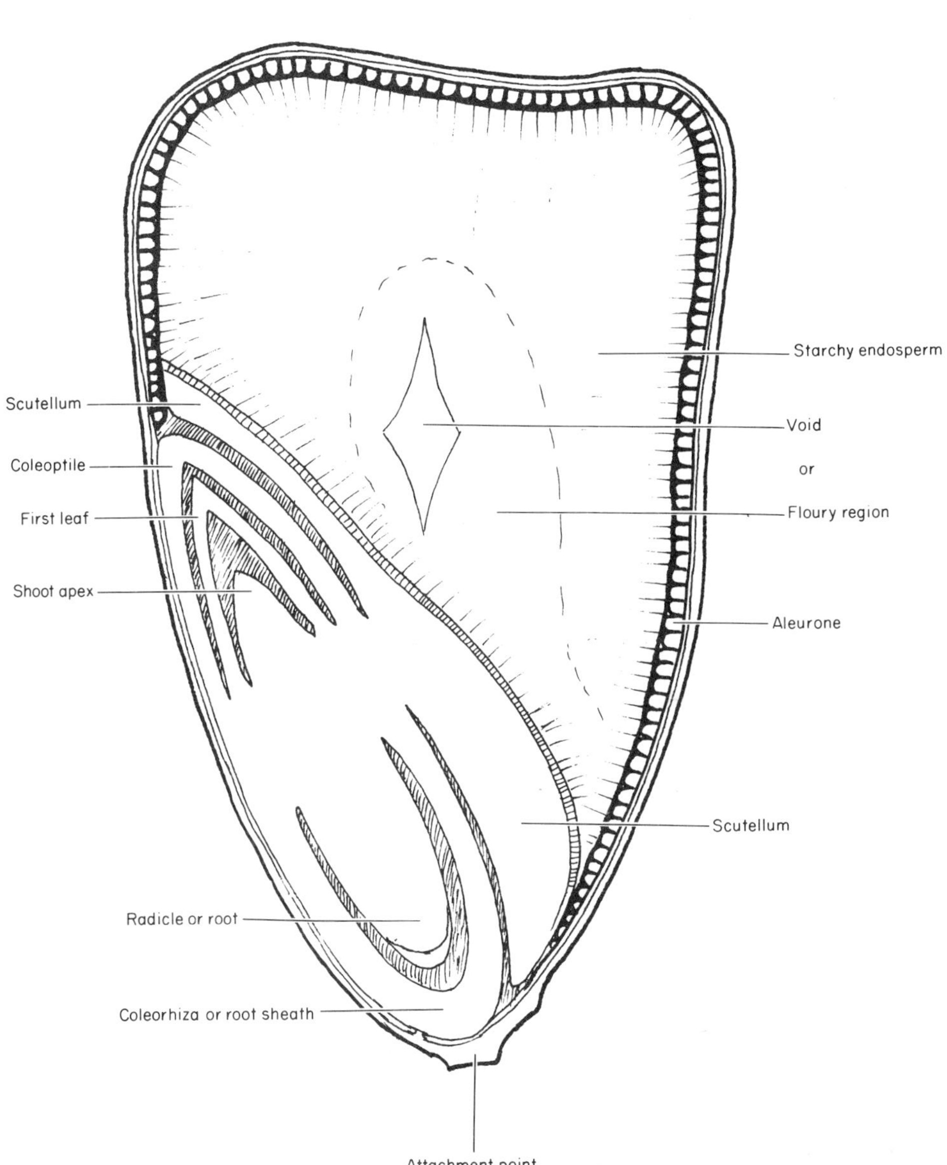

Fig. 1.3 Maize grain – median longitudinal section.

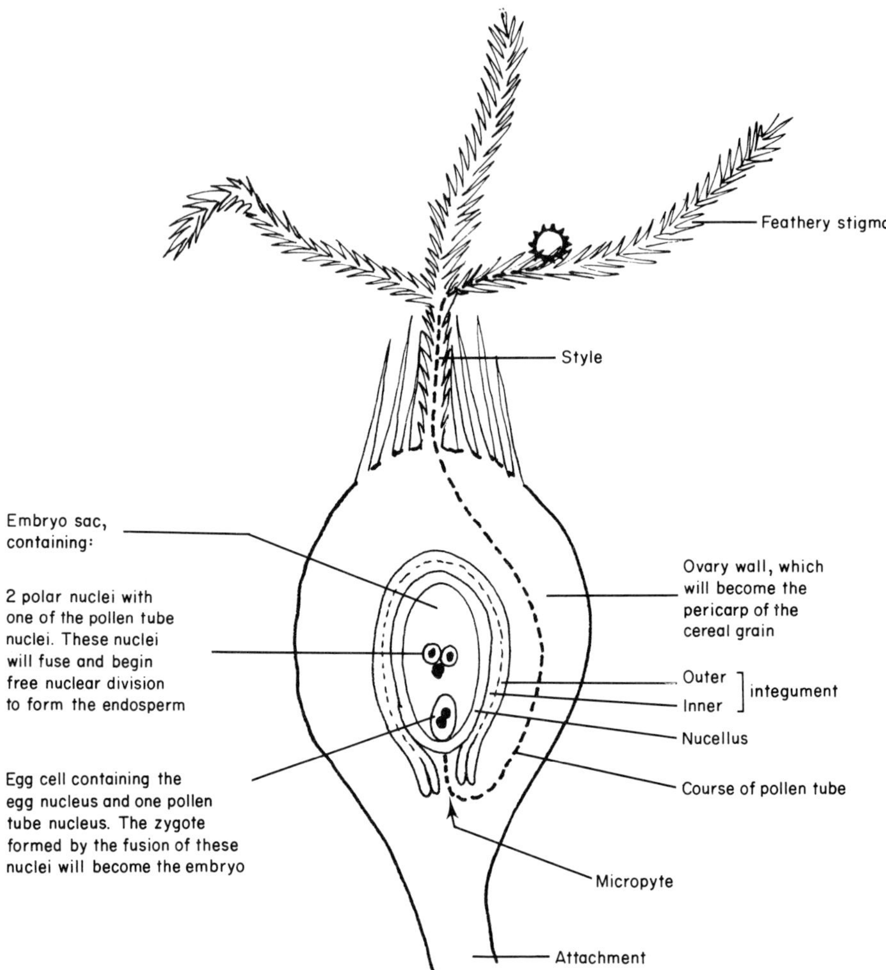

Fig. 1.4 Stylized cereal ovary at fertilization.

sperm initially form a syncytium, and then cell walls form between the nuclei. The cellular tissue appears to form a meristematic tissue, at least for wheat (Evers, 1970) and rye (Simmonds and Campbell, 1976). This tissue divides tangentially and radially to form files of cells perpendicular to the pericarp surface. At a relatively early stage (approximately one-third of the time period from fertilization to harvest ripeness) division stops and the outermost layer or layers of endosperm cells begin to thicken their walls and differentiate into the aleurone cells. By this stage, starch is being laid down in the inner endosperm cells.

Once the aleurone cell walls thicken a further increase in surface area of the endosperm cannot occur. In those grains with a crease, the crease provides a means of accommodating any expansion in the endosperm. In rice longitudinal furrows corresponding with ribs in the husk may permit an increase in endosperm volume, while an angular grain such as maize can become more or less convex. Dent maize, for example, has an identation in the distal end of the kernel. Shrinkage during drying, however, can result in a void forming in the endosperm of maize grain. Fig. 1.5 shows the pericarp and outer endosperm of the mature wheat grain.

The embryo (or germ) consisting of the embryo axis and the scutellum, has thin-walled cells. Those in the embryo axis are arranged to form the primordia of the root, shoot and leaves. In the Triticeae (which includes wheat, barley and rye) there is a rudimentary leaf, the epiblast, opposite the scutellum on the embryonic axis.

Thus, the cereal grain consists of three principal components: the pericarp, the endosperm and the embryo. The embryo (or germ) is divided into the axis, which will form the new plant, and the scutellum which serves initially as a food reserve for the embryo at germination; it also has a function in the digestion and absorption of the food reserves within the starchy endosperm. The starchy endosperm is surrounded by a differentiated layer or layers of cells, the aleurone, which is the outermost region of the endosperm tissue. The aleurone cells are a source of lipids and energy (the spherosomes) together with the phytin granules which are a rich source of phosphorus, and soluble, readily metabolized storage protein. Enzymes synthesized in the aleurone cells play a major part in the breakdown of reserves in the starchy endosperm.

III DISTRIBUTION OF LIPIDS IN THE CEREAL GRAIN

Detailed information about the nature and distribution of the lipids of many of the cereal species forms the major part of this book. The following is merely a comment on the functions and some of the properties of the tissues which make up the grain.

A Pericarp

Lipids in the pericarp will be principally components of cuticular layers, and will therefore be waxy and complex. The water relations of the grain are controlled by these layers. Other lipids present will be such membrane residues as remain when the pericarp cells senesce or collapse.

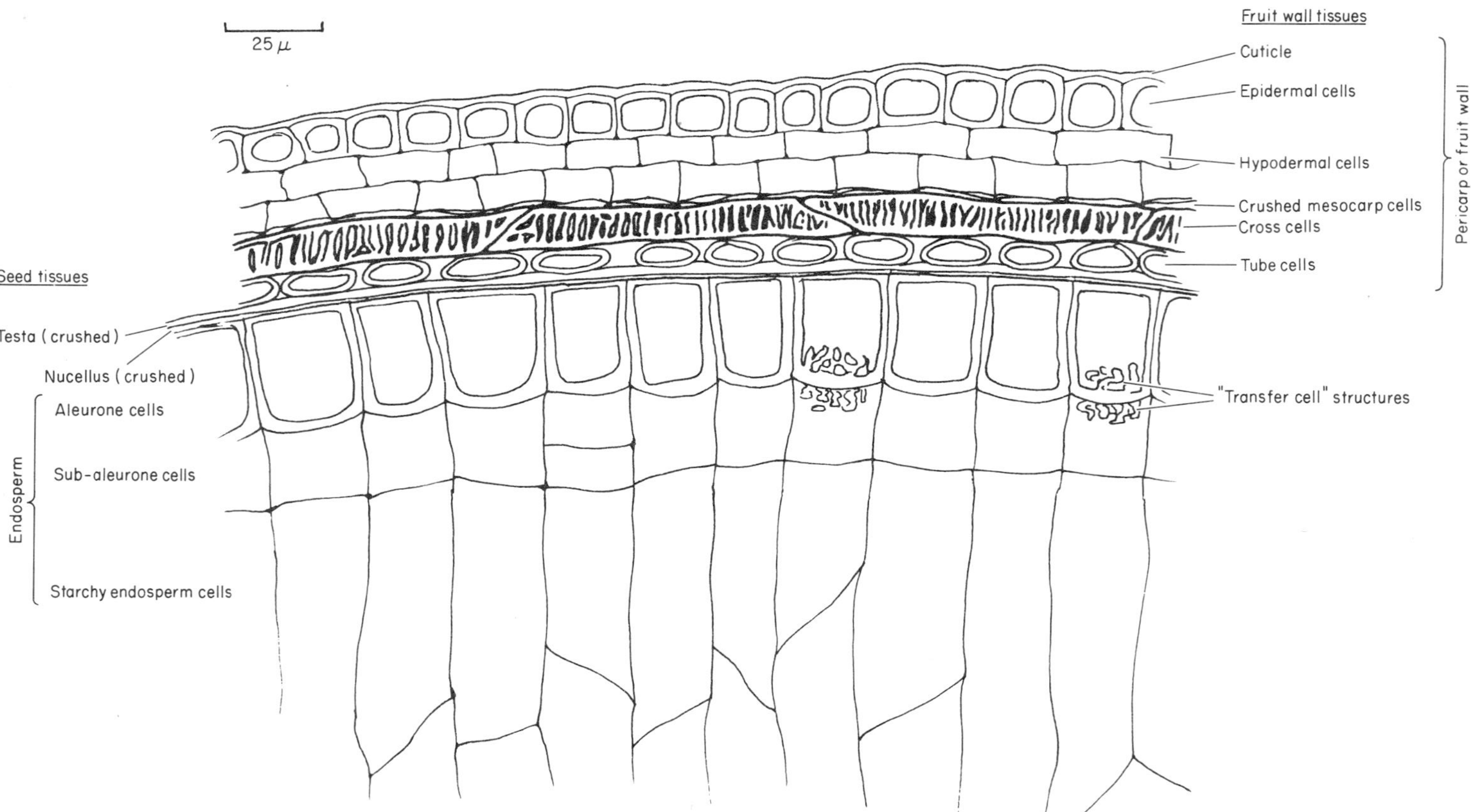

Fig. 1.5 Wheat grain – vertical section through the pericarp and outer endosperm (the section is across the long axis of the grain).

B Endosperm

The starchy endosperm contains membrane lipids, certain lipids associated with starch granules and some spherosomes. Some cereals, for example oats, have substantial quantities of lipid in the endosperm (see Chapter 16). The aleurone cells are rich in spherosomes, and contain membrane lipids. Rapid proliferation of endoplasmic reticulum can occur in aleurone cells, particularly those associated with the crease in the Triticeae. Plasma membrane proliferation can also occur: "transfer cell" wall and membrane elaborations occur, at least in wheat, during development, and, less frequently, at maturity.

C Embryo or Germ

Lipids are present in both the scutellum and the embryo axis. In maize (Chapters 17, 18) the embryo contains the largest portion of the total grain lipid. At germination, polar lipids in the embryo axis will contribute to the rapidly extending membrane systems as the axis elongates by the imbibition of water, and to the synthesis of endoplasmic reticulum and new membranes in the meristematic regions of the root and shoot. Numerous organelles will be forming; mitochondria and plastids, golgi systems and vesicles as cell walls are extended. Chapter 4 discusses this in more detail. Spherosomes in the scutellum form a further reserve of energy for the embryonic axis.

IV THE EFFECTS OF PROCESSING

Dry-milling is most extensively used in cereal processing, although wet-milling of maize is widely practised. Most milling processes attempt to do two things: to separate the components of the grain, and to reduce the grain to some kind of flour. The pericarp tissues are separated, usually together with the aleurone cells, as a "bran" or "offals" fraction from the millstream (Chapter 7) or by "polishing" as with rice (Chapter 15). The thick walls of the aleurone cells and the waxy nature of the pericarp lipids probably results in relatively little of the lipid from these tissues getting mixed into flour from the starchy endosperm, unless it is by contamination with bran fragments.

The softer, and more extensive tissues of the germ* create more of a

*In the other chapters of this book, the part of the grain consisting of the embryo axis and scutellum has been referred to as the "germ", rather than "embryo". This is done to avoid the confusion which has arisen from the use of the word "embryo" to denote, on the one hand, the embryo axis and, on the other, the embryo axis plus scutellum.

problem for the miller who wishes to separate a "germ" fraction from the other mill streams (Chapter 7). The difficulties of separating the scutellum from the endosperm, and the relatively large size of the germ with its high lipid content means that oils will be squeezed from the germ into that fraction of the flour which is expected to be derived solely from the starchy endosperm. The resulting presence of lipids from the embryo (germ) and, to a lesser extent, from the aleurone tissues as well as starchy endosperm lipids in flour is probably the principal reason for this book.

REFERENCES

Evers, A. D. (1970). *Ann. Bot.* **34**, 547–555.

Simmonds, D. H. and Campbell, W. P. (1976). In *Rye: Production, Chemistry and Technology* (W. Bushuk, ed.), pp. 63–110. American Association of Cereal Chemists, St. Paul, Minnesota.

2 Acyl Lipids in Cereals

W. R. MORRISON

University of Strathclyde, Glasgow, U.K.

I INTRODUCTION

A casual survey of the literature on acyl lipids in cereal grains gives the impression that the lipids are exceedingly complex and unsystematic. In fact, there is considerable order in the lipids of those cereals which have been studied most thoroughly and with an appreciation of the causes of interspecies variation, one can predict much about the lipids in the other cereals.

"Lipids in Cereal Technology"
ISBN 0-12-079020-3

Structural lipids (mainly glycolipids, GL, and phospholipids, PL) are located in membranes and organelles whose lipids are sometimes characteristic of particular tissues (e.g. phosphatidylglycerol containing *trans*-3-hexadecenoic acid and sulphoquinovosyldiglyceride in chloroplasts, diphosphatidylglycerol in mitochondria). Since cereal grains all have morphological variations of the same tissues and each tissue has a characteristic lipid composition, it is reasonable to expect similar patterns of lipid distributions especially within the subfamilies *Pooideae* (wheat, rye, triticale and barley) and *Panicoideae* (maize, sorghum and millets).

Variations superimposed on these common features due to lipid degradation, genetic mutations and effects of growing environment should be recognized as such and should not be regarded as evidence of more fundamental differences. Large variations in fatty acid composition of functional lipids should be accepted with great caution because a change to more saturated fatty acids could have a drastic effect on membrane permeability and plant viability. Such a change is therefore unlikely; a more probable explanation is often oxidative losses in analysis or other artefactual errors (Appendix 1).

If lipids isolated from cereals are examined in sufficient detail they will be found to contain most intermediates in the biosynthetic pathways leading to the principal end-products; the best examples are probably squalene, the pentacyclic triterpene alcohols, 4,4-dimethyl sterols and 4-methyl sterols which are the precursors of the 4-demethyl sterols (Morrison, 1978a). Most of the acyl lipid intermediates and some of the end-products are very minor components and are of academic interest only, but they are included in this review to give a more complete picture and update a previous comprehensive review of lipids in cereals (Morrison, 1978a).

II STRUCTURES OF ACYL LIPIDS

A Fatty Acids in Acyl Lipids

With very few exceptions, the fatty acids (FA) in cereal acyl lipids are, in order of abundance: linoleate (18 : 2, *n-6*; palmitate (16 : 0; oleate (18 : 1, *n-9*); linolenate (18 : 3, *n-3*) and stearate (18 : 0). Appreciable variations in the fatty acid composition of bulk lipids have been achieved in breeding maize, oats and barley (see Chapters 16 and 17). Other noteworthy variations in FA composition of individual lipids are mentioned later in this chapter.

B Acylglycerols and Free Fatty Acids (Non-polar Lipids, NL)

Triglycerides (triacylglycerols, TG), are deposited in spherosomes (oil droplets) bounded by a monolayer membrane (Jelsema *et al.*, 1977; Wanner and Theimer, 1978; Yatsu and Jacks, 1972) and this is the form in which plants usually store lipids. The highest TG levels occur in aleurone and scutellum tissue, but there are appreciable quantities in the cereal embryo, and in the starchy endosperm of oats (Morrison, 1978a; Youngs *et al.*, 1977). TG have also been reported in the starchy endosperm of wheat, maize and rice (Choudhury and Juliano, 1980a and b; Hargin and Morrison, 1980; Hargin *et al.*, 1980; Tan and Morrison, 1979a). In wheat they have the appearance of spherosomes and they are concentrated in the subaleurone region (Hargin *et al.*, 1980). The green tissues of immature pericarp contain appreciable quantities of spherosomes (see Chapter 1), but in mature pericarp the low levels of TG are merely residues of spherosomes which have survived lipolysis during pericarp senescence (Tan and Morrison, 1979b).

The stereospecific distribution of fatty acids in TG (Fig. 2.1), from single- or mixed-tissue sources, shows the usual pattern of more saturated acids of similar composition in positions 1 and 3, and more unsaturated acids in position 2 (Arunga and Morrison, 1971; Burini and Damiani, 1978; Miyazawa *et al.*, 1978; De La Roche *et al.*, 1971a–c). Molecular species are consistent with a 1-random, 2-random, 3-random* distribution pattern in TG insofar as they can be determined (Arunga and Morrison, 1971) and in the other major acyl lipids in wheat flour (Arunga and Morrison, 1971) and rice bran (Miyazawa *et al.*, 1977, 1978).

Diglycerides (diacylglycerols, DG) are intermediates in the biosynthesis of TG, glycosyldiglycerides (diacylglycosylglycerols) and phosphoglycerides. If care is taken to prevent equilibration to a mixture of isomers,

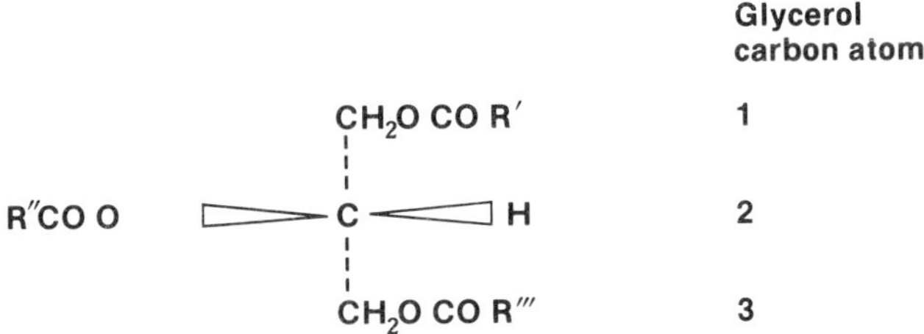

Fig. 2.1. Triglyceride (triacylglycerol, TG) showing stereo-specific numbering (*sn*) of carbon atoms in glycerol. Generally R″ is more unsaturated than R′ and R‴.

*The 3-position is acylated only in TG.

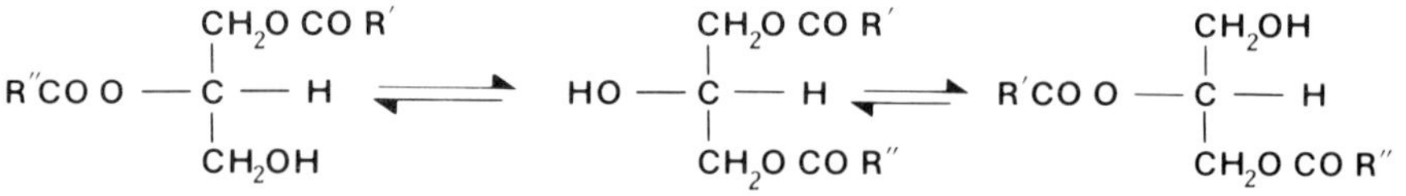

$$\begin{array}{ccccc} \mathrm{CH_2OCOR'} & & \mathrm{CH_2OCOR'} & & \mathrm{CH_2OH} \\ | & & | & & | \\ \mathrm{R''COO-C-H} & \rightleftharpoons & \mathrm{HO-C-H} & \rightleftharpoons & \mathrm{R'COO-C-H} \\ | & & | & & | \\ \mathrm{CH_2OH} & & \mathrm{CH_2OCOR''} & & \mathrm{CH_2OCOR''} \end{array}$$

Fig. 2.2 Diglycerides (diacylglycerols, DG), showing the particular case of isomerization of 1,2-diacyl-*sn*-glycerol (left) via 1,3-diacyl-*sn*-glycerol (centre) to 2,3-diacyl-*sn*-glycerol (right).

the DG in tissues of sound mature grain are often found to be devoid of *sn*-1,3-isomers on thin-layer chromatography (K. D. Hargin, S. L. Tan and W. R. Morrison, unpublished results) which suggests that they are *sn*-1, 2-DG biosynthesis intermediates. The DG in the lipids of fresh wheat flour (Fig. 2.2), which are derived from three separate parts of the wheat grain (see Chapter 7), are evidently isomerized *sn*-1, 2-DG (Arunga and Morrison, 1971).

DG are also formed by lipolysis of TG. This undoubtedly applies to DG in senesced pericarp and to DG in the aleurone and endosperm of mature maize and some rice. There is no evidence that DG could be formed by the action of glycosylhydrolases on glycosyldiglycerides or by the action of phospholipase-C on phosphodiglycerides.

Monoglycerides (monoacylglycerols, MG) are generally associated with partly degraded TG and are not present in significant amounts in sound mature tissues. The *sn*-1-MG and *sn*-3-MG isomers predominate at equilibrium (Fig. 2.3) and very often the low levels of *sn*-2-MG in lipid extracts separated by thin-layer chromatography are not quantified.

Free fatty acids (FFA) are always present in developing and mature wheat where they appear to be derived from normal intermediates of lipid metabolism. Lowest levels occur in wheat at maturity (Daftary and Pomeranz, 1965; Skarsaune *et al.*, 1970). Substantial quantities of FFA may be formed through hydrolysis of glycerolipids in developing maize (Tan and Morrison, 1979b), and as artefacts in the preparation of mature grain for analysis (Appendix 1). Free fatty acids are also formed by hydrolysis of lipids in milled products especially after a period of storage (Morrison, 1978a).

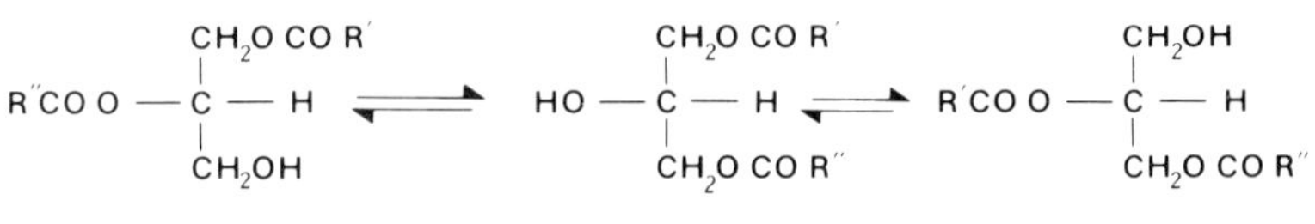

$$\begin{array}{ccccc} \mathrm{CH_2OCOR'} & & \mathrm{CH_2OCOR'} & & \mathrm{CH_2OH} \\ | & & | & & | \\ \mathrm{R''COO-C-H} & \rightleftharpoons & \mathrm{HO-C-H} & \rightleftharpoons & \mathrm{R'COO-C-H} \\ | & & | & & | \\ \mathrm{CH_2OH} & & \mathrm{CH_2OCOR''} & & \mathrm{CH_2OCOR''} \end{array}$$

Fig. 2.3 Monoglycerides (monoacylglycerols, MG), showing isomerization of 1-acyl-*sn*-glycerol to 2-acyl-*sn*-glycerol (centre) and 3-acyl-*sn*-glycerol (right).

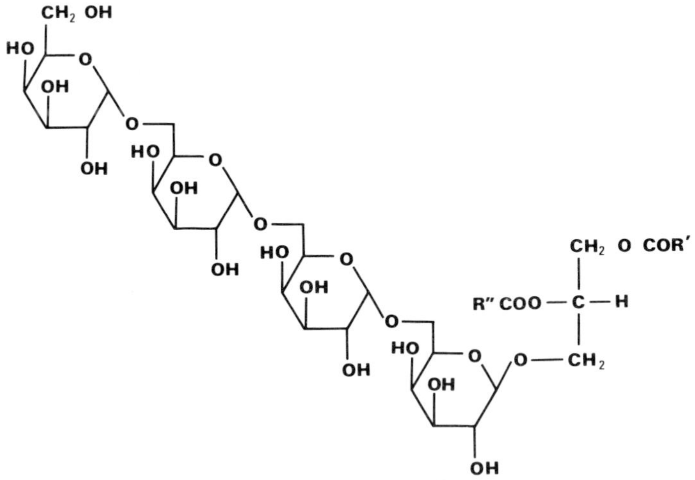

Gal—α(1→6)—Gal—α—(1→6)—Gal—α—(1→6)—Gal—β—(1→3)—[1,2—diacyl]—sn—glycerol

Fig. 2.4 Diacylgalatetraosyl-*sn*-glycerol (tetragalactosyldiglyceride). By removing successive terminal galactose residues the structure becomes diacylgalatriosyl-*sn*-glycerol (trigalactosyldiglyceride, TGDG), then diacylgalabiosyl-*sn*-glycerol (digalactosyldiglyceride, DGDG), and finally diacylgalactosyl-*sn*-glycerol (monogalactosyldiglyceride, MGDG). In the corresponding monoacyl lipids the fatty acyl residues will probably equilibrate mostly at the 1-position.

C Glycosylglycerides

The glycosylglycerides are quantitatively the major components of the GL in starchy endosperm and whole grains; they are the GL likely to be of greatest technological importance. The other GL are the sterylglycosides or glycosylsterols, the glycosylceramides, glycosylphosphoceramides and acylglycosyldiols (acyldiolglycosides) discussed in Sections IIE to IIH.

In the starchy endosperm of wheat and most other cereals, the principal sugar in the glycosylglycerides is galactose, and glucose is either a minor component or absent. The glycosylglyceride series consists of monoglycosyldiglyceride (MGDG) or 1,2-diacyl-3-O-β-*D*-glycopyranosyl-*sn*-glycerol (hexose = galactose or glucose), with subsequent galactose or glucose residues linked through α-(1 → 6) glycosidic bonds to form di-(DGDG), tri-(TGDG) and tetraglycosylglycerides (Fig. 2.4)* (Carter *et al.*, 1964; Fujino and Miyazawa, 1979; Fujino and Sakata, 1973; Miyazawa and Fujino, 1978a, b; Myhre, 1968). All permutations of glucose and galactose

*An alternative system of nomenclature is being introduced, whereby the galactosylglycerides would be diacylgalactosylglycerol (as before), diacylgalabiosylglycerol, diacylgalatriosylglycerol and diacylgalatetraosylglycerol respectively. Suitable abbreviations might then be GalDG, Gal_2DG, Gal_3DG and Gal_4DG, or simply GDG, G_2DG, G_3DG, and G_4DG if the hexose residues were not identified.

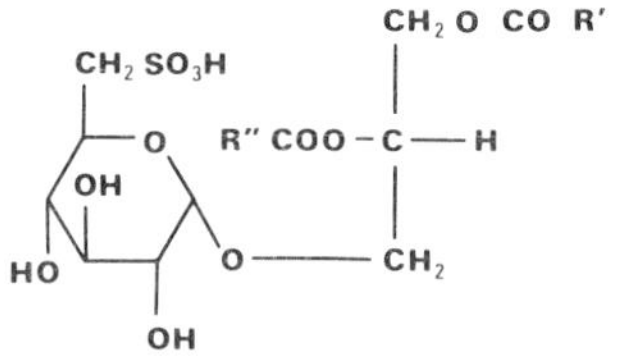

Fig. 2.5 6-sulpho-α-D-quinovopyranosyl-(1 → 3)-[1,2-diacyl]-*sn*-glycerol, (sulpholipid, sulphoquinovosyldiglyceride, SQDG).

in DGDG and TGDG have been reported (Fujino, 1978). The sugar in MGDG from Italian millet (*Setaria italica*) is exclusively glucose (Obara and Kihara, 1973).

Small amounts of galactosylmonoglycerides (MGMG and DGMG) are found in the endosperm of mature wheat where there is little evidence of lipase or lipolytic acyl hydrolase activity. In maturing maize endosperm there is extensive degradation of MGDG and DGDG but only low levels of MGMG and DGMG, indicating complete deacylation (Tan and Morrison, 1979b). A similar situation may prevail in some rice varieties.

Small amounts of GL occur with a fatty acid esterified at position 6 of the first sugar. The 6-O-acylmonogalactosyldiglyceride (AMGDG) and 6-O-acylmonogalactosylmonoglyceride (AMGMG) have been identified in wheat (Lin *et al.*, 1974; MacMurray and Morrison, 1970; Myhre, 1968). Such lipids are artefacts formed by enzyme catalysed acyl transfer in leaf homogenates (Heinz, 1967), but there is no evidence that they are laboratory artefacts in cereals.

Sulpholipid (sulphoquinovosyldiglyceride, SQDG-Fig. 2.5) is found in chloroplasts and etioplasts, and there might be some in other plastids such as amyloplasts (Fishwick and Wright, 1980). Small quantities of SQDG have been reliably identified in chloroplast-free tissues of maize (Weber, 1979) and rice (Kondo *et al.*, 1974).

D Phosphoglycerides (Phospholipids, PL)

Cereals contain the ubiquitous diacylphosphoglycerides: the major PL are phosphatidylcholine (PC), phosphatidylethanolamine (PE) and phosphatidylinositol (PI); the minor PL are phosphatidylglycerol (PG), phosphatidylserine (PS) and diphosphatidylglycerol (DPG) (Fig. 2.6). Phosphatidic acid (PA) is sometimes reported, but it is unimportant except in actively metabolizing tissue. There are no authenticated reports of PL with alkyl or alkenyl ether groups in place of acylester groups in cereals.

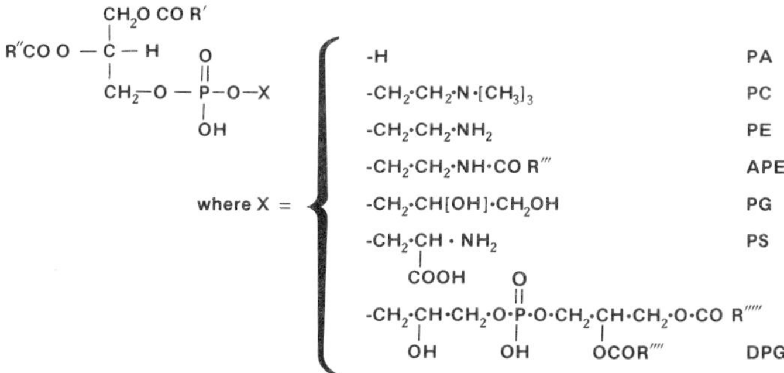

Fig. 2.6 Di-O-acylphosphoglycerides (phospholipids, PL). The general structure of 1,2-diacyl-*sn*-glycerol-3-phosphoryl-X is shown at the left; the groups at position X (right) give the PL commonly known as phosphatidic acid (PA), phosphatidylcholine (PC), phosphatidylethanolamine (PE), N-acylphosphatidylethanolamine (APE), phosphatidylglycerol (PG), phosphatidylserine (PS) and diphosphatidylglycerol (cardiolipin, DPG).

N-acylphosphatidylethanolamine (APE) and its lysoderivative (ALPE) are often misidentified or ignored, although they are major PL in the non-starch fraction of the endosperm of wheat and several other cereals. There is also one report of N-acylglycerylphosphorylethanolamine in wheat (Colborne and Laidman, 1975). In some situations these N-acyl lipids are undoubtedly artefacts; for example, they are found in wheat aleurone and germ when enzymes in these tissues have not been inactivated completely before lipid extraction (Hargin and Morrison, 1980). APE and ALPE in the germ, pericarp and tip cap of maize (Tan and Morrison, 1979a, Weber, 1979) may be artefacts, therefore. However, the highest levels of APE and ALPE are found in the starchy endosperm of wheat where they do not appear to be artefacts, and their obvious precursor, PE, has never been found in sufficient quantities for this to be likely.

Monoacylphosphoglycerides, or lysophospholipids (prefix L added to PL abbreviation), are usually regarded as degradation products of PL, and they are commonly found in lipids extracted from enzyme-active tissue (Appendix 1). They are also principal components of the internal starch lipids, together with some FFA (Becker and Acker 1976; Morrison, 1981; Morrison and Milligan, 1982). In wheat starch 89–94% of the internal starch lipids are LPC, LPE and LPG (Hargin and Morrison, 1980; Morrison, 1978b), and the very low levels of FFA show that they were not formed *in situ* by simple hydrolysis of PC, PE and PG. In maize and rice starch the proportions of lysoPL and FFA are more nearly equimolar, but they are not considered to be hydrolysis artefacts (Morrison and Milligan,

1982). Reports of LPS, LPI and PA in starch lipids (Becker and Acker, 1976; Thomas, 1979) need to be verified by appropriate structural analyses.

E Sterol Lipids

The sterols are described in Chapter 3, but some aspects of the sterol lipid classes should be mentioned. The principal forms are free sterol (S) and sterylester (SE). The subcellular distributions of S and SE are not known, but much of it may be associated with spherosomes. Sterylglycoside (SG) has been known for some time, but recently it has been shown (Fujino and Ohnishi, 1979a, b; Ohnishi and Fujino, 1978, 1980) that there is a homologous series of sterylglucosides (SG, SG_2, SG_3, SG_4 and SG_5). The structure of the sterylpentaglucoside is five β-(1 → 4) linked glucose residues with the terminal glucose linked through a β-(1 → 3) glycosidic bond to the sterol hydroxyl (Fig. 2.7). β-(1 → 3) glucose-glucose bonds have been detected in SG_2 and SG_3 (Fujino and Ohnishi, 1979a; Ohnishi and Fujino, 1978). There is one esterified sterylglucoside, 6-O-acylsterylglucoside (ESG and ASG) analogous to AMGDG and AMGMG (Kuroda *et al.*, 1977; MacMurray and Morrison, 1970; Myhre 1968). Sterylmannoside has also been found in brown rice (Sakata *et al.*, 1973).

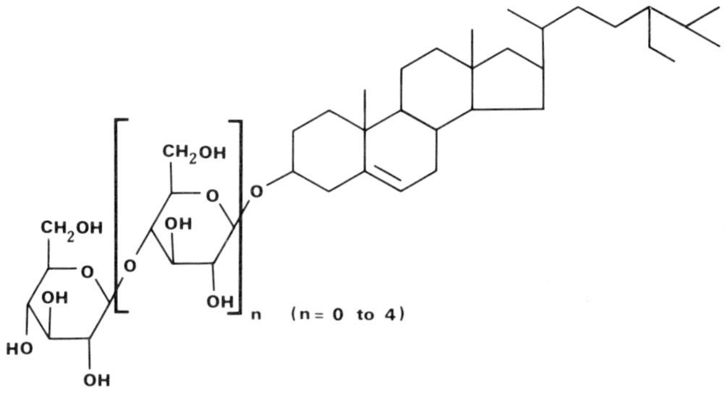

(n = 0 to 4)

Glc—β—(1→4)—Glc—β—(1→4)—Glc—β—(1→4)—Glc—β—(1→4)—Glc—β—(1→3)—sitosterol (n = 4)

Fig. 2.7 Cellopentaosylsitosterol (sitosterylpentaglucoside, SG_5). The structures of the lower glucosides are seen by removing successive terminal glucose residues, giving cellotetraosyl-, cellotriosyl-, cellobiosyl- and glucosylsitosterol (SG_4, SG_3, SG_2 and SG).

The fatty acids in endosperm SE have a high linoleate content in wheats lacking the D genome (*i.e.* AA diploids such as *T. monococcum* and AABB tetraploids such as *T. durum*) and in a few hexaploid breadwheats with recessive character for the Pln gene on the 7D chromosome (Torres and Garcia-Olmedo, 1974), but in the majority of breadwheats the endosperm SE has a high palmitate content (Morrison, 1978a). The fatty acids in ASG from bread flour are also comparatively saturated (MacMurray and Morrison, 1970). However, the fatty acids in SE and ASG from other parts of the hexaploid wheat grain always have a high linoleate content (Torres *et al.*, 1976; see also Chapter 7), as do those in rice bran (Kuroda *et al.*, 1977).

F Glycosphingolipids and Ceramide

The phosphorus-free sphingolipids consist of ceramide (CM) and glycosylceramides with one to four hexose residues (GCM, G_2CM, G_3CM, and G_4CM (Fujino and Ohnishi, 1976, 1982a, b; Fujino *et al.*, 1975). GCM usually has the structure β-D-glucopyranosyl $(1 \rightarrow 1)$ ceramide, and the higher glycosides have glucopyranosyl or mannopyranosyl residues lined through β-$(1 \rightarrow 4)$ glycosidic bonds to the first glucose or mannose (Fig. 2.8).

The fatty acids in these sphingolipids have typical long saturated or

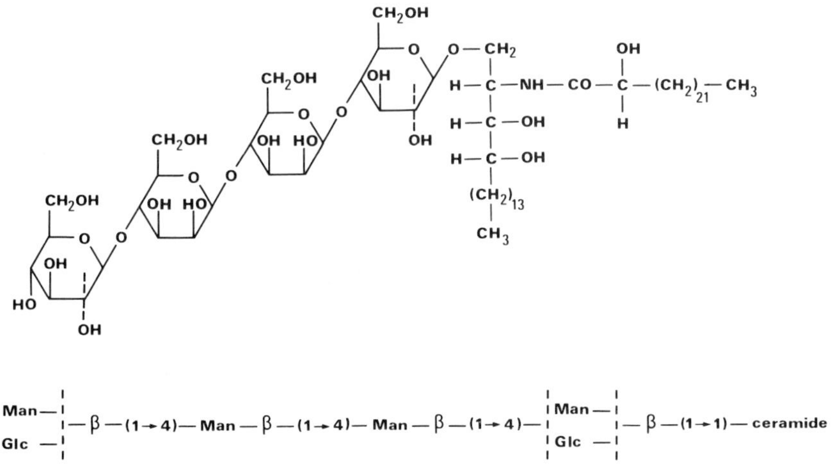

Man/Glc — β — (1→4) — Man — β — (1→4) — Man — β — (1→4) — Man/Glc — β — (1→1) — ceramide

Fig. 2.8 Tetraglycosylceramide (G_4CM). The first and the fourth sugar residues may be glucose or mannose, but the second and third sugars are always mannose. The structures of lower glycosylceramides are seen by removing successive terminal sugars until the monoglycosylceramide (cerebroside) is reached. In older literature these lipids are also termed ceramide (tetra-, tri-, di-, mono-) hexosides.

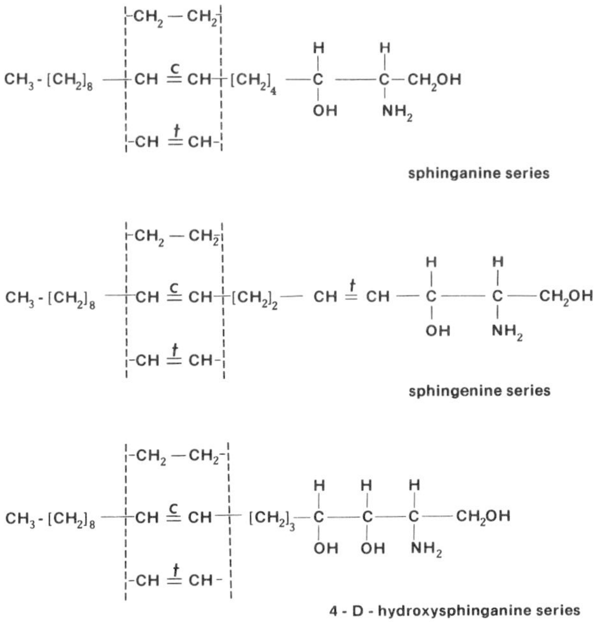

CH_3-$[CH_2]_8$ — {CH_2—CH_2 / CH $\overset{c}{=}$ CH / CH $\overset{t}{=}$ CH} — $[CH_2]_4$ — CH(OH) — CH(NH_2) — CH_2OH

sphinganine series

CH_3-$[CH_2]_8$ — {CH_2—CH_2 / CH $\overset{c}{=}$ CH / CH $\overset{t}{=}$ CH} — $[CH_2]_2$ — CH $\overset{t}{=}$ CH — CH(OH) — CH(NH_2) — CH_2OH

sphingenine series

CH_3-$[CH_2]_8$ — {CH_2—CH_2 / CH $\overset{c}{=}$ CH / CH $\overset{t}{=}$ CH} — $[CH_2]_3$ — CH(OH) — CH(OH) — CH(NH_2) — CH_2OH

4 - D - hydroxysphinganine series

Fig. 2.9 Long-chain bases in ceramides, consisting of three series of bases with three types of terminal hydrocarbon chain. These are classified as (top) the 1,3-dihydroxy-2-aminooctadecane or sphinganine bases, (centre) the 1,3-dihydroxy-2-amino-4-*trans*-octadecene or sphingenine bases, and (bottom) the 1,3,4-trihydroxy-2-aminooctadecane, 4-D-hydroxysphinganine, or phytosphinganine bases.

mono-unsaturated chains (14 : 0–28 : 0, 16 : 1–26 : 1) in normal, 2-hydroxy and (in CM only) 2,3-dihydroxy structures (Kondo *et al.*, 1975; Fujino and Ohnishi, 1976, 1982a).

The long-chain bases also show considerable complexity (Fig. 2.9) although they have almost exclusively 18-carbon chains. They consist of dihydroxy bases with saturated, or 8-*trans* or 8-*cis* unsaturated chains (d18 : 0, d18 : 1^{8t}, d18 : 1^{8c}), dihydroxy bases with the above structures and 4-*trans* unsaturation (d18 : 1^{4t}, d18 : $2^{4t,8c}$), and trihydroxy bases with similar outer chains (t18 : 0, t18 : 1^{8t}, t18 : 1^{8c}) (Fujino and Ohnishi, 1976, 1982a; Laine and Renkonen, 1973, 1974; MacMurray and Morrison, 1970).

G Glycophosphoceramides (Phytoglycolipids)

Glycophosphoceramides appear to be widely distributed in plants (Laine *et al.*, 1980) but there have been no reports of these lipids in cereals since

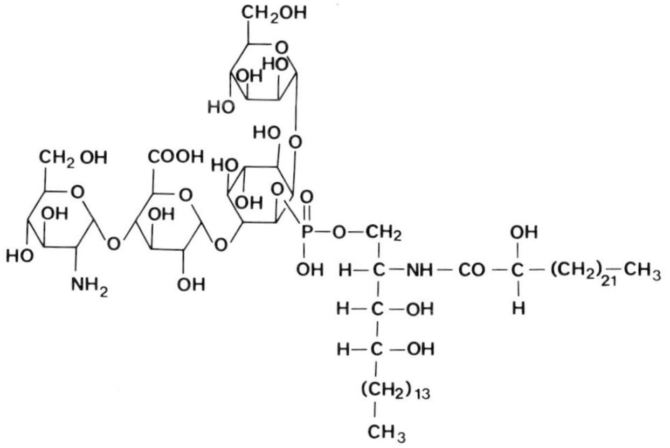

Man — α — (1→2) > Myo inositol — 1 — 0 — phosphoceramide
Glc NH_2 — α (1→4) — GlcUA — α — (1→6)

Fig. 2.10 The principal glycophosphoceramide (phytoglycolipid) in maize.

the pioneering work of Carter *et al.* (1969). Carter's group obtained evidence for a series of oligoglycophosphoceramides; the structure of the major component (Fig. 2.10) differs in some details from other plant glycophosphoceramides (Laine *et al.*, 1980; Hsieh *et al.*, 1981). In our laboratory no trace of polar glycophospholipids could be found in the total lipids (unwashed and unpurified) from wheat or maize germ (Hargin *et al.*, unpublished), and Fujino (1978) has noted that there is none in rice bran.

H Diol Lipids

Small amounts of lipids containing ethane, propane, butane and pentane diols instead of the usual glycerol (Fig. 2.11) have been found in immature wheat and maize (Bergel'son, 1973; Vaver *et al.*, 1969). The diacylesters are normally indistinguishable from TG, while

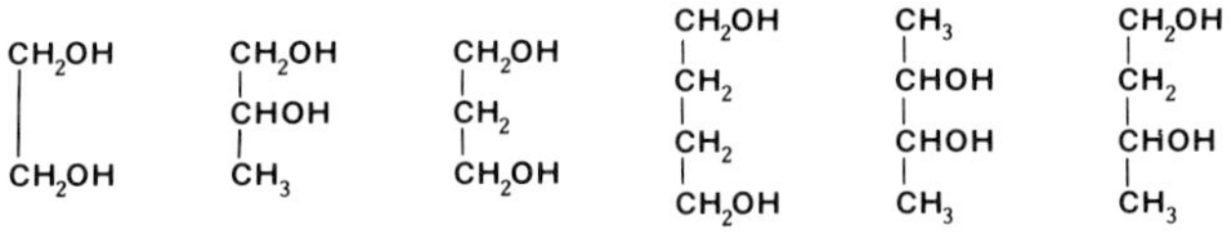

Fig. 2.11 Diols found in wheat and maize lipids analogous to triacylglycerol and diacylglycosylglycerols. The diols are (from the left) ethanediol, propane-1,2-diol, propane-1,3-diol, butane-1,4-diol, butane-2,3-diol and butane-1,3-diol.

1-O-β-D-glucopyranosyl-2-acylethanediol (in wheat) an 1-O-β-D-galactopyranosyl-2-acylethanediol (in maize) are found with MGDG (Vaver *et al.*, 1976, 1977).

I Tocopheryl and Tocotrienyl Esters

Berndorfer (1970) first reported tocopheryl esters among the tocopherols and tocotrienols in wheat germ oil, but the acyl moieties were not characterized. Kato *et al.*, (1981) have now found esterified tocopherols and tocotrienols in rice bran oil, soyabean oil and sesame oil – it is not clear whether these are artefacts or not.

J Xanthophyll Esters

Much of the xanthophyll (lutein) in wheat is esterified at one or both hydroxyls with the common C_{16} and C_{18} fatty acids (Farre-Rovira 1975; Farre-Rovira and Costes, 1974; Le Page and Sims, 1968). The proportions of free and esterified forms change in stored wheat flour (Farrington and Shearer, 1981), and some xanthophyll esters could be artefacts.

K Wax Esters

Wax esters have been reported in barley testa (Briggs, 1974), sorghum (Dalton and Mitchell, 1959), and rice bran oil (Ito *et al.*, 1981), and no doubt they occur in similar parts of all cereals. The short-chain wax esters in rice bran oil consist of methyl and ethyl esters of 16 : 0 and 18 : 1 FA, with lesser amounts of 14 : 0, 18 : 0 and 18 : 2 FA. The long-chain wax esters consist of even-numbered 20 : 0–36 : 0 straight-chain alcohols and 32 : 0, 34 : 0 and 36 : 0 branched-chain alcohols esterified with 16 : 0, 22 : 0 and 24 : 0 FA and small amounts of 14 : 0, 18 : 0, 18 : 1, 18 : 2 and 20 : 0 FA (Ito *et al.*, 1981).

III DISTRIBUTION OF ACYL LIPIDS

A Dissected Grain

1 *Maize*

The distributions of total extractable lipids, fatty acids, carotenoids, and tocols in maize have been known for some time (Morrison, 1978a), but quantification of the various acyl lipid classes was only achieved recently (Tan and Morrison, 1979a). Table 2.1 summarizes the data for amylomaize, normal maize, and waxy maize and Table 2.2 gives the com-

Table 2.1 Distribution of lipids (μg lipid $grain^{-1}$) in amylo-, LG11 hybrid, and waxy maize†

Lipid class	Pericarp	Tip cap	Germ	Endosperm			Whole kernel
				Non-starch + aleurone	Starch‡	Waxy‡	
SE	8–13	5–9	117–173	74–105	8–10	7	211–308
TG	17–76	40–81	7806–12 255	619–1671	17–27	76	8619–14 057
DG	3–8	3–8	161–751	25–118	6–7	8	205–887
FFA	1–69	11–18	64–259	286–1221	624–831	69	987–2329
MG	0–3	2	8–63	15–35	16–23	3	48–117
ASG	1–3	2–3	15–124	14–28	14–15	3	46–157
MGXG	1–2	1–2	27–65	§–26	23–41	2	58–129
DGXG	1–5	2–3	127–141	§–50	17–63	3	148–237
APE	0–1	0–1	23–36	1–6	0	0	30–39
ALPE	0–1	–	0–9	1–3	0	0	3–11
PG	–	0–1	7–8	1–11	0	0	9–20
PE	0–1	0–1	15–65	2–15	0	0	17–81
PC	0–1	0–2	140–178	3–58	0	0	167–201
PI	0–1	1	86–122	§–17	0	0	59–123
LPG	–	–	–	–	7–9	2	2–9
LPE	–	–	0–4	§–7	31–35	2	8–42
LPC	0–1	1	3–18	1–54	308–472	7	26–514
Other PL	1		16–124	1–22	7–17	0	37–156
Unsapon.	48–421	66–104	405–3393	88–198	0–54	19	618–4021
Total NL	34–49	6–7	8314–13 496	1215–3143	678–891	163	10 307–17 683
Total GL	2–7	2–9	170–320	25–90	73–119	10	343–510
Total PL	2–6	10–65	307–456	12–188	353–533	11	447–1012
Total	87–483	90–184	9196–17 633	1523–3509	1417–1523	203	12 180–23 226

†From Tan and Morrison (1979a).
‡Values for waxy maize starch given separately.
§= Not determined.

Table 2.2 Distribution of lipids in germ and endosperm of H51 maize (μg lipid grain^{-1})†

Lipid class	Germ	Endosperm‡
NL	7479	1008
ASG	3	76
MGDG	18	138
SG,CMH	111	293
MGMG		142
DGDG	24	320
SQDG,DGMG§	5	296
APE	8	5
ALPE	8	4
DPG	9	3
PG	20	6
PE	51	20
PC	252	154
PI	96	14
PS	4	–
LPE	1	7
LPC	2	81
Others	9	2
Total	8100	1561

†Calculated from data of Weber (1979).
‡Endosperm non-starch and aleurone lipids together, but excluding starch lipids.
§DGMG is identity of unknown suggested by the present author.

parable data calculated from the results of Weber's concurrent study (Weber, 1979). The data are interpreted here in the light of later studies of lipids in developing maize (Tan and Morrison, 1979b), lipids in maize starches (see B) and microscopic evidence (Chapter 1).

Pericarp tissue suffers extensive senescence during the later stages of grain filling, and its lipid, protein, and starch storage reserves are mostly consumed before the grain matures. Pericarp lipids (Table 2.1) are merely the remnants of spherosomes (TG, FFA, SE, DG) with traces of membrane GL and PL. High levels of unsaponifiable matter (sterols, aliphatic alcohols) are probably from surface wax lipids which are resistant to degradation. Tip cap lipids appear to be very similar.

Maize germ has an exceptionally high lipid content (39–47% in the four maizes described in Tables 2.1 and 2.2) and it is densely packed with spherosomes. Maize germ lipids are mostly TG, with 2.6–5.6% PL (characteristically PC > PI > PE) and traces of GL and N-acyl PL. Much of the unsaponifiable matter is free sterol. Glycosylglycerides, SE, DG and MG increase with germ lipid content during development, then decrease abruptly during the final stages of grain ripening and drying-out (Tan and Morrison, 1979b).

There have been no direct analyses of maize aleurone lipids, and only one analysis "by difference" (Tan and Morrison, 1979a). This showed that aleurone contains most of the endosperm TG, an appreciable quantity of FFA, and about 6% PL. Lipids will occur in spherosomes in maize aleurone, so it is to be expected that the aleurone lipids will resemble germ lipids, except for their higher FFA content.

The endosperm non-starch lipids are mainly FFA with small amounts of TG and other NL, and only traces of GL and PL (Table 2.1, Tan and Morrison, 1979a). Maximum levels of galactosylglycerides and PL are reached at an early stage of grain filling, and there is extensive degradation of these lipids during the later stages so that very little is left at maturity (Tan and Morrison, 1979b). Weber (1979) found much more GL (Table 2.2), but there is no indication whether this is normal or exceptional, or how the GL are distributed between aleurone and non-starch endosperm.

Starch lipids are discussed more fully in Section IIIB. The first detailed analyses (Table 2.1) show that waxy maize starch has very little lipids, but normal and amylomaize starches contain an appreciable proportion of the FFA and lysoPL in the whole grain.

2 *Wheat*

Hargin and Morrison (1980) determined the distribution of acyl lipids in four wheats (Table 2.3). There are many similarities between the lipids of wheat and maize, but there are also major differences due to the much smaller germ in wheat and the absence of lipid degradation in the developing wheat endosperm.

The pericarp lipids (in Atou wheat) are mainly partial glycerides, TG, and SE, and they appear to be remnants of spherosomes and membranes, as in maize pericarp.

There are roughly equal quantities of lipids in the aleurone and germ in each of the four wheats, and the acyl lipids are almost indistinguishable. They consist of spherosome lipids with a higher proportion of membrane PL than in maize (72–85% TG and other NL, 14–18% PL with PC > PI > PE + PG), and little or no detectable GL.

Endosperm non-starch lipids are found in oil droplets which may be spherosomes (Hargin *et al.*, 1980) and presumably in all membranes (including the amyloplast membrane) which are largely indistinguishable from the mass of dried-out storage protein in the mature endosperm. Most of the TG and other NL will be in the spherosomes, and the GL may be in amyloplast membranes (Fishwick and Wright, 1980). In all wheats

Table 2.3 Distribution of acyl lipids (μg lipid $grain^{-1}$) in four wheats†

Lipid class	Pericarp‡	Aleurone	Germ¶	Endosperm: Non-starch¶	Endosperm: Starch	Whole grain§
SE	5.6	3.4–11.0	7.5–9.8	4.3–17.0	0.7–2.3	20.0–25.8
TG	6.6	132.7–290.4	195.9–243.7	46.1–70.0	0.5–1.2	367.2–568.3
DG	14.8	1.9–11.8	0–13.0	9.7–17.8	0–0.9	18.2–72.9
FFA	9.1	5.9–14.4	2.5–5.7	7.7–20.5	3.9–8.3	3.6–110.5
MG	0.5	1.7–2.4	2.2–4.8 (MG + ASG)	6.9–26.5	0.9–1.2	13.8–30.5
ASG	1.6	4.8–6.5		3.9–13.2	0–2.8	9.4–16.3
MGDG	–	8.4–16.6 (MGDG–DGDG)	0–6.6 (MGDG–DGDG)	6.3–26.5	0.5–1.4	8.4–31.8
MGMG	–			0.9–7.4	0–2.6	4.7–14.0
DGDG	–			32.3–74.2	0–5.2	45.0–117.7
DGMG	–			3.9–24.0	0–8.9	13.1–42.7
APE	–	0–2.1	0–0.3	8.2–49.8	0	9.0–98.2
ALPE	–	0–1.7	0–0.9	6.6–29.7	0	6.8–35.6
DPG	–	0.3–2.3	0.6–1.7	–	0	< 7.6
PG	–	3.1–7.2 (PG + PE)	0.4–2.9	1.3–8.9 (PG + PE)	0	8.3–27.1 (PG + PE)
PE	–		4.3–6.9		0	
PC	–	19.8–32.2	20.3–28.1	5.0–19.8	0–1.3	21.2–79.6
PI	–	5.5–9.6	5.5–8.5	–	0	< 26.1
LPG	–	–	–	–	4.8–15.4	6.7–23.7
LPE	–	–	1.0–1.2	0.6–5.4	14.2–28.6	19.1–31.3
LPC	–	2.2–5.2	1.0–4.6	12.4–29.9	106.5–188.7	159.1–227.5
Other PL	–	0.2–1.1	–	1.1–3.5	1.2–29.4	6.0–22.1
Total NL	36.6	159.1–321.1	213.8–266.7	85.9–138.4	6.2–13.7	454.0–685.5
Total GL	2.9	13.2–21.6	0–9.5	44.1–111.7	1.6–17.2	96.2–175.4
Total PL	3.0	39.4–57.3	40.0–46.5	59.9–136.4	130.7–228.7	289.4–516.2
Total	42.5	220.1–386.8	269.8–319.0	237.0–386.5	138.5–255.8	916.1–1243.8

†Three hexaploid bread wheats and one tetraploid durum wheat (Hargin and Morrison, 1980).
‡Values for one hexaploid wheat (Atou) only.
§Experimental values, not the sum of values in other columns.
¶Colborne and Laidman (1975) reported PL compositions within these limits.
– = not determined.

DGDG > MGDG, and GL > N-acyl PL > diacylPL. The low levels of lysoPL are variable and may represent contaminating starch lipids.

Wheat starch lipids are almost entirely lyso-PL (LPC > LPE > LPG) but there always seems to be a small quantity of monoacyl NL (FFA, MG); often there are traces of what are considered to be non-starch lipids (TG, DG, GL). N-acylPL and diacylPL are invariably absent from purified wheat starch.

Each of the major parts of the grain contains a significant proportion of some class(es) of lipids in the whole grain. Germ and aleurone contain most of the SE, TG and diacylPL, nonstarch endosperm is the source of almost all the GL and N-acylPL, and starch provides the lysoPL.

3 *Other cereals*

There have been no detailed analyses of lipids in other cereals comparable with those in Tables 2.1 to 2.3. Zeringue and Feuge (1980) determined the acyl lipids in dissected parts of wheat, rye and triticale, but their work is open to criticism on several points (Morrison, 1982).

From the limited data in the literature (Morrison, 1978a), sorghum lipids should resemble maize lipids, and there should be considerable similarity between wheat, rye, triticale and (to a lesser extent) oat lipids. The distribution of lipids in rice may be like that in barley (which also has a multiple aleurone layer) or oats, but there is sometimes very little GL and PL in milled rice (Choudhury and Juliano, 1980a,b; Azudin and Morrison, unpublished results), which suggests a breakdown of endosperm non-starch lipids either during grain ripening or during post-harvest storage when ageing occurs.

Lipids in barley germ (Table 2.4) are very similar to those in wheat

Table 2.4 Distribution of lipids in germ of barley (μg lipid germ^{-1})†

Lipid class	Coleorhiza	Coleoptile	Scutellum	Whole germ
SE	6.6	2.7	1.7	11.0
TG	60.5	29.3	86.9	176.7
FFA	1.9	1.8	13.3	17.0
ASG	3.8	2.9	19.4	26.1
PE	4.8	2.4	5.0	12.2
PC	15.2	11.3	8.7	35.2
Total	92.8	50.4	135.0	278.2
(%PL)	(21.6)	(27.2)	(10.1)	(17.0)

†Calculated from data of Hølmer *et al.* (1973).

Table 2.5 Phospholipids in embryos of wild oats (*Avena fatua* L.)†

Phospholipid	μg 100 embryos^{-1}	
	Dormant	Non-dormant
PG	13	15
PE	27	22
PC	27	36
PI	1	2
PS	20	14

†Cuming and Osborne, 1978.

germ, with rather more TG in the spherosome-rich scutellum. Wild oat (*Avena fatua* L.) embryos have much less PL, even allowing for their smaller size, and their composition is atypical (Table 2.5).

B Starch Lipids

Starch lipids may be classified according to their source (Morrison, 1981). Non-starch lipids are spherosome and membrane lipids bound to surface protein; all are comparatively easily removed from the starch granules. Starch surface lipids are usually monoacyl non-starch lipids which have become bound to the carbohydrate surface of starch granules, possibly as amylose inclusion complexes. Starch internal lipids are inside the starch granules and they are generally believed to exist as inclusion complexes of amylose.

All cereal starches, except the waxy starches, contain internal lipids which are not easily extracted (Appendix 1). Fairly pure preparations of non-starch and starch lipids can be obtained by selective solvent extraction of wheat flour (Morrison *et al.*, 1975), but it is much better to work with highly purified undamaged starch granules to avoid the risk of contamination with non-starch lipids or premature extraction of starch internal lipids (Morrison, 1981; Morrison and Milligan, 1982). The discussion below refers only to starch internal lipids, although it is recognized that surface and non-starch lipids can modify the properties of starch significantly (Morrison, 1981).

Starch internal lipids are almost exclusively monoacyl lipids (Table 2.6). In wheat they are 86–94% lysoPL, 2–6% FFA, and 2–3% other lipids (Hargin and Morrison, 1980). Early analyses of wheat, barley, rye and oat starches give lysoPL as the major lipids accompanied by varying quantities of FFA and other NL (Becker and Acker, 1976). It may be deduced,

Table 2.6 Lipids in wheat, maize and rice starches (mg $100g^{-1}$)†

	Wheat‡		Maize§			
Lipid class	A-granules	B-granules	Normal	Amylo	Sugary	Rice¶
SE,TG,DG*	1–2	2–3	3	2	5	2–12
FFA	15–54	21–96	379	543	575	221–355
MG,ASG*	1–2	3–6	9	8	35	19–31
MGMG,DGMG*	4–12	8–16	13	23	49	20–40
LPG	30–41	46–62	7	16	15	38–48
LPE	41–75	56–99	20	36	33	86–112
LPC	451–734	564–734	262	419	383	453–513
Others	27–62	21–60	9	14	12	–

†W. R. Morrison, A. M. Coventry and M. N. Azudin, unpublished results.
‡Four samples each.
§One sample each.
¶Six samples, 16–22% amylose.
*Probably contaminating non-starch lipids.

therefore that the starches were not pure and that most of the FFA and NL were non-starch and surface lipids.

The internal lipids of maize and rice starches have more FFA than lysoPL. In a selection of maize genotypes containing up to 40% amylose, internal lipid content was found to be highly correlated with amylose content (Morrison and Milligan, 1982). There is a similar correlation in rice (Azudin and Morrison, unpublished results) and perhaps in the starches of other diploid cereals which exhibit large ranges in amylose content. However, there is no correlation between amylose and total lipids within the normal maize starches (20–28% amylose), although there is a strong negative correlation between the levels of FFA and lysoPL (Morrison and Milligan, 1982).

There seems no doubt that there is a connection between amylose and lipids, but the biochemistry is unknown, and it is already evident that subtle factors affect the balance of FFA and lysoPL.

REFERENCES

Arunga, R. O. and Morrison, W. R. (1971). *Lipids* **6**, 768–776.

Becker, G. and Acker, L. (1976). Die Lipide der Getreidestarken, *Handbuch der Starke in Einzeldarstellung* (M. Ulmann, ed.), Vol. VI-5 Paul Parey, Berlin and Hamburg.

Bergel'son, L. D. (1973). *Fette Seifen Anstrichm*. **76**, 89–96.

Berndorfer, E. K. (1970). *Elelmiszervizsgalati Kozlem.* **16**, 193–202.

Briggs, D. E. (1974). *Phytochemistry* **13**, 987–996.

Burini, G. and Damiani, P. (1978). *Tec. Molitoria* **29**, 97–108.

Carter, H. E., Ohno, K., Nojima, S., Tipton, C. L. and Stanacev, N. Z. (1964). *J. Lipid Res.* **2**, 215–222.

Carter, H. E., Strobach, D. R. and Hawthorne, J. N. (1969). *Biochemistry* **8**, 383–388.

Choudhury, N. H. and Juliano, B. O. (1980a). *Phytochemistry* **19**, 1063–1069.

Choudhury, N. H. and Juliano, B. O. (1980b). *Phytochemistry* **19**, 1385–1389.

Colborne, A. J. and Laidman, D. L. (1975). *Phytochemistry* **14**, 2639–2645.

Cuming, A. C. and Osborne, D. J. (1978). *Planta* **139**, 219–226.

Daftary, R. D. and Pomeranz Y. (1965). *J. Food Sci.* **30**, 577–582.

Dalton, J. L. and Mitchell, H. L. (1959). *J. Agric. Food Chem.* **7**, 570–573.

De la Roche, I. A., Alexander, D. E. and Weber, E. J. (1971a). *Crop Sci.* **11**, 856–859.

De la Roche, I. A., Weber, E. J. and Alexander, D. E. (1971c). *Lipids* **6**, 531–6. 871–874.

De la Roche, I. A., Weber, E. J. and Alexander, D. E. (1971c) *Lipids* **6**, 531–6.

Farre-Rovira, R. (1975). *Circ. Farm.* **33**, 399–417.

Farre-Rovira, R. and Costes, C. (1974). *Physiol. Veg.* **12**, 251–288.

Farrington, W. H. H. and Shearer, G. (1981). *J. Sci. Food Agric* **32**, 948–950.

Fishwick, M. J. and Wright, A. J. (1980). *Phytochemistry* **19**, 55–59.
Fujino, Y. (1978). *Cereal Chem.* **55**, 559–571.
Fujino, Y. and Miyazawa, T. (1979). *Biochim. Biophys. Acta* **572**, 442–451.
Fujino, Y. and Ohnishi, M. (1976). *Chem. Phys. Lipids* **17**, 275–289.
Fujino, Y. and Ohnishi, M. (1979a). *Biochim. Biophys. Acta* **574**, 94–102.
Fujino, Y. and Ohnishi, M. (1979b). *Proc. Japan. Acad.* **55**, ser. B, 243–246.
Fujino, Y. and Ohnishi, M. (1982a). *Proc. Japan. Acad.*, **58**, ser. B, 32–35.
Fujino, Y. and Ohnishi, M. (1982b). *Proc. Japan Acad.*, **58**, ser. B, 36–39.
Fujino, Y. and Sakata, S. (1973). *Cereal Chem.* **50**, 379–382.
Fujino, Y.; Sakata, S. and Nakano, M. (1974). *J. Food Sci.* **39**, 471–473.
Hargin, K. D. and Morrison, W. R. (1980). *J. Sci. Food Agric.* **31**, 877–888.
Hargin, K. D., Morrison, W. R. and Fulcher, R. G. (1980), *Cereal Chem.* **57**, 320–325.
Heinz, E. (1967). *Biochim. Biophys. Acta* **144**, 321–332.
Hølmer, G., Ory, R. L. and Høy, C.-E. (1973). *Lipids* **8**, 277–283.
Hsieh, T. C.-Y., Lester, R. L. and Laine, R. A. (1981). *J. Biol. Chem.* **256**, 7747–7755.
Ito, S., Suzuki, T. and Fujino, Y. (1981). *J. Agric. Chem. Soc. Japan* **55**, 247–253.
Jelsema, C. L., Morré, D. J.; Ruddat, M. and Turner, C. (1977). *Bot. Gaz*, **138**, 138–149.
Kato, A., Tanabe, K. and Yanoka, M. (1981). *Yukagaku* **30**, 515–516.
Kondo, Y., Ito, S. and Fujino, Y. (1974). *Agric. Biol. Chem.* **38**, 2549–2552.
Kondo, Y., Nakano, M. and Fujino, Y. (1975). *Agric. Biol. Chem.* **39**, 719–721.
Kuroda, N., Ohnishi, M. and Fujino, Y. (1977). *Cereal Chem.* **54**, 997–1006.
Laine, R. A. and Renkonen, O. (1973). *Biochemistry* **12**, 1106–1111.
Laine, R. A. and Renkonen, O. (1974). *Biochemistry* **13**, 2837–2843.
Laine, R. A., Hsieh, T. C.-Y. and Lester R. L. (1980). In *Cell Surface Glycolipids* (C. C. Sweeley ed.), Am. Chem. Soc. Symp. Ser. 128, pp. 65–78. American Chemical Society, Washington DC
LePage, M. and Sims, R. P. A. (1968). *Cereal Chem.* **45**, 600–604.
Lin, M. J. Y., Youngs, V. L. and D'Appolionia, B. L. (1974). *Cereal Chem.* **51**, 17–33.
MacMurray, T. and Morrison W. R. (1970). *J. Sci. Food Agric.* **21**, 520–528.
Miyazawa, T. and Fujino, Y. (1978a). *Agric. Biol. Chem.* **42**, 1979–1980.
Miyazawa, T. and Fujino, Y. (1978b). *J. Agric. Chem. Soc. Japan* **52**, 37–43.
Miyazawa, T., Yoshino, Y. and Fujino, Y. (1977). *J. Sci. Food Agric.* **28**, 889–894.
Miyazawa, T., Tazawa, H. and Fujino, Y. (1978). *Cereal Chem.* **55**, 138–145.
Morrison, W. R. (1978a) *Adv. Cereal Sci. Technol.* **2**, 221–348.
Morrison, W. R. (1978b). *J. Sci. Food Agric.* **29**, 365–371.
Morrison, W. R. (1981). *Starke* **33**, 408–410.
Morrison, W. R. (1982). *J. Am. Oil Chem. Soc.* **59**, 102.
Morrison, W. R., Mann, D. L., Wong, S. and Coventry, A. M. (1975). *J. Sci. Food Agric.* **26**, 507–521.
Morrison, W. R. and Milligan, T. P. (1982). In *Maize: Recent Advances in Chemistry and Technology* (G. E. Inglett, ed.). pp. 1–18. Academic Press, London and New York.
Myhre, D. V. (1968). *Can. J. Chem.* **46**, 3071–3077.
Obara, T. and Kihara, H. (1973). *J. Agric. Chem. Soc. Japan* **47**, 231–236.
Ohnishi, M. and Fujino, Y. (1978). *Agric. Biol. Chem.* **42**, 2423–2425.
Ohnishi, M. and Fujino, Y. (1980). *Agric. Biol. Chem.* **44**, 333–338.

Sakata, S., Ito, S. and Fujino, Y. (1973). *J. Agric. Chem. Soc. Japan* **47**, 125–128.
Skarsaune, S., Youngs, V. L. and Gilles, K. (1970). *Cereal Chem.* **47**, 533–544.
Tan, S. L. and Morrison, W. R. (1979a). *J. Am. Oil Chem. Soc.* **56**, 531–535.
Tan, S. L. and Morrison, W. R. (1979b). *J. Am. Oil Chem. Soc.* **56**, 759–764.
Thomas, W. (1979). *Stärke* **31**, 54–57.
Torres, J. V. and Garcia-Olmedo, F. (1974). *Plant Sci. Lett.* **3**, 213–217.
Torres, J. V., Carbonero, P. and Garcia-Olmedo, F. (1976). *Phytochemistry* **15**, 677–680.
Vaver, V. A., Prokazova, N. V., Starkhova, G. D. and Bergel'son, L. D. (1969). *Dokl. Akad. Nauk SSSR* **188**, 227–229.
Vaver, V. A., Stoyanova, V. G., Geiko, N. S., Nechaev, A. P., Todriya, K. G. and Bergel'son, L. D. (1976). *Bioorg. Khim.* **2**, 530–534.
Vaver, V. A., Todria, K. G., Prokazova, N. V., Rozynov, B. V. and Bergel'son, L. D. (1977). *Biochim. Biophys. Acta* **486**, 60–69.
Wanner, G. and Theimer, R.R. (1978). *Planta* **140**, 163–169.
Weber, E. J. (1979). *J. Am. Oil Chem. Soc.* **56**, 637–641.
Yatsu, L. Y. and Jacks, T. J. (1972). *Plant Physiol.* **49**, 937–943.
Youngs, V. L., Püskülcü, M. and Smith, R. R. (1977). *Cereal Chem.* **54**, 803–812.
Zeringue, H. J. and Feuge, R. O. (1980). *J. Am. Oil Chem. Soc.* **57**, 373–376.

3 Non-Saponifiable Lipids in Cereals

P. J. BARNES

The Lord Rank Research Centre, High Wycombe, Bucks, U.K.

I INTRODUCTION

As already described in Chapter 2, the acyl lipids of cereal grains consist of free fatty acids and fatty acids esterified to alcohols, predominantly glycerol or glycerol derivatives. Other alcohols, including sterols, carotenoids and tocopherols may also be esterified to fatty acids and the esters can thus be considered as acyl lipids. However, the non-acyl parts of these lipids often occur in the unesterified form and thus the sterols, triterpenols, carotenoids and tocopherols together with the hydrocarbons are more conveniently placed in a separate group for the purposes of this book.

"Lipids in Cereal Technology"
ISBN 0-12-079020-3

The non-saponifiable lipids were included in a review by Morrison (1978) in which the lipids in individual cereals were discussed separately. In this chapter the lipids in cereals will be discussed together in sections covering sterols and triterpenols, hydrocarbons, carotenoids and tocols. The carotenoids and tocols are the non-saponifiable lipids of greatest technological importance in cereals.

A number of papers have been published concerning the non-saponifiable lipids of commercial wheat germ oil, maize germ oil (corn oil) and rice bran oil; later chapters will be concerned with these oils and their compositions. The information will not be described in detail in this discussion.

II STEROLS AND TRITERPENOLS

Most of the published information on the composition of sterols and triterpenols in cereal grains concerns whole grains and mill products such as flour, semolina, bran and germ; there are very few reported analyses of these lipids in dissected grain fractions.

A Total Sterols

1 Whole Grain

Sitosterol (Fig. 3.1; B-3) is the major sterol in most cereal grains and grain products (Morrison, 1978). It represents at least 55% of the sterols in non-saponifiable fractions from whole grains of wheat, rye, oats, rice and maize (Knights, 1967) and 50% of the total sterols of sorghum grain (Palmer and Bowden, 1975a) (Table 3.1). The second major component in the sterols of whole grain is campesterol (B-2), except for oat which contains a relatively high proportion of Δ^5-avenasterol (B-5). Although stigmasterol (B-6) and Δ^5- and Δ^7- avenasterols (B-5; C-5) were not detected in all the cereals in these studies of whole grain extracts, they are present in rice bran and in commercial maize germ oil and wheat germ oil (Chapters 15, 17, and 19).

The work by Knights (1967) confirmed much earlier reports that the wheat grain contained a relatively large proportion of saturated sterols (stanols) and also demonstrated their presence in rye and maize grain. The stanols in wheat grain consisted of 47% C_{28} and 53% C_{29} 3β-hydroxy-5α-stanols (A-2, A-3). Later studies using a range of *Triticum* species indicated the presence of even greater proportions of stanols, the C_{28} representing 13.5 to 22.5% and the C_{29} 6.5 to 22.5% of the grain sterols (Berrie

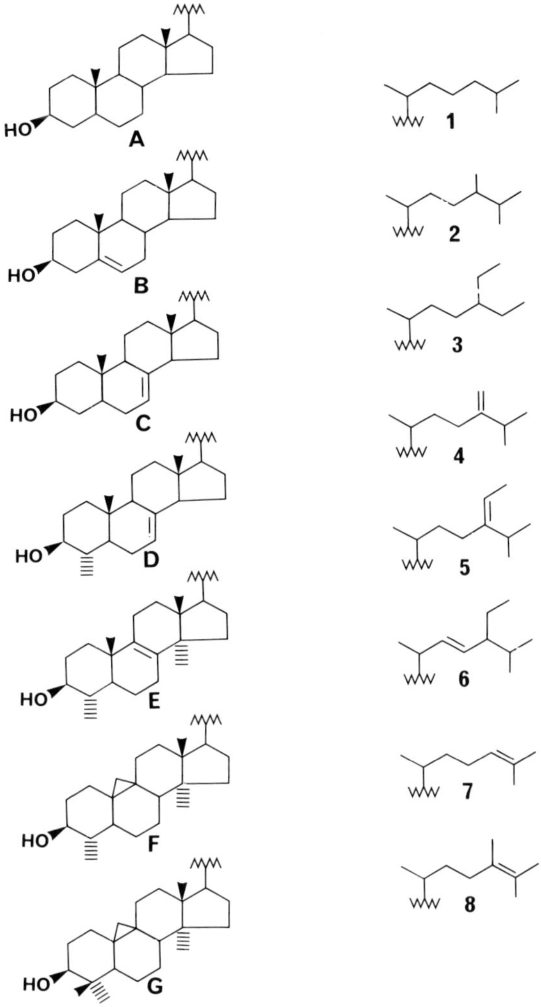

Fig. 3.1 The sterols found in cereal grains can be represented by combinations of the sterol nuclei and side-chains as follows:
4-demethyl sterols: A-1 – 5∝-cholestan-3β-ol. A-2 – 24-methyl-5∝-cholestan-3β-ol. A-3 – sitostanol, 24-ethyl-5∝-cholestan-3β-ol. B-1 – cholesterol. B-2 – campesterol. B-3 – sitosterol. B-4 – 24-methylene cholesterol. B-5 – Δ^5-avenasterol; isofucosterol; 24-ethylidene cholesterol. B-6 – stigmasterol. C-1 – Δ^7-cholesten-3β-ol. C-2 – Δ^7-campesterol. C-3 Δ^7-stigmasterol. C-4 24-methylenecholest-7-en-3β-ol. C-5 – Δ^7-avenasterol.
4-methyl-sterols: D-2 – 24-methyl lophenol. D-3 – 24-ethyl lophenol. D-4 – gramisterol; 24-methylene lophenol. D-5 – citrostadienol; 24-ethylidene lophenol. E-2 – 24-dihydro obtusifoliol. E-4 – obtusifoliol. F-4 – cycloeucalenol.
4,4-dimethyl sterols: G-1 – cycloartanol. G-4 – 24-methylene cycloartanol. G-7 – cycloartenol. G-8 – cyclobranol.

Table 3.1 Composition of sterols from cereal grains

Sterol	% of total sterols extracted					
	wheat†	rye†	oat†	rice†	maize†	sorghum‡
Sitosterol	55.0	59.0	60.0	55.5	65.0	49.6
Campesterol	21.0	20.0	6.4	22.0	21.0	27.8
Stigmasterol	–	tr	–	11.0	5	15.0
Δ^5-avenasterol	–	8	20.8	–	–	5.0
Δ^7-avenasterol	–	–	6.5	–	–	2.2
Cholesterol	tr¶	0.5	2.5	2.5	tr	0.4
5α-stanols	18	9	–	–	9	
Others	5	5	tr	5	tr	

†Knights (1967).
‡Palmer and Bowden (1975a).
¶tr indicates trace quantities detected.

and Knights, 1972). In a more detailed examination of the sterols of the oat grain by gas chromatography-mass spectrometry (gc-ms), Knights and Laurie (1967) detected the corresponding C_{27} derivative (A-1) in addition to the C_{28} and C_{29} stanols. MacMurray and Morrison (1970) have confined the presence of stanols in wheat flour by gc-ms and it is surprising that so little work has been published concerning these components. Sitostanol (A-3) was identified in an extract of triticale grain (Dominguez and Rodriguez-Bores, 1972) and sitostanol palmitate was claimed to be present in wheat (Garcia-Olmedo, 1965). Kemp *et al.* (1967) detected three stanols in the mixture of sterols from the endosperm of germinating maize grain.

It is noteworthy that stanols have not been reported in rice bran or in the oils of rice bran, wheat germ and maize germ; this suggests the possibility that stanols may be present in substantial quantities in the endosperm only.

The gc-ms investigations by Knights and Laurie (1967) not only increased our knowledge of the 4-demethyl sterols of the oat grain but also identified several sterols not previously known to occur in higher plants. All the 4-demethyl sterols shown in Fig 3.1 were found in the non-saponifiable fraction from oat grain. They comprise two series based on cholest-5-en-3β-ol (B-1) and cholest-7-en-3β-ol (C-1), unsubstituted at C-24 or with 24-methyl, 24-ethyl, 24-methylene or 24-ethylidene substitution. In addition, the three stanols and stigmasterol were detected. In sorghum grain, 24-methylenecholest-5-en-3β-ol (B-4) and 24-methylene-cholest-7-en-3-β-ol (C-4) occur, together with the sterols listed in Table 3.1 (Palmer and Bowden, 1975b).

The analysis of sorghum grain sterols also showed the presence of 4-methyl sterols and pentacyclic triterpenols but 4,4-dimethyl sterols could not be detected. Gramisterol, (D-4), citrostadienol (D-5), 24-methyl-

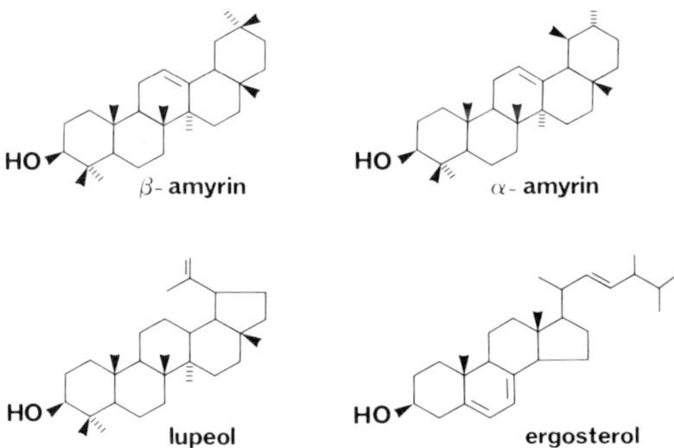

Fig. 3.2 Triterpenols and ergosterol.

lophenol (D-2), 24-ethyl-lophenol (D-3), obtusifoliol (E-4) and 24-dihydro-obtusifoliol (E-2) were identified in the 4-methyl sterol fraction (Fig. 3.1) and β-amyrin, δ-amyrin and lupeol (Fig. 3.2) in the triterpenols (Bowden and Palmer, 1975; Palmer and Bowden, 1975b). Gramisterol and citrostadienol were the major 4-methyl sterols and lupeol was the major triterpenol (Palmer and Bowden 1975a). The composition of the corresponding fractions from whole grain of other cereals has not been reported. However, detailed information is available for rice bran oil and rice bran, maize germ oil and wheat germ oil (Chapters 15, 17, 19); the most notable difference compared with sorghum grain is the presence of the 4,4-dimethyl sterols, 24-methylene cycloartanol (G-4), cycloartanol (G-1), cycloartenol (G-7) and cyclobranol (G-8) in the oils and bran, accompanied by α-amyrin in the germ oils (Fig. 3.2).

2 *Dissected Grain Fractions*

The distribution of lipids in cereal grains cannot be determined accurately by analysis of commercial mill products such as germ and bran because these are invariably contaminated with tissues from other parts of the grain. However, very little information has been published on the sterol composition of dissected grain fractions. Torres *et al*. (1976) analysed the 4-demethyl sterols of endosperm, bran, embryo and scutellum from one sample each of *Triticum aestivum* and *T. durum*. The percentage composition of four sterols was given for each tissue (Table 3.2). The proportions of stigmasterol and cholesterol are very high compared with results for wheat germ oil (Itoh *et al*., 1973), wheat grain (Knights, 1967) and flours (see below). The proportion of campesterol was lower in the endosperm than in the other tissues and the concentration of total 4-demethyl sterols was least in the endosperm and greatest in the germ.

In sorghum, three-quarters of the total sterols of the grain were found in

Table 3.2 Composition of sterols in dissected wheat grain fractions†

Sterol	Proportion of total sterols in fraction (%)			
	endosperm	bran	embryo	scutellum‡
Triticum aestivum				
Sitosterol	67.0	67.2	63.3	81.9
Campesterol	4.6	10.6	16.1	14.5
Stigmasterol	14.6	10.6	9.4	2.3
Cholesterol	13.3	11.6	11.2	1.3
Triticum durum				
Sitosterol	75.0	65.3	55.2	69.6
Campesterol	3.8	10.9	12.8	17.4
Stigmasterol	13.6	10.2	16.2	10.4
Cholesterol	7.6	13.6	15.8	2.6

†Torres *et al.* (1976).
‡Scutellum from grain germinated for 6 days.

the germ but the triterpenols were equally distributed between germ and endosperm ("endosperm" presumably consisted of all non-germ tissue but the distinction was not clearly made) (Palmer and Bowden, 1975a). The percentage composition of the sterols and triterpenols from the two grain fractions were similar.

The lipids from the testa and pigment strand of barley contain at least five sterols of which two, sitosterol and campesterol, were identified (Briggs, 1974). Both free and esterified sterols were present.

3 *Wheat Flour and Semolina*

The distribution of grain lipids during flour milling is described in Chapter 7. Wheat flour contains not only the endosperm lipids but also a contribution from germ and aleurone; the extent is governed by the grade of flour.

An investigation by gas-liquid chromatography (glc) of the sterols in flours and millstreams from *T. aestivum* and semolina from *T. durum* indicate that sitosterol and campesterol were the major components and suggest that cholesterol was also present (Berry *et al.*, 1968). An earlier report had also indicated sitosterol to be the main sterol in wheat flour by glc analysis (McKillican, 1964). MacMurray and Morrison (1970) used gc-ms to confirm that the sterols in wheat flour were sitosterol and campesterol together with the C_{28} and C_{29} 3β-hydroxy-5α-stanols, as found for whole wheat grain by Knights (1967). About two-thirds of the total flour sterol was sitosterol while campesterol and C_{29} stanol represented about one-quarter and the remainder was C_{28} stanol; cholesterol was not detected.

The results of glc analysis of the sterols of three wheat flours have recently been reported (Amelotti *et al.*, 1980). No account was taken of the 5α-stanols but Δ^7-avenasterol, Δ^7-stigmasterol (C-3) and minor amounts of stigmasterol (B-6) and Δ^7-campesterol (C-2) were claimed to be present.

B Distribution of Sterols among the Sterol Lipids

MacMurray and Morrison (1970) showed by gc-ms that the sterol compositions of the free sterols, steryl glycosides (SG) and acyl steryl glycosides (ASG) from wheat flour were similar; the steryl esters (SE) contained a greater proportion of sitosterol and less campesterol (Table 3.3). Other reports have confirmed that sitosterol and campesterol are the major sterols of the free sterol fraction, SE, SG and ASG of wheat flour, millstreams and durum semolina (Berry *et al.*, 1968); Myhre, 1968; Lin *et al.*, 1974).

In sorghum grain lipids the 4-demethyl sterols were found in the free sterols, SE and SG but 4-methyl sterols were restricted to the free sterol fraction and the triterpenols could only be found in the SE (Palmer and Bowden, 1975a). The sterol compositions of free sterols, SE, SG and ASG from grains of *Eleusine coracana* (finger millet) were similar to each other, with 80–84% sitosterol and 15–19% stigmasterol (Mahadevappa and Raina, 1978).

Kuroda *et al.* (1977) separated the sterol lipid fractions from rice bran and determined the composition of 4-demethyl sterols, 4-methyl sterols and 4,4-dimethyl sterols. Unlike sorghum grain, the rice bran SE contained all three types of sterol but in both cases the glycosides contained only 4-demethyl sterols. Sitosterol, campesterol, stigmasterol and Δ^5-avenasterol were the major components of the 4-demethyl sterols, while gramisterol and citrostadienol predominated in the 4-methyl sterols and 24-methylene cycloartanol and cycloartenol in the 4,4-dimethyl sterols.

Table 3.3 Composition of sterols in wheat flour sterol lipids†

	Proportion of total sterols in fraction (%)			
Sterol	free	SE	SG	ASG
Sitosterol	63.2	75.2	63.8	65.6
Campesterol	16.8	4.8	17.2	14.4
C_{28} stanol	6.4	5.9	5.7	6.0
C_{29} stanol	13.6	14.1	14.3	14.0

†MacMurray and Morrison (1970).

C Ergosterol in Cereal Grains and Flour

The presence of ergosterol (Fig. 3.2) in wheat germ oil was reported in 1935 (Drummond *et al.*) but was not confirmed by later studies using gc-ms (Itoh *et al.*, 1973). Stoyanova *et al.* (1975) found ergosterol to be a major component (together with sitosterol and campesterol) in the total sterol fraction of wheat grain but ergosterol has not been detected by most other workers; it is possible that ergosterol found in cereal grains is of fungal origin.

Seitz *et al.* (1977) studied ergosterol content of wheat, maize and sorghum grain as a means of estimating the extent of fungal contamination, based on the fact that it is the major component of the sterols in most fungi. At very low levels of fungal contamination only traces of ergosterol could be detected but the concentration increased with increasing levels of preharvest fungal invasion.

D Sterol Biosynthesis

Although sterol biosynthesis is not discussed in detail here, a brief consideration of the later stages of the biosynthetic pathway in plants will explain the presence of most of the sterols found in cereal grains.

In plants, the first cyclic product from squalene oxide is cycloartenol

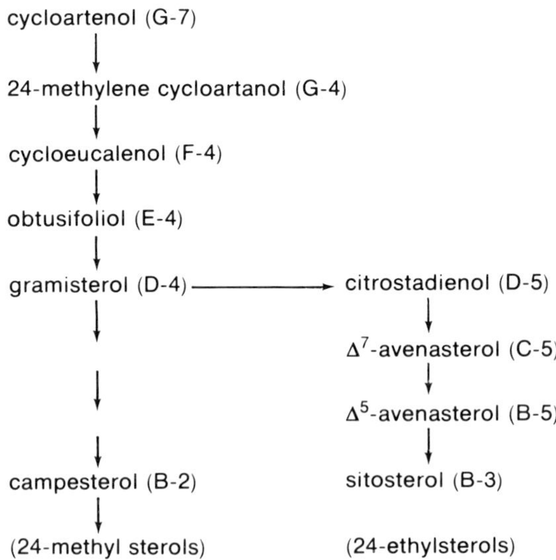

Fig. 3.3 Pathways for the biosynthesis of sterols in plants (modified from Goodwin, 1980).

from which the other 4,4-dimethyl sterols, 4-methyl sterols and 4-demethyl sterols are derived. Fig. 3.3 shows the possible pathway by which these transformations may occur and illustrates the interrelationships of the sterols (Goodwin, 1980). Side-chain modification proceeds through the 24-methylene to the 24-methyl sterols or by a second methylation producing the 24-ethylidene group and then the 24-ethyl sterols. Nuclear transformations include the opening of the cyclopropane ring, demethylations at C-4 and C-14 and the $\Delta^8 \rightarrow \Delta^7 \rightarrow \Delta^5$ steps. It can be seen that the range of sterols found in cereal grains represents not only a variety of end-products but also a set of intermediates common to the sterol biosynthetic pathway in plants.

III TOCOLS

Tocols are responsible for the vitamin E activity of plant tissues. Their structures comprise a phenolic ring and a side-chain which may be saturated or unsaturated. Here, tocol is used to encompass both those compounds having a saturated side-chain – the tocopherols – and those having a triunsaturated side-chain – the tocotrienols (Fig. 3.4). In each series four patterns of methyl substitution of the phenolic ring are found giving rise to

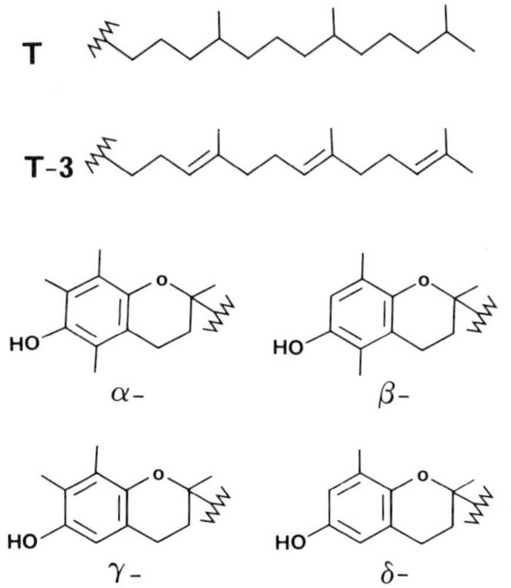

T

T-3

HO O α-

HO O β-

HO O γ-

HO O δ-

Fig. 3.4 Tocols found in cereal grains. The tocopherols have a saturated side-chain (T) and the tocotrienols have a triunsaturated side-chain (T-3). Methyl substitution of the phenolic ring gives rise to four members of each tocol series.

Table 3.4 Nomenclature and abbreviations for naturally occurring tocols†

Substitution	Tocopherol (T)	Tocotrienol (T-3)
5,7,8-trimethyl	α-T	α-T-3 (ζ-T)
5,8-dimethyl	β-T	β-T-3 (ε-T)
7,8-dimethyl	γ-T	γ-T-3 (η-T)
8-monomethyl	δ-T	δ-T-3

†Based on Pennock *et al.*, 1964.

the eight tocols listed in Table 3.4. Tocodienols and tocomonoenols have not been reported in ungerminated cereal grains.

α-T has the highest vitamin E activity followed by β-T and α-T-3 (Brubacher and Wiss, 1972). Estimates of the vitamin E content of foodstuffs and animal feeds are based frequently on the concentration of α-T, however, these estimates may be inaccurate in cereal grains which contain large proportions of other tocols in addition to α-T.

The protection of membrane lipids against oxidation is one of the suggested mechanisms of vitamin E activity (McCay and King, 1980). An antioxidant effect of tocols is also observed in edible oils and fats but the relative activities of the different tocols depend upon temperature and type of fat (Bourgeois, 1981).

A Tocol Composition of Whole Cereal Grains

The mature seeds and fruits of plants contain relatively large proportions of other tocols in addition to α-T compared with vegetative tissues. Various combinations of all eight of the tocopherols and tocotrienols are found among cereal grains. In grains that contain significant proportions of β-T and β-T-3, γ-T and γ-T-3 are not usually detectable (e.g. wheat, rye and triticale). Conversely, the β-tocols are not found normally in significant quantities when the corresponding γ-tocols are present (e.g. maize and rice). Barley grain is a notable exception and contains both β- and γ-tocols. It is possible that the β-tocols (5,8-dimethyl derivatives) and γ-tocols (7,8-dimethyl derivatives) represent components of alternative biosynthetic routes to tocopherols and tocotrienols.

1 *Barley and Oat*

Barley grain contains all eight tocols listed in Table 3.5. The presence of the four tocopherols together with α-, β- and γ-tocotrienols has been confirmed by gas chromatography-mass spectrometry (gc-ms) (Govind Rao

Table 3.5 Tocol composition of the grain of barley, oat, wheat, rye and triticale

	Composition of tocols (mg kg^{-1})							
	α-T	β-T	γ-T	δ-T	α-T-3	β-T-3	γ-T-3	δ-T-3
Barley								
Slover *et al.*, 1969a	2	0.4	0.3	0.1	11	3	2	
Barnes, 1983‡	4.6	0.2	0.7	0.2	13.0	2.7	8.4	0.7
Oat†								
Lasztity *et al.*, 1980	5.5	0.9	2.2	0.4	12.8	1.0	2.7	1.0
Slover *et al.*, 1969a	5	0.9	–	–	11	2	–	
Barnes, 1983‡	4.3	0.8	–	–	11.8	1.9	–	–
Wheat								
Barnes, 1983‡	9.7	3.9	–	–	2.4	19.0	–	–
Slover *et al.*, 1969a	10	7	–	–	4	28	–	
Slover *et al.*, 1969‡	11.9	6.2	–	–	5.5	33.3		
Rye								
Barnes, 1983‡	13.1	3.9	–	–	14.8	11.8	–	–
Slover *et al.*, 1969a	16	4	–	–	15	8	–	
Triticale								
Barnes, 1983‡	9.1	3.0	–	–	10.3	15.1	–	–

†See also Chapter 16.
‡Results expressed on dry weight basis; all other results either wet weight or not specified.

and Perkins, 1972) and the presence of δ-T-3 is indicated by the results of high-performance liquid chromatographic analysis (hplc) (Barnes, 1983). The quantitative data presented in Table 3.5 show that α-T-3 is the major tocol in the barley grain; this is also confirmed by other studies (Green, 1958; Mason and Jones, 1958; Chow *et al.*, 1969; Govind Rao and Perkins, 1972).

Only α-T, β-T, α-T-3 and β-T-3 have been detected in oat grain extracts using gc-ms (Govind Rao and Perkins, 1972) and hplc (Barnes, 1983); a similar composition has been determined using gas-liquid chromatography (glc) (Slover *et al.*, 1969a) (Table 3.5). However, it has been claimed that γ- and δ-tocols are present in addition to α- and β-tocols as determined by using thin-layer chromatography (tlc) (Chow *et al.*, 1969; Lásztity *et al.*, 1980). It is generally agreed that α-T-3 is the major tocol together with a substantial proportion of α-T. Oat grain, therefore, primarily contains the α- and β-tocols; further investigations are needed to determine definitely whether or not significant quantities of the γ- and δ-tocols are present. Further details of oat grain tocols are discussed in Chapter 16.

2 *Wheat*

Only the α- and β-tocols have been detected in the wheat grain (Table 3.5), a finding that confirms the results of other studies (Green, 1958; Mason and Jones, 1958; Hall and Laidman, 1968a; Suckewer and Rakowska, 1974; Thompson *et al.*, 1972; Thompson and Hatina, 1979). Wheat flours (Slover *et al.*, 1969b; Toepfer *et al.*, 1972), dissected grain fractions (Hall and Laidman, 1968a; Morrison *et al.*, 1982), germinating wheat grain (Hall and Laidman, 1968b) and developing wheat grain have also been shown not to contain γ- and δ-tocols.

There are two reports concerning the presence of relatively large proportions of γ- and δ-tocopherol in the tocol fractions from wheat grains (Telegdy-Kovats *et al.*, 1970; Davis *et al.*, 1980). In both cases, β-T-3 was not reported although in other studies this tocotrienol has been found to be the major tocol in the wheat grain. It may be that tocols were misidentified in these two investigations and most of the available information supports the conclusion that wheat grain contains only α- and β-tocols.

In contrast to barley and oat, the major tocols in wheat grain are α-T and β-T-3, and these are present at a higher concentration in the wheat grain than in other cereal grains (Table 3.5).

3 *Rye and Triticale*

Rye, in common with wheat, contains only the α- and β-tocols but α-T-3 is present at a higher concentration than in wheat (Table 3.5). Triticale grain has a tocol composition similar to that of rye.

4 *Maize, Rice, Sorghum and Millets*

The tocols of the maize grain consist of α-T, γ-T, α-T-3 and γ-T-3 (Table 3.6 and Chapter 17). The results in Table 3.6 illustrate the wide variation in the content and relative proportions of the tocols. Even greater variation has been demonstrated among maize inbreds (see Chapter 17).

Extracts from rice grain have been found to contain α-T, γ-T, δ-T, γ-T-3 and δ-T-3 (Table 3.6) with the major components being α-T, γ-T and γ-T-3. In rice bran oil, β-T and α-T-3 have also been detected (Kato *et al.*, 1981; Tanabe *et al.*, 1982).

Until recently there had been no data reported for the tocol compositions of the grain of sorghum or identified millets. Analysis by hplc has indicated that γ-T is the major tocol in sorghum and that the millets as a group do not all share the same tocol composition (Table 3.6). The major tocol is γ-T in bulrush, foxtail and finger millet, and β-T-3 in common millet. In two previous reports of millet tocol composition the identity of

Table 3.6 Tocol composition of the grain of maize, rice, sorghum and millets

	Composition of tocols (mg kg^{-1})							
	α-T	β-T	γ-T	δ-T	α-T-3	β-T-3	γ-T-3	δ-T-3
Maize†								
Grams *et al.*, 1970‡	22.1		56.0	trace	6.4		9.9	
Beringer and Dompert, 1976‡	11.1		76.3	–	1.4		3.2	
Barnes, 1983‡	0.2	–	17.5	0.3	0.3	–	3.3	–
Rice								
Slover, 1971	3		3	0.4	trace		5	
Barnes, 1983‡	0.6	–	0.8	0.2	–	–	2.6	0.2
Sorghum‡,¶	0.8	–	11.5	–	–	–	–	–
Bulrush millet‡,¶ (*Pennisetum americanum*)	1.3	–	55.4	0.8	–	–	5.3	2.0
Foxtail millet‡,¶ (*Setaria italica*)	1.9	0.4	27.8	–	0.4	–	0.6	–
Finger millet‡,¶ (*Eleusine coracana*)	3.2	0.3	17.6	–	0.5	–	–	–
Common millet‡,¶ (*Panicum miliaceum*)	–	0.4	–	3.5	–	15.0	–	–

†See also Chapter 17.
‡Results expressed on dry weight basis; all other results either wet weight or not specified.
¶Barnes, 1983.

the millet was not given; American workers found γ-T to be the major tocol (Slover *et al.*, 1969a) whereas in a Russian study a composition of α-T (33%), β-T (27%), γ-T (5%) δ-T (5%) and γ-T-3 (31%) was reported (Seit-Ablaeva *et al.*, 1973).

B Distribution of Tocols in the Cereal Grain

It has long been recognized that tocols are not uniformly distributed throughout the tissues of the wheat grain. The analysis of total tocols in mill fractions has revealed a higher content in the germ fraction than in flour and bran (Binnington and Andrews, 1941). Russell Eggitt and Ward (1955) showed that tocopherols are concentrated mainly in the germ fraction from flour milling while tocotrienols are present at much higher concentrations in the bran fractions.

Germ fractions from wheat, rye, triticale, barley and oat, prepared using

Table 3.7 Concentration of tocols in dissected wheat grain fractions†

Fraction	Concentration of tocols (mg kg^{-1} dry weight)				
	α-T	β-T	α-T-3	β-T-3	Total
Germ	255.6 (98.0)‡	114.4 (97.3)	< 2 –	< 2 –	370.0
Pericarp, testa, aleurone	0.5 (1.2)	< 0.4 –	10.0 (81.2)	68.6 (49.8)	79.1
Starchy endosperm	0.07 (0.8)	0.10 (2.7)	0.45 (18.8)	13.5 (50.2)	14.1

†Morrison *et al.*, 1982.
‡Figures in parentheses represent percentage distribution among the fractions of the whole grain.

an impact mill and purified manually, have been shown to contain tocopherols by hplc, but none of the four tocotrienols has been detected (Barnes and Taylor, 1981). With the exception of barley germ, which also contained γ-T, only α-T and β-T have been found in the germ from these cereal grains.

The distribution of tocols among the dissected tissues of wheat grain has been investigated recently using hplc with fluorescence detection, a selective and sensitive method capable of detecting all eight tocols (Morrison *et al.*, 1982). Only α- and β-tocols are present in wheat grain tissues (Table 3.7). α-T and β-T are the only tocols detectable in the germ and the amounts of these compounds in germ represents more than 97% of the total α-T and β-T in the grain. β-T-3 is distributed equally between the starchy endosperm and bran tissues although the concentration is higher in the latter. In contrast, α-T-3 is found predominantly in the bran tissues, where it accounts for over 80% of the total grain α-T-3.

These tocopherols can be used as markers for germ lipid in flour because the concentrations of α-T and β-T in the endosperm and bran are very low. The application of tocol measurement to detect germ and bran lipids in flour millstreams is further discussed in Chapter 7.

Although maize grain differs from wheat in having γ-tocols in place of β-tocols, the pattern of distribution of tocopherols and tocotrienols in the grain tissues has been shown to be similar to that in wheat (Grams *et al.*, 1970). The germ contains the major proportion of the tocopherols while the tocotrienols are found in the whole endosperm (Table 3.8). The inclusion of the aleurone with the starchy endosperm in the maize dissection accounts for the absence of tocotrienols from the pericarp fraction.

During the development of wheat grains, tocotrienols are not detectable

Table 3.8 Concentrations of tocols in dissected maize grain fractions†

	Concentration of tocols (mg kg^{-1} dry weight)					
	α-T	γ-T	δ-T	α-T-3	γ-T-3	Total
Germ	129–194 (95.8–95.9)‡	332–388 (94.8–94.6)	7.3–10.4 (100)	–	–	470–589
Pericarp and tip cap	5.1–7.5 (3.7–3.8)	17.2–20.4 (4.9–5.0)	–	trace	trace	25.5–32.3
Endosperm	0.5–0.7 (0.4)	1.0–1.9 (0.2–0.4)	–	4.5–8.7 (~100)	3.8–18.9 (~100)	11.2–30.2

†Grams *et al.*, 1970.
‡Figures in parentheses represent percentage distribution among the fractions of the whole grain.

in the dissected germ at any stage but are found in whole grain extracts from 14 days after flowering (Barnes, 1983); α-T and β-T are detectable in the germ from 35 days after flowering. On germination, only α-T is synthesized in the embryo axis (Hall and Laidman, 1968b) and it appears that the endosperm may be the only tissue during the life cycle of the wheat plant in which tocotrienols represent a major proportion of the tocols (Green, 1958).

IV CAROTENOIDS

Carotenoids are polyisoprenoid compounds with structures based on the acyclic hydrocarbon lycopene (Fig. 3.5; A-1-1′). Carotenoid hydrocarbons are known as carotenes and are the biosynthetic precursors of the oxygenated derivatives called "xanthophylls".

A Structure and Properties of Carotenoids

Carotenoids may be acyclic or have one or two cyclid end groups (Fig. 3.5). Oxygenated derivatives include alcohols, epoxy and alkoxy compounds, aldehydes, ketones and acids (Spurgeon and Porter, 1980). Although the xanthophylls of chloroplasts are present as the free alcohols, those in fruits and seeds are mainly esterified to long-chain fatty acids and esterification also occurs in senescent plant tissues.

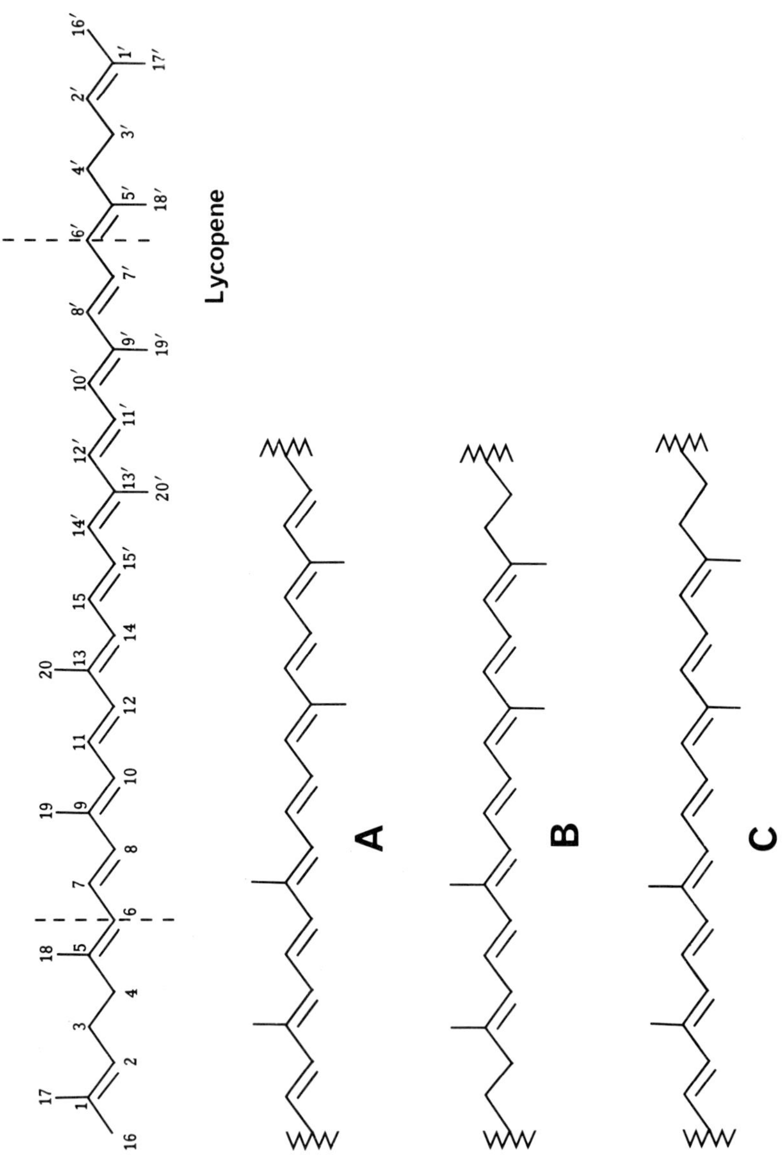

Lycopene

A

B

C

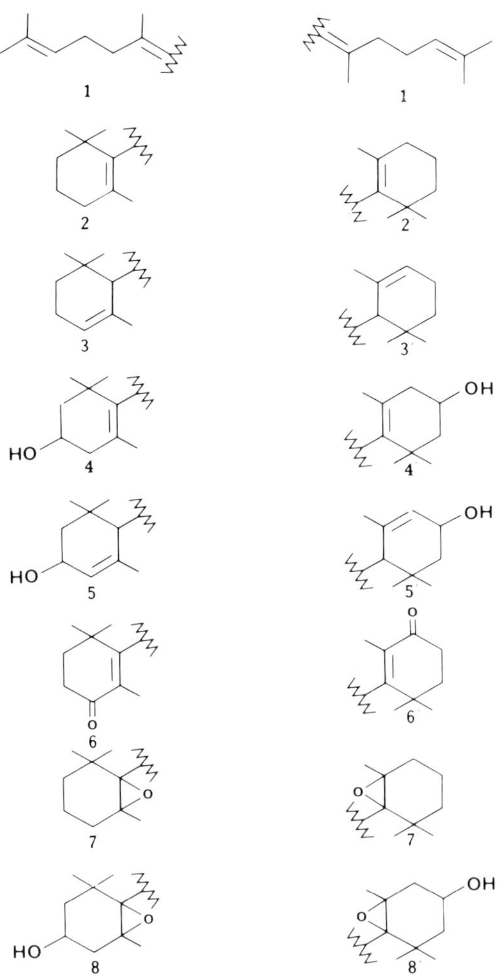

Fig. 3.5 The carotenoids found in cereal grains are represented by combinations of the polyene chains (denoted A, B and C) with the end groups shown. A-1-1′ – lycopene. B-1-1′ – ζ-carotene. A-2-1′ – γ-carotene. C-2-1′ – β-zeacarotene. A-2-2′ – β-carotene. A-2-3′ – α-carotene. A-4-2′ – cryptoxanthin. A-4-4′ – zeaxanthin. A-4-5′ – lutein. A-6-6′ – canthaxanthin. A-7-7′ – β-carotene-5,6,5′,6′-diepoxide. A-8-3′ – lutein-5,6-epoxide. A-8-4′ – antheraxanthin. A-8-8′ – violaxanthin.

The characteristic properties of the carotenoids reflect the presence of the conjugated polyene chain which accounts for the colour and the sensitivity to light, heat, oxygen and acid. Colour due to carotenoids is an important factor in the use of cereal grains in food manufacture, especially in durum wheat used to make pasta. (The significance of carotenoids in pasta manufacture is discussed in Chapter 13). Bleaching agents are used in some countries to remove the yellow carotenoid colour in wheat flour to increase the whiteness of the flour and bread. Even when chemical agents are not used, carotenoids undergo co-oxidation linked to the lipoxygenase-catalysed oxidation of polyunsaturated fatty acids during dough-mixing (Chapters 6 and 10).

Certain carotenoids are precursors of vitamin A (retinol). Although they have no intrinsic activity, α-, β- and γ-carotene are converted into vitamin A in the intestinal mucosa and liver.

B Occurrence of Carotenoids in Cereal Grains

Much of the research on cereal grain carotenoids has been done on wheat as a result of the involvement of carotenoids in durum wheat and pasta quality. Chapter 13 discusses this work. This chapter discusses only the general aspects of carotenoids in cereals.

1 *Whole Grain*

Lutein (A-4-5′) is the major carotenoid of wheat grain and has been detected in most studies of the carotenoids of grains of other cereals (Morrison, 1978). β-Carotene (A-2-2′) is also commonly found but is not a major component except in oil extracted from barley grain where it is present in concentrations from 52 to 85 mg kg^{-1} of oil together with xanthophylls (46 to 65 mg kg^{-1}) and minor quantities of α-carotene (A-2-3′) and a β-carotene isomer (Demchenko, 1969). Xanthophylls are the major carotenoids in other cereal grains besides wheat, as demonstrated by the presence of zeaxanthin (A-4-4′) and lutein in maize (Quackenbush *et al*., 1963; Grogan and Blessin, 1968; Cabulea, 1971; Washuttl, 1974; Goodwin, 1973, 1976) and in yellow sorghum (Wall and Blessin, 1969).

There are a number of other reports of the carotenoids detected in cereal grains, usually with no indication of the relative abundance of the individual compounds. A wide range of carotenoids is claimed to be present in wheat grain including lutein, β-carotene, β-apocarotenal, lutein-5,6-epoxide (A-8-3′) cryptoxanthin (A-4-2′), zeaxanthin, antheraxanthin (A-8-4′), canthaxanthin (A-6-6′), taraxanthin, triticoxanthin, and flavoxanthin. Barley grain xanthophylls consist of lutein, cryptoxanthin, zeaxanthin and violaxanthin (A-8-8′) (Nakayama, 1962). Little information is available for rye, oats, rice and millets but Russian studies indicate the presence of β-carotene, lutein, lutein-5,6-epoxide, taraxanthin and poly-*cis*-lycopene B in rye grain (Klyushkina *et al*., 1969, 1970; Nechaev and

Table 3.9 Composition of carotenoids in maize grain

Carotenoid	Composition of carotenoids (mg kg^{-1})		
	Cabulea (1971)	Quackenbush *et al.* (1963)	Grogan and Blessin (1968)
Zeaxanthin	6.1–15.8	0.8–27.4	0.1–33.1
Lutein	7.0–14.9	2.0–33.1	0.1–16.1
Zeaxanthin and lutein esters		0.3– 4.1	
Cryptoxanthin	0.9– 2.3	0.3– 5.1	0 – 6.9
Fisoxanthin	0.5– 2.0		
Zeinoxanthin		0 – 7.8	0.1– 5.1
Polyoxy compounds		0.4– 3.0	0.1– 7.5
α-carotene	0.1– 0.6		} 1.5– 4.8
β-carotene	0.4– 1.7	0.1– 5.4	
β-zeacarotene	< 0.6	0.1– 4.7	

Sandler, 1975); lutein and taraxanthin in rice (Sandler *et al.*, 1969); lutein-5,6-epoxide and taraxanthin in oat and β-carotene. β-carotene-5,6,5′,6′-diepoxide (A-7-7′), lutein, antheraxanthin and zeaxanthin in a millet (Nechaev and Sandler 1975).

Table 3.9 lists data for the quantitative composition of carotenoids in maize grain and illustrates the minor carotenoids found in addition to lutein and zeaxanthin. In contrast to these results, Goodwin (1973, 1976) found the major carotenoids in maize grain to be β-carotene, cryptoxanthin and zeaxanthin with minor quantities of α-carotene, ζ-carotene (B-1-1′), γ-carotene (A-2-1′) and β-zeacarotene (C-2-1′).

The maize samples examined in the studies summarized in Table 3.9 show a total carotenoid range of from < 1.0 to 58 mg kg^{-1}. Most other cereals have total carotenoid contents at the lower end of this range, with wheat containing from 1.8 to 5.8 mg kg^{-1} (Fortmann and Joiner, 1978); the range of 43 to 99 mg kg^{-1} of lipid extract found for wheat by Nechaev and Sandler (1975) suggests lower values. Nakayama (1962) found 82 μg per 1000 grains of barley, which would indicate approximately 1 mg kg^{-1} in grain and is in accord with values of 100 to 150 mg kg^{-1} in barley oil (Demchenko, 1969). Data for the carotenoid content of the lipid fraction from rye, oat, rice and a millet suggest that the carotenoid content of the grain would be less than 100 mg kg^{-1} (Klyushkina, 1969, 1970; Nechaev and Sandler, 1975; Sandler *et al.*, 1969). Common sorghums contain about 1.5 mg kg^{-1} but yellow sorghums have higher carotenoid contents in the region of 25 to 30 mg kg^{-1} grain.

2 *Grain Fractions*

In dissected maize fractions 74–86% of the carotenoids are found in the vitreous endosperm, 9–23% in the floury endosperm, 2–4% in the germ and 1% in the bran (Blessin *et al*., 1963). The total carotenoid concentration in wheat is greater in the embryo (4.1 to 11.0 mg kg^{-1}) than in the endosperm (1.6 to 2.2 mg kg^{-1}) or bran (0.9 to 2.2 mg kg^{-1}) (Chen and Geddess, 1945). In bread wheat, from one-third to two-thirds of the xanthophyll in the endosperm and bran is esterified to fatty acids while the xanthophyll in the embryo is mainly unesterified. In durum wheat, the endosperm xanthophylls also are predominantly unesterified.

V HYDROCARBONS

Hydrocarbons are minor components of whole cereal grains and flours and there is no evidence for involvement of endogenous long-chain hydrocarbons in the major quality characteristics of cereals. Added hydrocarbons have been shown to influence the loaf volume of bread baked from wheat flour, but the concentrations were much higher than those of the flour hydrocarbons (Elton and Fisher, 1968).

Youngs and Gilles (1970) found the average total hydrocarbon content of wheat flour and durum semolina to be 360 mg kg^{-1} dry weight. Of the total, 65% was composed of *n*-hydrocarbons and 3% was squalene; the remainder included *iso*-, *anteiso*- and 1-cyclohexyl derivatives. The major components were C_9 to C_{12}, C_{17} and C_{25}. In contrast, Lorenz and Maga (1972) found the main components of the *n*-hydrocarbons of wheat and triticale flours to be C_{16} to C_{20} but they did not analyse chain lengths greater than C_{22}.

The hydrocarbon fraction from wheat bran from commercial and laboratory milling consists of predominantly *n*-alkanes together with *n*-alkanes and *iso*-, *anteiso*-, 1-cyclohexyl and phenolic derivatives (Youngs and Gilles, 1970; Martelli and Frattini, 1979). Bran has a hydrocarbon distribution which is shifted towards greater chain length; in comparison with the corresponding flour, the major components are C_{25}, C_{27} and C_{29} (Youngs and Gilles, 1970). The predominance of the longer chain length, odd carbon number *n*-alkanes in the hydrocarbon fraction from the outer-layers of cereal grains is confirmed by studies of the composition of barley testa, rice bran and sorghum grain wax (Briggs, 1974; Ito *et al*., 1981; Dalton and Mitchell, 1959). The principal carbon numbers of *n*-alkanes are C_{29} and C_{31}. The hydrocarbon composition of wheat germ oil is discussed in Chapter 19.

It is difficult to determine to what extent the hydrocarbons found in extracts from cereal grains, and especially from cereal products, are derived from the biosynthetic activities of the plant unless certain precautions are taken. The hydrocarbon content is very low and significant contamination could arise during processes such as grain-drying and milling. It has been shown that polycyclic aromatic hydrocarbons are present at low concentrations in harvested grain of wheat and maize and that the concentration increases during grain-drying as a result of absorption from the combustion gases (Hutt *et al.*, 1978).

REFERENCES

Amelotti, Gr., Galbusera, P. and Montorfano, P. (1980). *Riv. Ital. Sost. Grasse* **57**, 577–581.

Barnes, P. J. (1983). In *Proceedings of the 7th World Cereal and Bread Congress*, Prague, 1982. (J. Holas, ed.), in press. Elsevier, Amsterdam.

Barnes, P. J. and Taylor, P. W. (1981). *Phytochemistry* **20**, 1753–1754.

Beringer, H. and Dompert, W. U. (1976). *Fette Seifen Anstr.* **78**, 228–231.

Berrie, A. M. M. and Knights, B. A. (1972). *Phytochemistry*, **11**, 2363–2365.

Berry, C. P., Youngs, V. L. and Gilles, K. A. (1968). *Cereal Chem.* **45**, 616–626.

Binnington, D. S. and Andrews, J. S. (1941). *Cereal Chem.* **18**, 678–686.

Blessin, C. W. Brecher, J. D. and Dimler, J. R. (1963). *Cereal Chem.* **40**, 582–586.

Bourgeois, C. F. (1981). *Rev. Franç. Corps. Gras.* **28**, 353–356.

Bowden, B. N. and Palmer, M. A. (1975). *Phytochemistry* **14**, 1140–1141.

Briggs, D. E. (1974). *Phytochemistry* **13**, 987–996.

Brubacher, G. and Wiss, O. (1972). In *The Vitamins: Chemistry, Physiology, Pathology, Methods* (W. H. Sebrell and R. S. Harris, eds), Vol. 5, 2nd ed, pp. 255–258. Academic Press, London and New York.

Cabulea, I. (1971) *Eucarpia* **5**, 85–91.

Chen, K. T. and Geddes, W. F. (1945). "Studies on the wheat pigments". M.Sc. thesis, University of Minnesota, St. Paul, Minnesota.

Chow, C. K., Draper, H. H. and Csallany, A. S. (1969). *Analyt. Biochem.* **32**, 81–90.

Dalton, J. L. and Mitchell, H. L. (1959). *J. Agr. Food Chem.* **7**, 570–573.

Davis, K. R., Litteneker, N., LeTourneau, D., Cain, R. F., Peters, L. J. and McGinnis, J. (1980). *Cereal Chem.* **57**, 178–184.

Demchenko, A. I. (1969). *Izv. Vyssh. Ucheb. Zaved. Pishch. Tekhnol.* **5**, 18–20.

Dominguez, X. A. and Rodriguez-Bores, F. (1972). *Phytochemistry* **11**, 2655.

Drummond, J. C., Singer, E. and MacWalter, R. J. (1935). *Biochem. J.* **29**, 456–471.

Elton, G. A. H. and Fisher, N. (1968). *J. Sci. Food Agric.* **19**, 178–181.

Fortman, K. L. and Joiner, R. R. (1971). In *Wheat: Chemistry and Technology* (Y. Pomeranz, ed.), 3rd edn, pp. 493–522. American Association of Cereal Chemists, St. Paul, Minnesota.

Garcia-Olmedo, F. (1965). *Bol. Inst. Nac. Invest. Agron.* **25**, 409–416.

Goodwin, T. W. (1973). In *Phytochemistry* (L. P. Miller, ed.), Vol. 1, pp. 112–142. Van Nostrand-Reinhold, Princeton, N.J.

Goodwin, T. W. (1976). In *Chemistry and Biochemistry of Plant Pigments* (T. W. Goodwin, ed.), 2nd ed., Vol. 1, pp. 225–261. Academic Press, London and New York.
Goodwin, T. W. (1980). In *The Biochemistry of Plants* (P. K. Stumpf, ed.), Vol. 4, pp. 485–507. Academic Press, London and New York.
Govind Rao, M. K. and Perkins, E. G. (1972). *J. Agr. Food Chem.* **20**, 240–245.
Grams, G. W., Blessin, C. W. and Inglett, G. E. (1970). *J. Am. Oil Chem. Soc.* **47**, 337–339.
Green, J. (1958). *J. Sci. Food Agric.* **9**, 801–812.
Grogan, C. O. and Blessin, C. W. (1968). *Crop Sci.* **8**, 730–732.
Hall, G. S. and Laidman, D. L. (1968a). *Biochem. J.* **108**, 465–473.
Hall, G. S. and Laidman, D. L. (1968b). *Biochem. J.* **108**, 475–482.
Hutt, W., Meiering, A., Oelschlager and Winkler, E. (1978). *Can. Agric. Eng.* **20**, 103–107.
Ito, S., Suzuki, T. and Fujino, Y. (1981). *Nippon Nogei Kagaku Kaishi*, **55**, 247–253.
Itoh, T., Tamura, T. and Matsumoto, T. (1973). *J. Am. Oil Chem. Soc.* **50**, 122–125.
Kato, A., Tanabe, K. and Yamaoka, M. (1981). *Yukagaku* **30**, 515–516.
Kemp, R. J., Goad, L. J. and Mercer, E. I. (1967). *Phytochemistry* **6**, 1609–1615.
Klyushkina, Yu. F., Denisenki, Ya. I. and Nechaev, A. P. (1969). *Maslo-Zhir. Prom.* **35**, 8–9.
Klyushkina, Yu. F., Denisenki, Ya. I., Nechaev, A. P. and Yanotovskii, M. T. (1970). *Prikl. Biokhim. Mikrobiol.* **6**, 95–98.
Knights, B. A. (1967). In *The Gas Liquid Chromatography of Steroids* (J. K. Grant, ed.), pp. 211–221. Cambridge University Press, Cambridge.
Knights, B. A. and Laurie, W. (1967). *Phytochemistry* **6**, 407–416.
Kuroda, N., Ohnishi, M. and Fujino, Y. (1977). *Cereal Chem.* **54**, 997–1006.
Lásztity, R., Berndorfer-Kraszner, É. and Huszár, M. (1980). In *Cereals for Food and Beverages* (G. E. Inglett, and L. Munck, eds), pp. 429–445. Academic Press, London and New York.
Lin, M. J. Y., Youngs, V. L. and d'Appolonia, B. L. (1974). *Cereal Chem.* **51**, 17–33.
Lorenz, K. and Maga, J. (1972). *J. Agr. Food Chem.* **20**, 769–772.
MacMurray, T. A. and Morrison, W. R. (1970). *J. Sci. Food Agric.* **21**, 520–528.
Mahadevappa, V. G. and Raina, P. L. (1978). *J. Am. Oil Chem. Soc.* **55**, 647–648.
Martelli, A. and Frattini, C. (1979). *Relata Tech.* **11**, 23–29.
Mason, E. L. and Jones, W. L. (1958). *J. Sci. Food Agric.* **9**, 524–527.
McCay, P. B. and King, M. M. (1980). *Basic Clin. Nutr.* **1**, 239–317.
McKillican, M. E. (1964). *J. Am. Oil Chem. Soc.* **41**, 554–557.
Morrison, W. R. (1978). *Adv. Cereal Sci. Technol.* **2**, 221–348.
Morrison, W. R., Coventry, A. M. and Barnes, P. J. (1982). *J. Sci. Food Agric.* **33**, 925–933.
Myhre, D. V. (1968). *Can. J. Chem.* **46**, 3071–3077.
Nakayama, T. O. M. (1962). *Proc. Ann. Mtg. Am. Soc. Brew. Chem.*
Nechaev, A. P. and Sandler, Zh. Ya. (1975). *Lipidy Zerna*. Kolos, Moscow.
Palmer, M. A. and Bowden, B. N. (1975a). *Phytochemistry* **14**, 1813–1815.
Palmer, M. A. and Bowden, B. N. (1975b). *Phytochemistry* **14**, 2049–2053.
Quackenbush, F. W., Firch, J. G., Brunson, A. M. and House, L. R. (1963). *Cereal Chem.* **40**, 250–259.

Pennock, J. F., Hemming, F. W. and Kerr, J. D. (1964). *Biochem. Biophys. Res. Commun.* **17**, 542–548.

Russell Eggitt, P. W. and Ward, L. D. (1955). *J. Sci. Food Agric.* **6**, 329–336.

Sandler, Zh. Ya., Denisenko, Ya. I., Nechaev, A. P. and Yanotovskii, M. T. (1969). *Isz. Vyssh. Ucheb. Zaved. Pishch. Tekhnol.* **3**, 23–26.

Seit-Ablaeva, S. K., Nechaev, A. P., Denisenko, Ya. I. and Yanotovskii, M. P. (1973). *Prikl. Biokhim. Mikrobiol* **9**, 737–739.

Seitz, L. M., Mohr, H. E., Burroughs, R. and Sauer, D. G. (1977). *Cereal Chem.* **54**, 1207–1217.

Slover, H. T. (1971). *Lipids* **6**, 291–296.

Slover, H. T., Lehmann, J. and Valis, R. J. (1969a). *J. Am. Oil Chem. Soc.* **48**, 417–420.

Slover, H. T., Lehmann, J. and Valis, R. J. (1969b). *Cereal Chem.* **46**, 635–641.

Spurgeon, S. L. and Porter, J. W. (1980). In *The Biochemistry of Plants* (P. K. Stumpf, ed.), Vol. 4, pp. 419–483. Academic Press, New York and London.

Stoyanova, V. G., Geiko, N. S., Segal, G. M. and Nechaev, A. P. (1975). *Khim. Priv. Soedin.* **11**, 357–359.

Suckewer, A. and Rakowska, M. (1974). *Roczn. Pzh.* **25**, 159–169.

Tanabe, K., Yamaoka, M., Tanaka, A., Kato, A. and Junko, A. (1982). *Yukagaku* **31**, 205–208.

Telegdy-Kovats, L., Berndorfer-Kraszner, É. and Hunyadvari, E. (1970) *Elelmezesi Ipar* **24**, 302–305.

Thompson, J. N., Erdody, P. and Maxwell, W. B. (1972). *Analyt. Biochem.* **50**, 267–280.

Thompson, J. N. and Hatina, G. (1979). *Liquid Chromatog.* **2**, 327–344.

Toepfer, E. W., Polansky, M. M., Enheart, J. F., Slover, H. T., Morris, E. R., Hepburn, F. N. and Quackenbush, F. W. (1972). *Cereal Chem.* **49**, 173–186.

Torres, J. V., Carbonero, P. and Garcia-Olmedo, F. (1976). *Phytochemistry* **15**, 677–680.

Wall, J. S. and Blessin, C. W. (1969). *Cereal Sci. Today* **14**, 264–266, 268–270, 276.

Washuttl, J. (1974). *Qual. Plant.* **23**, 369–372.

Youngs, V. L. and Gilles, K. A. (1970). *Cereal Chem.* **47**, 317–323.

4 Lipid Metabolism in Germinating Cereals

N. A. CLARKE, M. C. WILKINSON and D. L. LAIDMAN
University College of North Wales, Bangor, Gwynedd, U.K.

I INTRODUCTION

Lipids fulfil two important roles in germinating cereals. The first role is played by the triacylglycerols (triglycerides), and to a lesser extent by the diacylglycerols (diglycerides), which are used as energy resources. The principal energy reserve of cereals is starch, so the importance of the triacylglycerols as a source of energy is often not fully appreciated. The starch reserve is, however, restricted to the starchy endosperm. Although this starch is hydrolyzed during germination and the products sustain the growing seedling, the principal energy stores within the embryo tissues are triacylglycerols. It is generally believed that these triacylglycerol reserves

Abbreviations: NAD+ oxidized nicotinamide adenine dinucleotide. NADH reduced nicotinamide adenine dinucleotide. Fp oxidized flavoprotein. FpH_2 reduced flavoprotein. ATP adenosine triphosphate.

"Lipids in Cereal Technology"
ISBN 0-12-079020-3

are mobilized and that they are an important means of sustaining the embryo during early germination before the arrival of sugars from the starchy endosperm. Similarly, the principal energy reserve of the aleurone tissue is triacylglycerol. Despite the fact that it surrounds the starchy endosperm, and therefore has copious amounts of sugars made available to it, the known evidence indicates that the aleurone tissue utilizes triacylglycerols as an important energy source during germination.

The second important role of lipids is played by the polar lipids which are structural components of cellular membranes. In common with other seeds, the cells of quiescent cereal grains contain poorly developed membrane structures especially in mitochondria and endoplasmic reticulum. During germination these structures undergo extensive developments which are important aspects of the germination process. Several developmental processes are controlled by hormones and many of them involve active metabolism of the membrane lipids.

Many gaps still exist in our understanding of lipid biochemistry during germination in cereal grains. Some of these can be filled, at least in a speculative manner, by referring to studies of other species, particularly the oleaginous seeds. These studies will be referred to from time to time to construct a comprehensive account of the control and integration of lipid metabolism in the germinating cereal grain. Our account will be concerned mainly with the aleurone tissue since more quantitative information is available for this tissue than for the others. The germ tissues will also be discussed where information is available. Together the aleurone and germ tissues contain the greatest concentrations of stored lipid and they undergo the greatest ultrastructural changes during germination. The lipids of the starchy endosperm are considered less important in the germination process since they are minor energy reserves in that tissue. Moreover, the tissue itself is dead and does not contribute to the cellular biochemistry of germination – it is essentially an inanimate nutrient store for the developing seedling.

II THE ULTRASTRUCTURE OF CEREAL GRAINS

To properly explain the process of lipid metabolism in the germinating grain, it is necessary to describe first the various sub-cellular structures involved. As well as organelles such as mitochondria, which are found ubiquitously in plant cells, the cells of germinating grains contain other structures which are typical of storage tissues and of seed storage tissues in particular. These special structures include the spherosomes and glyoxy-

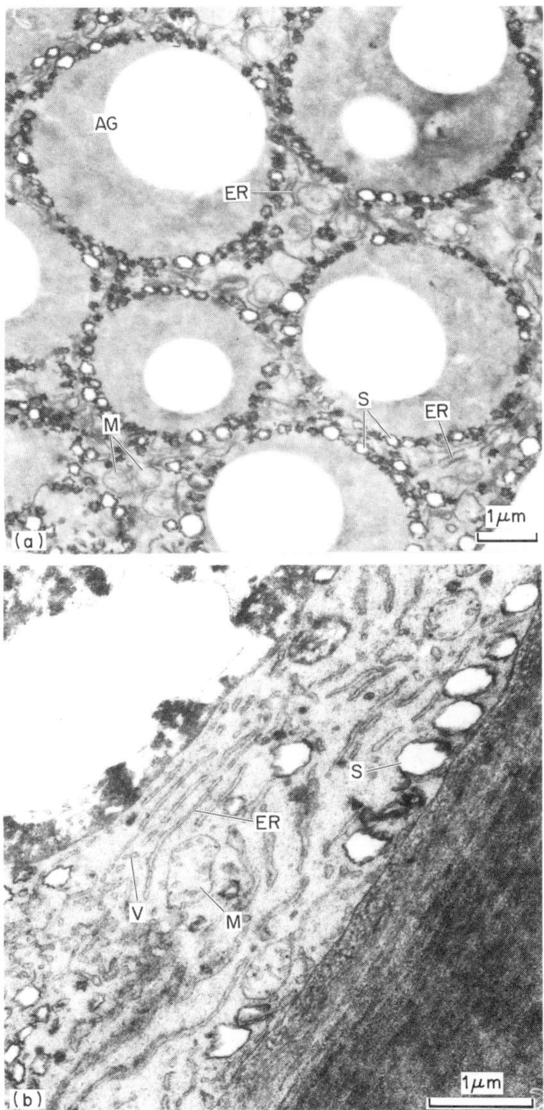

Fig. 4.1 Electron micrographs of an aleurone cell from an ungerminated wheat grain (a) and from a grain germinated for 4 days (b). In the ungerminated grain, aleurone grains (AG), with spherosomes (S) attached, fill most of the cytoplasm. Endoplasmic reticulum (ER) is sparse and present as short profiles. The mitochondria (M) are poorly developed. In the germinated grain, long cisternae of endoplasmic reticulum (ER) are present and numerous vesicles (V), which have formed from the endoplasmic reticulum, are visible. The mitochondria (M) have well developed cristae (from Colborne *et al.*, 1976).

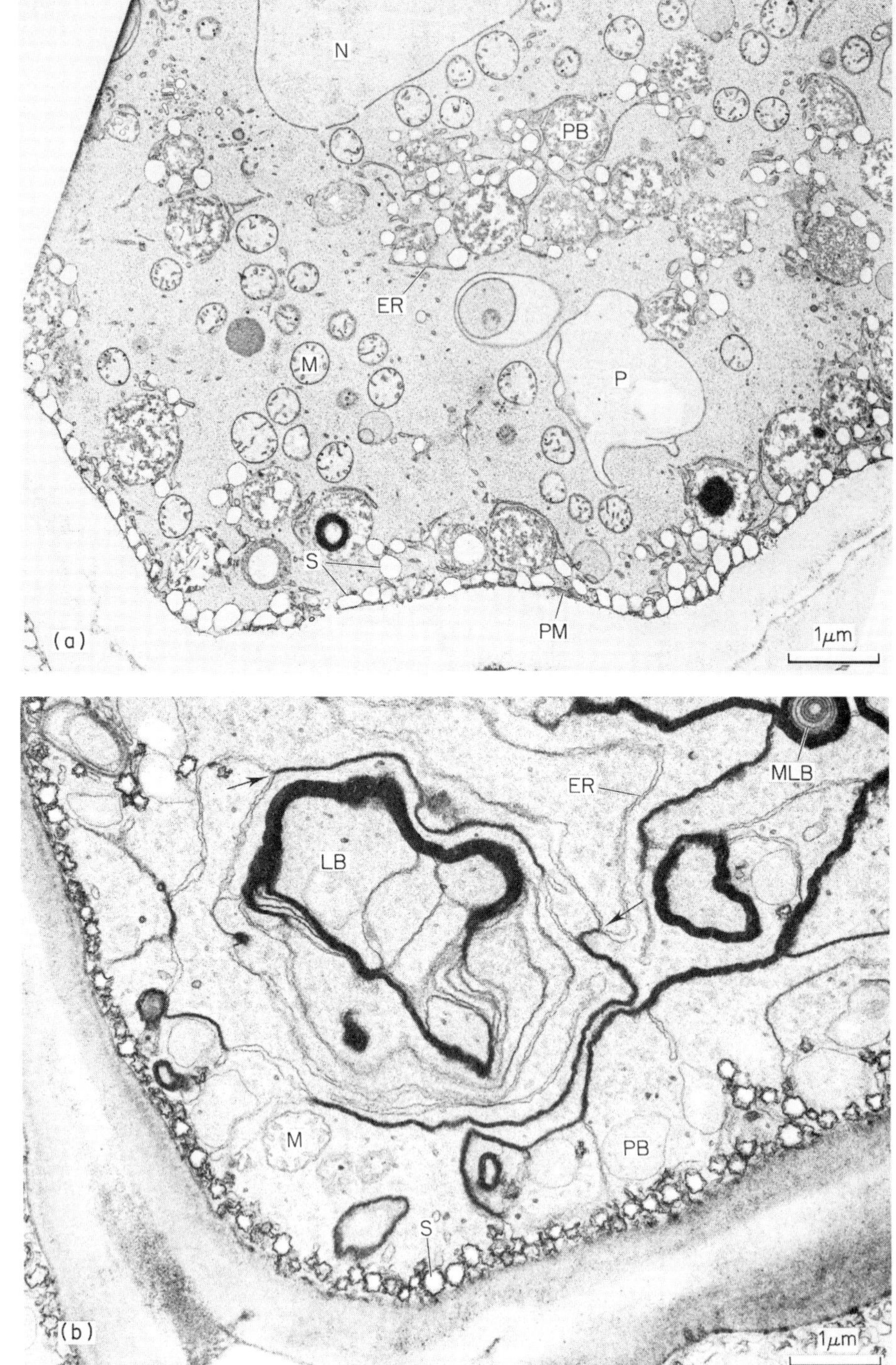
N
PB
ER
M
P
S
PM
(a)
1μm
ER
MLB
LB
M
PB
S
(b)
1μm

somes. Together with the aleurone grains that give seed cells a characteristic appearance in the electron microscope. As an example of such cells the electron micrographs in Fig. 4.1 show the aleurone cell of wheat. Similar electron micrographs have been published for the aleurone cells of barley (Buttrose, 1963; Paleg and Hyde, 1964; Van der Eb and Nieuwdorp, 1967; Jones, 1969a) and maize (Kyle and Styles, 1977) and for the scutellum of barley (Nieuwdorp, 1963; Swift and O'Brien, 1972). The cells of the embryo axis tissues have also been studied (Berjack and Villiers, 1970; Deltour and Bronchart, 1971; Hallam *et al.*, 1972; Mollenhauer *et al.*, 1978). These studies show the embryo axis cells to contain similar structures to those in aleurone and scutellum cells, but their organization within the cells is different (Fig. 4.2). (Please refer to Figs. 4.1 and 4.2 while reading the following accounts.)

A The Ultrastructure of the Quiescent Grain

1 *Aleurone Tissue*

The quiescent aleurone cell contains a large nucleus which is up to 25 μm in diameter in wheat and barley. The cytoplasm around the nucleus fills the rest of the volume of the cell and there is no central vacuole (cf. vegetative plant cells which have a large central vacuole). Up to 60% of the volume of the cytoplasm is filled by aleurone grains (Buckhout *et al.*, 1981) of about 3–4 μm diameter. Aleurone grains are surrounded by a single membrane and they contain a matrix of protein. Within the protein matrix are two types of electron-dense inclusions, one consisting of phytin and the other probably of glycoprotein. It has been reported that the phytin inclusion may also contain lipid (Jacobsen *et al.*, 1971), but this has not been confirmed and it may be an artefact produced during the sectioning of the

---

Fig. 4.2 Electron micrographs of an embryo cell from a maize grain imbibed for 2 h (a) and from a grain germinated for 18 h (b). In the 2 h imbibed grain, protein bodies (PB) can be seen dispersed through the cytoplasm, a few having dense inclusions. Spherosomes (S) line the plasma membrane (PM) and some are free in the cytoplasm, though rarely associated with protein bodies (cf. Fig. 1a). Endoplasmic reticulum (ER) is present as short profiles. Many mitochondria (M) are visible. The nucleus (N) and a plastid (P) are also shown. In the germinated grain, long cisternae of endoplasmic reticulum (ER) are present. Protein bodies (PB) are still present. The protein bodies are believed to form myelin-like bodies (MLB) from their dense inclusions. They in turn develop into lamellar bodies from which dense membrane structures "peel off". These structures split at specific points to form two endoplasmic reticulum profiles (open arrows). Spherosomes (S) and mitochondria (M) can be seen. (Fig. 2a from Mollenhauer *et al.* (1978), Fig. 2b courtesy of H. H. Mollenhauer, unpublished data).

tissue for microscopy. Recent studies of the protein bodies in developing barley, maize and wheat endosperms have led to the hypothesis that they are derived from the endoplasmic reticulum (Miflin *et al.*, 1981). Since aleurone grains in the aleurone tissue are morphologically equivalent to the protein bodies in the starchy endosperm, it is very likely that they are also formed from endoplasmic reticulum during grain development.

The surface of the aleurone grains is covered by spherosomes which are tightly pressed to the grain surface. Other spherosomes similarly line the inner surface of the plasma membrane. Spherosomes in different plant species have been variously termed "oil bodies", "oleosomes" and "premembrane bodies" (Bewley and Black, 1978). Their mixed nomenclature reflects the uncertainty concerning their structure and cellular origin. They consist of a triacylglycerol droplet of about 0.3 μm in diameter and investigators have been uncertain whether they are surrounded by a phospholipid bilayer membrane or by a monolayer half-membrane, or indeed whether they have a limiting membrane at all (Frey-Wyssling *et al.*, 1963; Schwarzenbach, 1971; Sorokin, 1967). Yatsu and Jacks (1972) have proposed that lipid bodies are bound by a half unit membrane and that this monolayer of phospholipid contains occluded proteins. This was also found to be true in wheat aleurone tissue (Jelsema *et al.*, 1977). The proteins did not, however, include lipid metabolizing enzymes such as hydrolases. Recently it has been suggested that during spherosome development, triacylglycerols accumulate at specific sites within the lipophilic middle layer of one of the unit membranes of the endoplasmic reticulum (Wanner and Theimer, 1978; Wanner *et al.*, 1981). The resulting expansion of the membrane would lead to the formation of spherical structures which could either break away as separate spherosomes or remain attached to the endoplasmic reticulum. The latter possibility may be the situation in the cereal aleurone cell where the spherosome half-membrane appears to be continuous with the outer half of the aleurone grain membrane (Buttrose, 1971). This situation suggests that the spherosomes and the aleurone grain might have a similar origin in the endoplasmic reticulum of the cell. The aleurone grain-spherosome complexes fill most of the cytoplasm of the aleurone cell (Buckhout *et al.*, 1981).

The small amount of space which remains between the aleurone grain-spherosome structures in the aleurone cell contains the remaining organelles and cytoplasmic membranes. Mitochondria and plastids with poorly developed internal membrane structures are present (Buttrose, 1971; Doig *et al.*, 1975b). The plastids are non-pigmented and they have been termed "leucoplasts". In several species of angiosperms a number of workers have shown plastids to be the principal site for the synthesis of long-chain fatty acids. Plastids have also been shown to produce some mono-, di- and triacylglycerols (see Wanner *et al.*, 1981). Dilations of the lipid layer and intimate associations of the plastid membrane with spherosomes indicate direct transport of lipids from the plastid to the spherosome (Wanner *et al.*, 1981). Unfortunately none of this work has been done specifically on cereals, but we can expect to find a similar relationship between plastids and spherosomes during the ontogeny of spherosomes in developing cereal grains. Thus, alternative mechanisms involving either the endoplasmic reticulum alone, or the endoplasmic reticulum together with plastids, might exist for the formation of spherosomes.

The endoplasmic reticulum is also poorly developed in the quiescent aleurone cell. It is sparse and present in the form of short profiles or crescents (Jones, 1969a; Paleg and Hyde, 1964; Colborne *et al.*, 1976). The failure of Van der Eb and Nieuwdorp (1967) to find endoplasmic reticulum in the barley aleurone cell can probably be discounted since it is so scarce. Endoplasmic reticulum is believed to be the major site of phospholipid biosynthesis (Getz *et al.*, 1968) and also of triacylglycerols (see Wanner *et al.*, 1981). Golgi bodies or dictyosomes are also sparse. They were not found by Van der Eb and Nieuwdorp, 1967 or Colborne *et al.*, 1976, but they have been observed in the quiescent aleurone cells of both maize (Kyle and Styles, 1977) and barley (Jones, 1969a). Glyoxysomes which have a single limiting bilayer membrane have been observed in the aleurone tissues of wheat (Doig *et al.*, 1975a) and barley (Buttrose, 1971). They are few in number and small in size. In the germinating grain, glyoxysomes contain the enzymes of fatty acid β-oxidation and the glyoxylate cycle (Doig *et al.*, 1975a). They are, therefore, important sites of lipid metabolism during germination.

2 *Starchy Endosperm*

It has already been shown that the starchy endosperm is significantly different from the other tissues of the cereal grain because it is a non-living food source for the cereal grain during germination. It does, however, appear to possess spherosomes, but at a lower concentration than in the aleurone or scutellar tissues (Hargin *et al.*, 1980).

3 *Scutellum*

The epithelial cells of barley scutella have been studied (Nieuwdorp, 1963; Swift and O'Brien, 1972), and they are very similar in appearance to those of aleurone cells. Phytin globoids are present within the protein bodies.

4 *Embryo Axis*

Studies on the embryo axis cells show significant differences compared with aleurone cells. The presence of protein bodies is in doubt. Mollenhauer *et al*. (1978) have identified protein bodies in maize embryo cells. Hallam *et al.* (1972) and Berjack and Villiers (1970) did not observe protein bodies. It is probable that the organelles referred to as electron-lucent bodies and lysosome-like bodies respectively in these papers are, in fact, protein bodies. Some cells contain phytin globoids within their protein bodies. Spherosomes are predominantly seen around the plasma membrane (Hallam *et al.*, 1972). They are occasionally seen free in the cytoplasm, but they are rarely seen in intimate association with the protein bodies (Mollenhauer *et al.*, 1978) (see Fig. 4.2).

The situation regarding protein bodies, phytin inclusions and spherosomes in the embryo axis and scutellum has recently been classified (Bechtel and Pomeranz, 1978; Okamoto *et al.*, 1982). In particular Bechtel and Pomeranz have classified the cells of the quiescent rice embryo into three types:

(1) Cells containing protein bodies with phytin inclusions and spherosomes dispersed through the cytosol.

(2) Cells containing protein bodies which lack phytin inclusions and spherosomes concentrated around the cell periphery.

(3) Cells lacking protein bodies, but containing spherosomes around the cell periphery.

B. Ultrastructural Changes during Germination

1 *Aleurone Tissue*

In reviewing the literature on this subject it is important to appreciate that different procedures have been used by different investigators. In particular, different germination temperatures have been employed so that ultrastructural changes have occurred at different rates. To rationalize this situation, Jones (1981b) has proposed that a biochemical event such as the appearance and time-course of α-amylase synthesis would be a more useful indicator of the progress of germination. Where possible we will cross-refer to such biochemical events in the following account of ultrastructural changes and the time-scale will be referred to only in general terms. According to the available information, the following account applies equally to wheat (Colborne *et al*., 1976), barley (Van der Eb and Nieuwdorp, 1967; Vigil and Ruddat, 1973) and oats (Sevinate-Pinto *et al*., 1981).

Events in the aleurone cell can be conveniently divided into three phases. During the first 1–2 days of germination at 20–25 °C, long cisternae of endoplasmic reticulum develop near to the nuclear membrane. The cisternae form into parallel stacks and become extensively covered with ribosomes. Mitochondria develop more extensive cristae and appear more numerous. Plastids also appear to be more abundant and are closely associated with the developing endoplasmic reticulum. Some reservations should be placed on these observations however; while there is no doubt about the structural developments of the mitochondria and plastids, the evidence for their increases in number is far from convincing. Some investigators have observed structural continuities between spherosomes and the developing endoplasmic reticulum (Sevinate-Pinto *et al.*, 1981).

After 1–2 days, a second phase of activity takes place which overlaps the first probably by several hours. The ends of the cisternae of the endoplasmic reticulum fill with electron-dense material, and the swollen ends bud off to form vesicles (see Fig. 4.1b). Although conclusive evidence is still lacking, information from several sources indicates that these vesicles migrate, possibly via Golgi bodies, to the plasma membrane. They disgorge their contents by exocytosis through the plasma membrane and electron-dense material can be seen to accumulate nearby in the cell wall. Golgi bodies become more numerous and vesicles are observed budding off from the peripheries of their cisternae. This situation persists for several days and it coincides with the time when α-amylase is synthesized and secreted from the aleurone tissue into the starchy endosperm. In addition to these events, glyoxysomes increase in numbers and the spherosomes decrease in both number and size as their triacylglycerol reserves are mobilized and metabolized. The phytin and protein reserves in the aleurone grains are also mobilized during this time. The aleurone grains themselves swell and eventually coalesce with one another to form large vacuoles. This process leads to the third and final phase where the cells become extensively vacuolated, their membranes degenerate and the tissue dies.

2 *Scutellum*

The situation in the scutellum is very similar to that in the aleurone tissue, at least in barley (Nieuwdorp and Buys, 1964). In the scutellum however, the formation of endoplasmic reticulum occurs earlier than in the aleurone cell, presumably because the former imbibes water and commences germination more quickly since it is at the exposed end of the grain. After imbibition, spherosomes decrease in size and in number and often appear to be connected to the endoplamsic reticulum. In contrast to the aleurone tissue, the endoplasmic reticulum of the scutellum does not proliferate vesicles, but numerous Golgi bodies do. Glyoxysomes were not observed by Nieuwdorp and Buys in their electron micrographs of the barley scutellum. This apparent oversight was undoubtedly because glyoxysomes had not been discovered at that time (see Beevers, 1969), and we can expect active glyoxysomes to be present in the scutella of all germinating cereals. After several days of germination the cells of the barley scutellum become extensively vacuolated and their membrane structures degenerate similar to the situation described for the aleurone tissue (like the aleurone tissue, the scutellum has a short functional life during germination; it then senesces and dies).

3 *Embryo Axis*

The embryo axis is a more difficult tissue to study because of the variety of cell types present and their different developmental ages. In general, endoplasmic reticulum develops during the first day of germination in maize (Berjack and Villiers, 1970: Mollenhauer *et al*., 1978) and rye (Hallam 1972). The developing cisternae are closely associated with protein bodies. Organelles, including mitochondria, undergo increased membrane development. After the first day of germination, Golgi bodies can be seen vesiculating but the cisternae of the endoplasmic reticulum do not vesiculate extensively. The cells become vacuolated. Spherosomes along the plasma membrane migrate through the cytoplasm and are progressively reduced until they are no longer visible. Some of these features can be seen in Fig. 4.2b.

4 *Continuities between Cellular Membrane Structures*

Many of the investigators of germinating seed tissues have remarked upon the close association between the developing cisternae of the endoplasmic reticulum and several cellular structures including protein bodies, spherosomes, leucoplasts and the nuclear membrane. It has been natural for these investigators to extend their observations to infer functional relationships between these structures and the formation of endoplasmic reticulum. Thus, Mollenhauer *et al*. (1978) proposed that endoplasmic reticulum in the germinating maize embryo is formed from lamellar bodies which are in turn derived from myelin bodies and protein bodies (see Fig. 4.2b). Sevinate-Pinto *et al.* (1981) have recently published impressive micrographs showing structural continuity between spherosomes and endoplasmic reticular-like cisternae in the aleurone cells of germinating oats. They propose that the spherosomes give rise to a system of neoformed membranes which in turn develop into endoplasmic reticulum. Laidman (1982) also speculated that the close association between leucoplasts and developing endoplasmic reticulum in wheat aleurone cells might indicate functional relationships between the two. In a more general review of the endomembrane system in plants, however, Morré (1975) concluded that the endoplasmic reticulum is mainly self-generating but it may be partly derived from the nuclear membrane. Clearly, these proposals must be considered with some reservations for the present and this aspect of germination is an important topic for further research.

C Control of the Ultrastructural Changes in the Aleurone Tissue

This topic has been critically reviewed elsewhere (Laidman, 1980; 1982). Therefore, only a summary will be presented here.

In terms of the three phases of ultrastructural development described previously, it is now generally agreed that the formation of endoplasmic reticulum (Phase 1) is initiated by the imbibition of water by the tissue (Jones, 1980b; Buckhout *et al.*, 1981). Development of the cristae of the mitochondria is initiated similarly (Doig *et al.*, 1975b). Thus, one of the first responses of the aleurone tissue to imbibition is to establish the cellular ultrastructure which will be needed for germination. The embryo tissues respond similarly to imbibition.)

The second phase of development, where vesicles bud off from the cisternae of the endoplasmic reticulum and Golgi bodies show increased activity, does not occur in grains where the embryo has been removed (Colborne *et al.*, 1976; Obata and Suzuki, 1976). The development of glyoxysomes also does not occur (Doig *et al.*, 1975a). These findings suggest that the embryo is responsible for controlling some aspects of development in the aleurone tissue of the germinating grain. Experiments using the hormone gibberellic acid have confirmed this situation. Researchers have known for more than 20 years that gibberellic acid can replace the embryo in inducing the synthesis of α-amylase in the aleurone tissue (Yomo, 1960; Paleg, 1960). Moreover, gibberellic acid is known to be secreted from the embryo into the endosperm during germination. More recently it has been found that gibberellic acid also initiates the vesiculation of the endoplasmic reticulum and the formation of increased numbers of active Golgi bodies and glyoxysomes in both wheat (Doig *et al.*, 1975a; Colborne *et al.*, 1976) and barley (Paleg and Hyde, 1964; Jones, 1969c; Vigil and Ruddat, 1973) aleurone tissue. The hormone action may also lead to the formation of additional endoplasmic reticulum, over and above that formed during the first phase of ultrastructural development following imbibition.

Although information for the cereal grain is limited, studies of many other species have indicated that the nuclear membrane, endoplasmic reticulum, Golgi bodies, glyoxysomes and the plasmalemma belong to an integrated membrane system which has been called "the endomembrane system"; the aleurone grains and spherosomes also belong to this system in the cereal aleurone tissue. It has been proposed that most of the membrane material in the system is synthesized in the endosplasmic reticulum and that the other components are formed by the flow of membrane from the endoplasmic reticulum. The flow probably occurs via the small vesicles which bud off from the cisternae of the endoplasmic reticulum. Tomos and Laidman (1979) have implicated the endomembrane system in the gibberellic acid response mechanism in the aleurone tissue during germination. Reviewing the available evidence they proposed that the hormone is responsible for initiating the flow of membrane material from the endoplasmic reticulum into vesicles. These vesicles are believed to be the vehicles by which freshly synthesized α-amylase is carried, perhaps via Golgi bodies, to the plasma membrane. It was also proposed that gibberellic acid initiates endomembrane flow from the endoplasmic reticulum into glyoxysomes. Fig. 4.3 summarizes these proposals which have been presented in greater detail elsewhere (Tomos and Laidman, 1979; Laidman, 1980).

In this paper, the scheme in Fig. 4.3 is important for two reasons. First, it implicates the hormone gibberellic acid in the regulation of glyoxysome formation. It has been mentioned previously that glyoxysomes are the site of fatty acid degradation in the aleurone tissue during germination. Second, the scheme illustrates the importance of controlled endomembrane flow to the secretion of α-amylase and other hydrolases from the aleurone

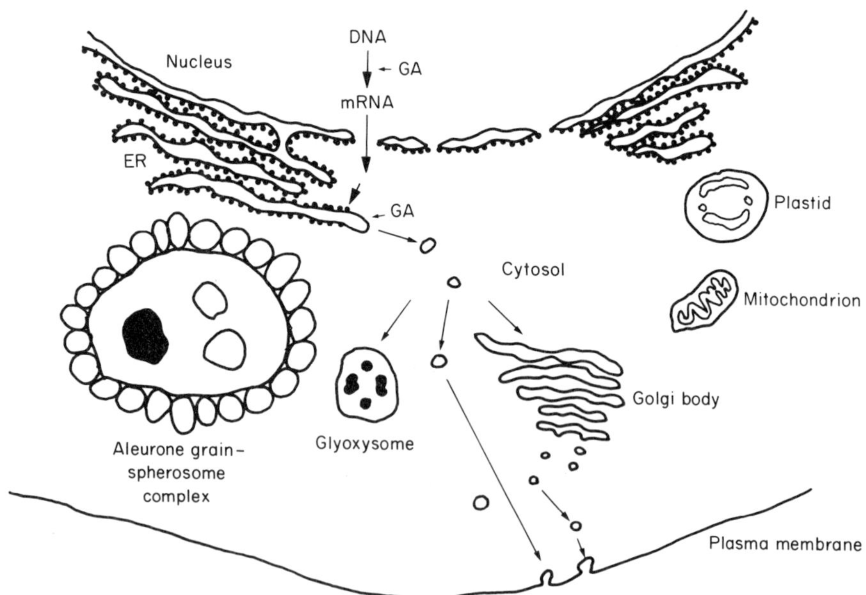

Fig. 4.3 Suggested routes of endomembrane flow in the aleurone cell. GA → marks the possible sites at which the gibberellin response is manifest. These sites are at the point of DNA transcription and at the cisternae of the endoplasmic reticulum where vesicles are formed (from Tomos and Laidman, 1979).

cell during germination. We will return to both of these aspects in later sections on the biochemistry of lipid metabolism in the aleurone tissue.

Although spherosomes are also believed to be part of the endomembrane system, and despite the fact that several investigators have commented upon the structural continuities between spherosomes and endoplasmic reticulum (see above), no evidence is available regarding the control of spherosome mobilization during germination. In particular, there is no evidence for the involvement of gibberellic acid in the control of either triacylglycerol breakdown in spherosomes or the spherosome-endoplasmic reticulum interactions. The regulation of these ultrastructural aspects of germination awaits detailed investigations.

III LIPID METABOLISM DURING GERMINATION

A Triacylglycerol Mobilization

The reserves of triacylglycerols together with the smaller quantities of diacylglycerols, which are present in the spherosomes, are mobilized during germination. Most of the reserves are believed to be mobilized by the actions of lipases to yield glycerol and free fatty acids.

1 *Starchy Endosperm*

Mobilization of triacylglycerol reserves in the starchy endosperm starts within 12 h after the beginning of imbibition, coinciding with the appearance of neutral lipase activity (Tavener and Laidman, 1972a,b). Figure 4.4

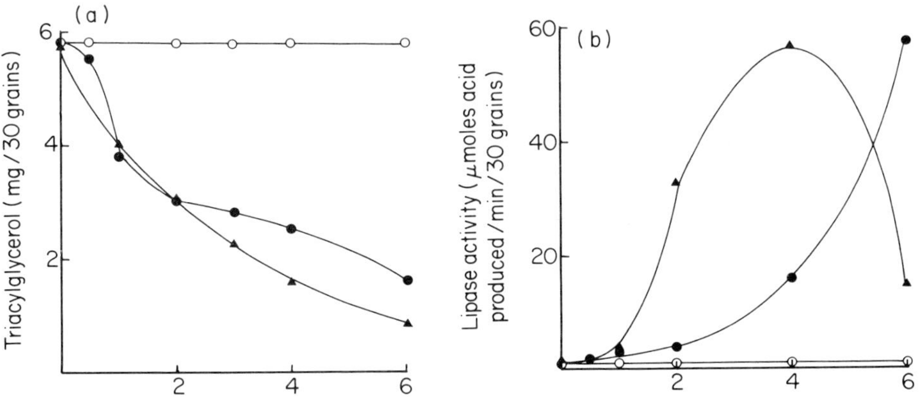

Fig. 4.4 Triacylglycerol metabolism (a) and lipase activity (b) in the starchy endosperm of wheat. ●, in germinated whole grains; ○, in incubated embryoectomized grains; ▲, in embryoectomized grains incubated with 1 mM hydroxylamine (from Tavener and Laidman, 1972a,b).

shows the time courses of these activities. Tavener and Laidman investigated the control of this mobilization. They found that removal of the embryo prevented the appearance of neutral lipase activity and the lipolysis of triacylglycerols when the embryoectomized grains were incubated under germination conditions. This indicated that a factor emanating from the embryo is responsible for initiating triacylglycerol mobilization in the starchy endosperm. Incubation of the embryoectomized grains with diffusates from the embryo confirmed this view. It was not possible to replace the embryo factor with gibberellic acid, but applying substrate concentrations of certain nitrogen-containing compounds, particularly hydroxylamine or glutamine, mimicked the embryo in bringing about both lipase activity and lipolysis in the starchy endosperm. The source of the active lipase remains unknown. It may originate from the aleurone tissue. Alternatively, the lipase induced by hydroxylamine or glutamine may not be the same enzyme as that in the germinating grain, and the latter could come from the epithelial cells of the scutellum which have been implicated as an important source of other hydrolases for the mobilization of endosperm nutrients (Briggs, 1972; Okamoto and Akazawa, 1980). A more tentative possibility is the suggestion that fungi within the dead endosperm tissue may provide the source of lipase (Tavener and Laidman, 1972b). Similarly, the fate of the lipolysis products in the starchy endosperm is unknown.

2 *Aleurone Tissue*

Mobilization of triacylglycerol reserves in the aleurone tissue of wheat also begins within 12 h after the start of imbibition, and it continues steadily for several days. A sharp rise in neutral lipase activitiy is associated with this mobilization. Figure 4.5 shows the time courses of these activities through germination. When they were investigating the control aspects of the mobilization, Tavener and Laidman (1972a,b) found that the removal of the embryo had a profound effect on the process. Thus, when embryoectomized grains were incubated under germination conditions, the triacylglycerol level in the aleurone tissue decreased by about 20% during the first 24 h (Fig. 4.5a). During this time the rate of breakdown exceeded that in the germinating whole grain; then no further breakdown occurred. Removing the starchy endosperm from the aleurone tissue abolished the mobilization entirely. These results suggest that two pools of triacylglycerol exist within the aleurone tissue. One of them, constituting about 20% of the total triacylglycerol reserve, is mobilized in response to a factor(s) emanating from the starchy endosperm, while mobilization of the second and larger pool is controlled by a factor(s) from the embryo. The role of the

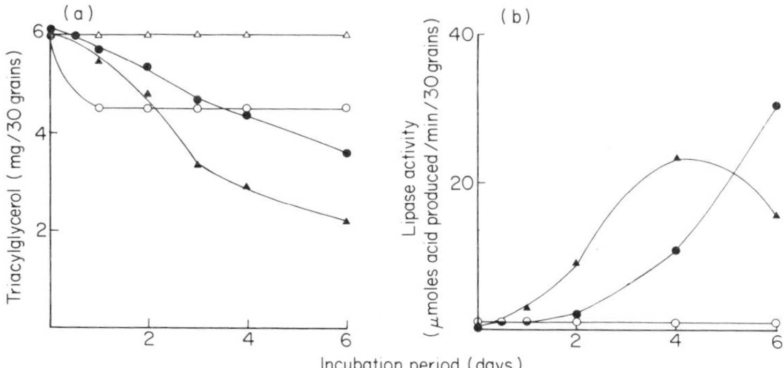

Fig. 4.5 Triacylglycerol metabolism (a) and lipase activity (b) in the aleurone tissue of wheat. ●, in germinated whole grains; ○, in incubated embryoectomized grains or aleurone tissue incubated with 1 nM kinetin; △, in incubated aleurone tissue; ▲, in embryoectomized grains incubated with 1 mM hydroxylamine + 10 μM indole acetic acid (from Tavener and Laidman, 1972a,b).

starchy endosperm could be mimicked exactly by applying physiological concentrations of cytokinins (kinetin or benzyladenine) to incubating aleurone layers. Although the cytokinin caused triacylglycerol to be mobilized, it did not cause neutral lipase to appear in the aleurone tissue (Fig. 4.5b). The pathway of this mobilization is not known. Mobilization of the second pool could be brought about by adding an embryo diffusate to embryoectomized grains. The embryo factor could not be replaced by gibberellic acid (*cf.* the induction of carbohydrate metabolism by this hormone). Conversely, applying physiological concentrations of either hydroxylamine or glutamine together with indole acetic acid to embryoectomized grains initiated the mobilization of this larger pool. The same two factors also caused the appearance of neutral lipase activity (Fig. 4.5b) and they retarded the metabolism of the smaller endosperm-controlled pool. It has been suggested that the action of hydroxylamine/glutamine and indole acetic acid on the aleurone tissue is dependent upon RNA and protein synthesis (Tavener and Laidman, 1972b).

Although the above effects of hydroxylamine, glutamine and indole acetic acid on the reserve tissues of the wheat grain are very interesting, doubts have been raised as to their physiological significance. Glutamine, in particular, seems unlikely to be the active agent *in vivo* (Chittenden *et al.* 1978), despite the fact that it accumulates in the reserve tissues during germination. Although indole acetic acid has been found in the shoots of germinating embryos and cytokinins have been isolated from maize endosperm and germinating barley, neither has been proved to be present and active in the wheat system under discussion. Hydroxylamine is an even less likely candidate for the role of natural factor. Similarly, the data of Tavener and Laidman leave us with an incomplete account of the enzymology of triacylglycerol mobilization. They concentrated upon neutral lipase activities in the wheat tissues. We know, however, that many seeds contain acid lipases and Jelsema *et al.* (1977) have reported an acid lipase located in the aleurone grain fraction from wheat aleurone tissue. The site of lipolysis during germination is also poorly understood. There has always been the assumption that the mobilization of triacylglycerols takes place exclusively in the spherosomes. To date, however, no spherosome-bound lipase has been identified. Moreover, Moreau *et al.* (1980) have described a monoacylglycerol-specific alkaline lipase located in the glyoxysomes of germinating castor bean. This suggests that partially hydrolysed glycerides may be transferred from spherosomes to glyoxysomes.

3 *Scutellum*

In the scutellum of wheat, triacylglycerol mobilization does not start until the third day of germination (Tavener and Laidman, 1972a) and it coincides with the appearance of neutral lipase activity in the tissue (Tavener and Laidman, 1972b). The time courses of these activities are shown in Fig. 4.6. Interestingly, this mobilization in the scutellum commences when the rate of triacylglycerol mobilization in the embryo axis is decreasing and at the time when hydrolysis of nutrient reserves in the starchy endosperm is just beginning. The significance of these temporal relationships is unclear, however. In contrast to wheat, the barley scutellum mobilizes its triacyl-

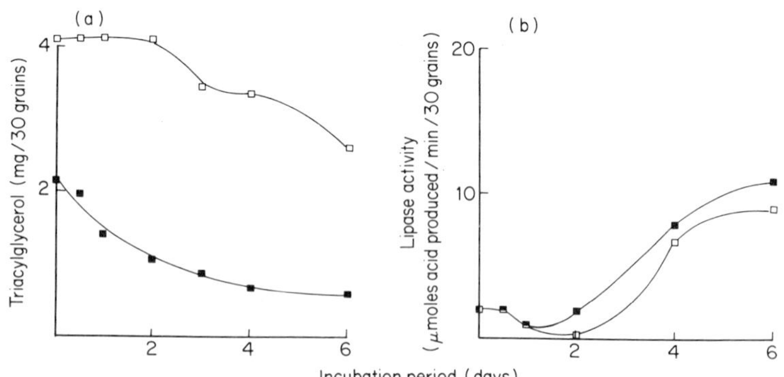

Fig. 4.6 Triacylglycerol metabolism (a) and lipase activity (b) in the embryo tissues of germinated wheat. ■, embryo axis; □, scutellum (from Tavener and Laidman, 1972a,b).

glycerol reserves within 24–28 hours following the commencement of imbibition (MacLeod and Palmer, 1966). It has been suggested that the triacylglycerol is converted to starch. Although a neutral lipase which may initiate this metabolism has been identified in the barley embryo (MacLeod and White, 1962), the β-oxidation and gluconeogenic enzymes required to convert the fatty acid products into sugars have not been reported at such an early stage in germination. Glyoxylate cycle enzyme activities do, however, appear during the first day of germination in maize scutellum (Longo and Longo, 1970) indicating that glyoxysomes are present and that early gluconeogenesis may occur in the tissue.

4 *Embryo Axis*

The embryo axis of wheat commences moblization of its triacylglycerol reserves about 12 h after the beginning of imbibition (Tavener and Laidman, 1972a). Mobilization proceeds rapidly at first, but it continues more slowly after the second day (Fig. 4.6a). Interestingly, this period of 2–3 days corresponds approximately to the length of time for which isolated embryos can support themselves in nutrient-free media. In the intact germinating grain, therefore, the triacylglycerol reserves present in the embryo axis are probably acting as an energy source for that tissue during early germination and before other nutrients are mobilized from the starchy endosperm. The exact nature of triacylglycerol metabolism is not known. Neutral lipase activity in the tissue does not increase significantly until the third day of germination (Fig. 4.6b) (Tavener and Laidman, 1972b). The earlier metabolism of triacylglycerol must, therefore, be catalysed either by the low levels of neutral lipase present in the quiescent tissue or by another unidentified lipase possibly an acid lipase.

B Glycerol and Fatty Acid Metabolism

It is generally assumed that the glycerol produced from lipolysis is phosphorylated by glycerol kinase and that the glycerol phosphate is incorporated into the intermediary metabolism of the cell.

Our understanding of the biochemistry of fatty acid metabolism during germination is much better than that of lipolysis. Much of the basic information on this subject comes from studies of the germinating castor bean, but sufficient data is now available from cereals to speculate that the two groups of plants are similar in this respect.

1 *Non-cereal Oleaginous Seeds*

Beevers and his colleagues have shown that fatty acid catabolism in the castor bean endosperm takes place in glyoxysomes (Breidenbach and

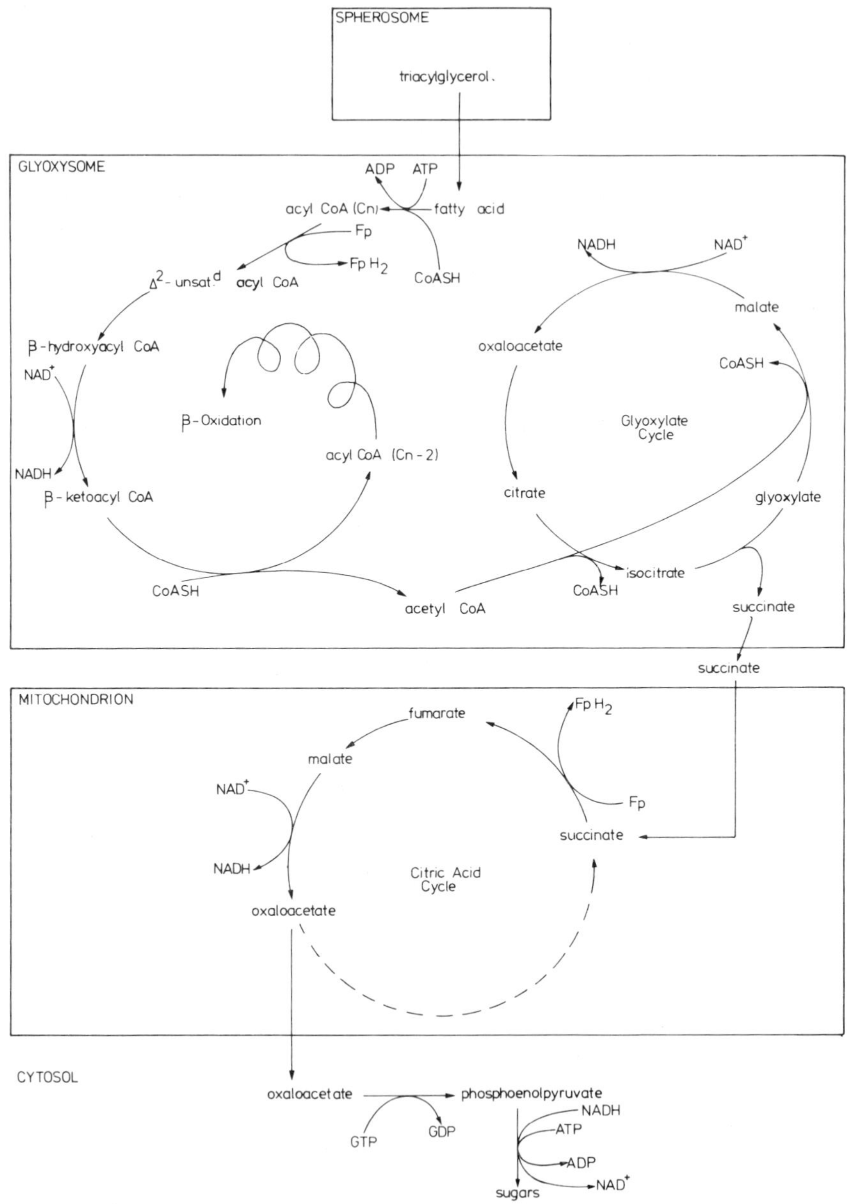

Fig. 4.7 Fatty acid metabolism in germinating oleaginous seed tissues (see text for detailed description).

Beevers, 1967; Cooper and Beevers, 1969) within which the enzymes of both β-oxidation and the glyoxylate cycle are located (incidentally, it is not known how the fatty acids are transported to the glyoxysomes from the spherosomes). The pathways of fatty acid metabolism are illustrated in Fig. 4.7. In a repeating cycle of reactions, β-oxidation breaks down fatty acyl CoA to acetyl CoA. The oxidation is coupled to the reduction of NAD^+ to NADH and Fp to FpH_2. Acetyl CoA cannot escape from the glyoxysome and it is incorporated directly into the glyoxylate cycle. This cycle of reactions consumes acetyl CoA and produces succinate which is the end-product of fatty acid metabolism in the glyoxysome. The succinate passes through the limiting membrane of the glyoxysome into the cytoplasm. Like β-oxidation, the glyoxylate cycle is an oxidative sequence of reactions and it is coupled to the reduction of NAD^+ to NADH. The succinate released from the glyoxysome is taken up into the mitochondrion where it is converted to oxaloacetate by enzymes of the citric acid cycle. This sequence of reactions is also oxidative in nature since it is coupled to the reduction of Fp and NAD^+. The NADH produced by the glyoxysomal and mitochondrial metabolism of fatty acids is potentially available for the reduction of other metabolites or for the production of ATP by oxidative phosphorylation. The oxaloacetate produced by the mitochondria is passed out into the cytoplasm where it is converted to sugars via phosphoenol pyruvate and reversed glycolysis. This reversed glycolysis sequence is reductive in nature and requires energy in the form of NADH and nucleoside triphosphates. The germinating seed thus contains a most elegant mechanism to convert stored fats into amphibolic intermediates which can be used either for energy or carbohydrate production. Compartition of the mechanism between different parts of the cell, which function co-operatively, allows the metabolism to be directed and regulated. (See articles by Bewley and Black (1978) and Lord and Roberts (1980) for more comprehensive accounts of the glyoxysome and its biochemistry.)

2 *Cereals*

Our view that fatty acids are metabolized in a similar way in cereals is based upon the identification of β-oxidation activity and the unique enzymes of the glyoxylate cycle (isocitratase and malate synthetase) in cereal tissues. These enzyme activities are not present in quiescent grains, but they have been found in the scutellum of germinating maize (Oaks and Beevers, 1964) and the aleurone tissues of germinating barley (Jones, 1972) and wheat (Doig and Laidman, 1972; Doig *et al*., 1975a). Our view is further supported by the demonstration using electron microscopy which shows that the aleurone tissue contains glyoxysomes (see Section IIB).

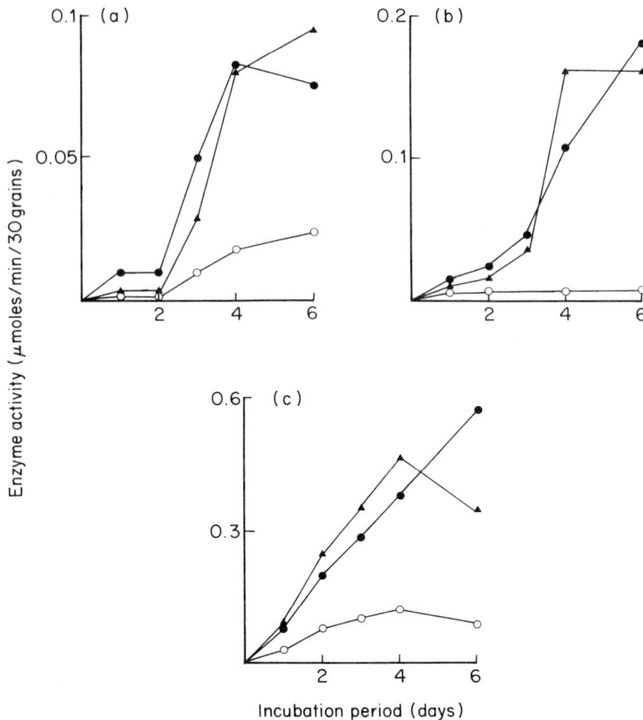

Fig. 4.8 β-Oxidation (a), isocitratase (b) and malate synthetase, (c) activities in the aleurone tissue of wheat. ●, in germinated whole grains; ○, in incubated embryoectomized grains; △, in embryoectomized grains incubated with 10 μM gibberellic acid (from Doig *et al.*, 1975a).

Studies using embroyoectomized grains of barley and wheat have shown that the appearance of both fatty acid β-oxidation and glyoxylate cycle enzyme activities in the aleurone tissue is controlled by gibberellic acid from the embryo. Grains which have had the embryos removed do not develop these enzyme activities when they are incubated under germination conditions, but the addition of gibberellic acid to the incubation medium leads to the appearance of all the activities (Jones, 1972; Doig and Laidman, 1972; Doig *et al.*, 1975a). Figure 4.8 shows the data for wheat. These results together with those from ultrastructural studies show that the formation of glyoxysomes and their contained enzymes in the aleurone tissue is induced by the hormone gibberellic acid during germination. Studies using metabolic inhibitors suggest that the induction is dependent upon both RNA and protein synthesis (Doig *et al.*, 1975a).

C Phospholipid Metabolism

The account in Section IIB has described how development of the endoplasmic reticulum is an important early event in germinating seeds. It is useful, therefore, to look at the biochemistry of this process by studying the synthesis of phospholipids which are important components of biological membranes. Both the aleurone tissue and the embryo have been studied in this respect. They will be discussed in that order in this section.

1 *Aleurone Tissue*

The aleurone tissue of the cereal grain has a secretory function during germination and the endomembrane system plays an important part in the secretory mechanism. Consequently, much attention has been paid to phospholipid metabolism in the tissue and particular interest has been taken in the control of phospholipid synthesis and turnover. Much confusion has arisen, however, about the action of gibberellic acid on the phospholipid metabolism, and the literature contains several conflicting reports. It is not the purpose here to go through all of the contradictory data on the subject. This has, in any case, been critically reviewed elsewhere, and the reader should refer to articles by Tomos and Laidman (1979) and Laidman (1980, 1982) for this information. Hopefully, these reviews have clarified some of the issues so that we can now present an accurate account of events.

During the first 24 h of germination the level of total phospholipids in the wheat aleurone tissue increases by about 50% (Fig. 4.9) (Varty and

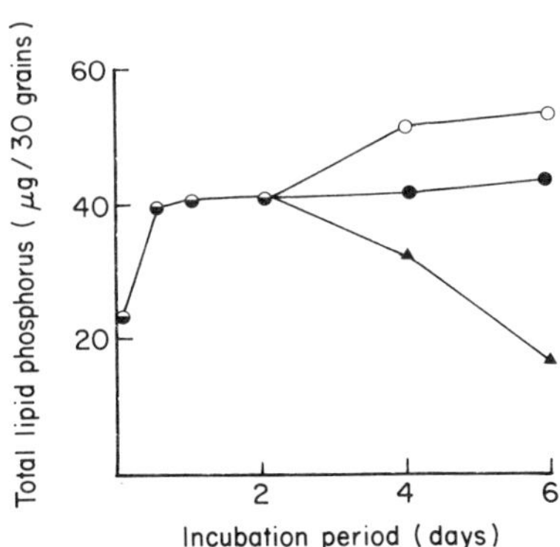

Fig. 4.9 Total phospholipid levels in the aleurone tissue of wheat. ●, in germinated whole grains; ○, in incubated embryoectomized grains; ▲, in embryoectomized grains incubated with 1 μM gibberellic acid (from Varty and Laidman, 1976).

Laidman, 1976). This level then remains constant up to the sixth day of germination. During the period of increasing phospholipid levels, the aleurone tissue can actively incorporate radio-labelled precursors into phospholipids, and the radio-labelled phospholipids which are produced are located in the microsomal fraction of the aleurone cell. The increase in phospholipid levels during the first 24 h coincides with the first phase of membrane development in the tissue when endoplasmic reticulum is being rapidly formed (see Section IIB1). Moreover, the increase is not dependent upon the presence of either the embryo or gibberellic acid, just as the formation of the endoplasmic reticulum was independent of these factors. The biochemical and the ultrastructural information thus agree that membrane synthesis during this phase of germination is not dependent upon gibberellic acid. A similar situation has recently been conclusively demonstrated in the aleurone tissue of barley. In a comprehensive sub-cellular study, where the endoplasmic reticulum was estimated following ultracentrifugation through a sucrose density gradient, Jones (1980a,b) showed that the membrane formation is induced by imbibition, although gibberellic acid may slightly enhance the rate of formation during the first 24 h.

There are some doubts regarding the origin of the phospholipid in the newly formed endoplasmic reticulum. Early investigations of the barley aleurone tissue suggested that membrane formation might take place without the net synthesis of phospholipids (Koehler and Varner, 1973; Firn and Kende, 1974; Vredevoogd and Ruddat, 1974) and it was proposed that spherosomes might be the source of phospholipid. Such a situation has considerable attractions since it would provide a biochemical function for the ultra-structural connections that have been observed between the spherosomes and the endoplasmic reticulum (see Section IIB4).

Spherosomes, however, are composed predominantly of triacylglycerol and adequate supplies of phospholipids would not be found there. If the spherosomes are to be considered as the source of new membrane, it becomes necessary to propose the conversion of triacylglycerol to phospholipid, presumably by limited lipase and CDP: 1,2-diacylglycerol phosphocholine transferase activities. Recent experiments in our laboratory have, however, shown that, in wheat, radioactive acetate, glycerol, phosphate and choline are actively and specifically incorporated into the fatty acyl, glycerol, phosphate and choline moieties of the phospholipid molecule (Mirbahar, 1982). The enzymes of the CDP: choline biosynthetic pathway to phosphatidylcholine have also been identified in barley aleurone tissue (Johnson and Kende, 1971; Ben-Tal and Varner, 1974). Therefore, at least part of the phospholipid pool can be synthesized *in toto*, and the ultrastructural connections between spherosomes and the endoplasmic reticulum may be part of the process of triacylgylcerol mobilization rather than a membrane synthesizing system. Interestingly, Vick and Beevers (1978) have found that the fatty acyl moieties of newly synthesized phospholipids in germinating castor beans are synthesized in plastids and incorporated into phospholipids in the endoplasmic reticulum. In the cereal aleurone tissue this would conveniently explain the close ulstrastuctural association observed between plastids and the developing endoplasmic reticulum. Harwood and his colleagues, however, have presented evidence (Harwood, 1979) that germinating seeds do not necessarily synthesize fatty acids *in toto* but may chain-elongate pre-existing acids, thus explaining the incorporation of radioactive acetate. The origin of the endoplasmic reticulum in germinating seeds thus remains an exciting area of study.

The second phase of ultrastructural development in the aleurone tissue, which starts during the second day of germination, is characterized by the gibberellic acid-induced flow of membrane material from endoplasmic reticulum into secretory vesicles and glyoxysomes (see Section IIB1). The action of gibberellic acid also leads to changes in phospholipid metabolism during this phase. Embryoectomized grains slowly accumulate phospholipids between the second and sixth day of incubation (Fig. 4.9). This contrasts with the situation in the intact germinating grain where the level of phospholipids remains constant. Incubation of the embryoectomized grains with gibberellic acid causes a breakdown of phospholipid between the second and sixth day of incubation. The breakdown of phosphatidylcholine accounts for more than 80% of the total phospholipid lost during this period. Thus, gibberellic acid does not have the same effect as the embryo has on phospholipid levels. This difference is in contrast to their similar regulatory effects on several other aspects of aleurone tissue metabolism including α-amylase production and the synthesis of glyoxysome enzymes. A similar difference between the action of the embryo and that of gibberellic acid has been observed in the aleurone tissue of barley (Koehler and Varner, 1973; Firn and Kende, 1974). To explain the apparent discrepancy between the action of the embryo and that of gibberellic acid, Varty and Laidman (1976) envisaged the existence of another regulatory factor emanating from the embryo which can modulate the gibberellic acid action. They obtained evidence for the presence of such a factor in diffusates from isolated wheat embryos, but the nature of the factor remains unknown.

R. B. Mirbahar, working in our laboratory, has recently studied the effects of gibberellic acid on phospholipid turnover in the aleurone tissue (Mirbahar, 1982; Mirbahar and Laidman, 1981a,b). He incubated embryo-ectomized grains for 24 h in the presence of [^{14}C]glycerol to

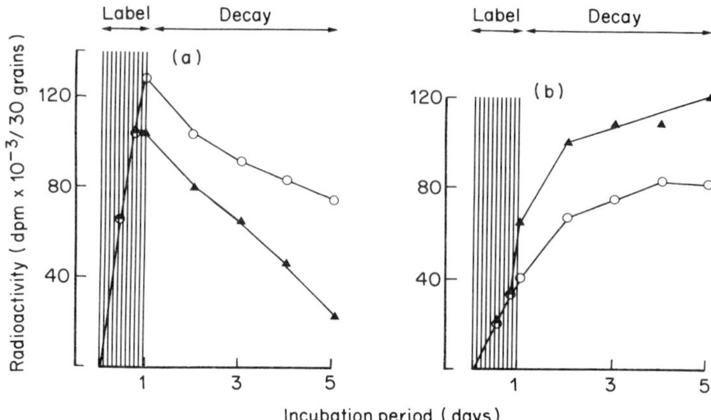

Fig. 4.10 Incorporation and turnover of [^{14}C]glycerol in total phospholipids (a) and triacylglycerols (b) in the aleurone tissue of wheat. ▲, in incubated embryoectomized grains; ○, in embryoectomized grains incubated with 1 μM gibberellic acid (from Mirbahar and Laidman, 1981a).

radiolabel the aleurone tissue pools of phospholipids. This 24 h period corresponds to the first phase of ultrastructural development in the tissue when phospholipids are being actively synthesized. At the end of the labelling period the radioactive precursor was removed and the embryoectomized grains were incubated for further periods up to 96 h, corresponding to the second phase of ultrastructural development. Samples were harvested at various times and the radioactivity in each phosphatide was measured.

The decay of radioactivity and of specific radioactivity provided estimations of the breakdown and turnover of the phosphatides. The experiment was carried out either in the absence or presence of gibberellic acid and the results are shown in Fig. 4.10 and Table 4.1. The expected pattern of labelling was obtained in which radioactivity accumulated in the phospholipids during the first incubation and was lost from them during the second incubation (Fig. 4.10a). As the radioactivity was lost from the phospholipids it appeared quantitatively in the triacyglycerol fraction (Fig. 4.10b) indicating a stoichiometric conversion of phospholipids into triacylglycerols. The stoichiometry was maintained for at least 72 h of the second incubation. Gibberellic acid enhanced the rate of phospholipid conversion into triacylglycerol. Moreover, the effect of the hormone is evident even before the end of the first incubation, and the rate of conversion was increased by about three-fold between 18 and 24 h following hormone application.

Table 4.1 Effects of gibberellic acid on phosphatide levels and the turnover of [^{14}C] glycerol in phosphatides†

Incubation period (h)	Phosphatide level (μmoles/30 seeds)		Radioactivity (10^{-3} × dpm/30 seeds)		Sp. Radioactivity (10^{-3} × dpm/μmole)	
	−GA	+GA	−GA	+GA	−GA	+GA
Phosphatidyl choline						
12	0.54	0.54	40.8	40.8	–	–
18	–	–	54.7	55.9	–	–
24	0.61	0.63	72.6	53.7	119	85
48	0.67	0.63	53.4	37.1	80	59
72	0.73	0.56	46.4	28.7	64	51
96	0.73	0.48	40.7	19.0	56	40
120	0.74	0.40	35.2	10.4	48	26
Phosphatidyl inositol						
12	0.23	0.23	7.5	7.5	–	–
18	–	–	12.7	13.1	–	–
24	0.25	0.24	18.0	16.3	72	68
48	0.24	0.24	18.0	15.0	75	63
72	0.23	0.22	16.0	12.8	70	58
96	0.25	0.24	15.7	9.3	63	39
120	0.25	0.23	15.5	4.0	62	17
Phosphatidyl ethanolamine						
12	0.10	0.10	4.7	4.9	–	–
18	–	–	8.5	9.1	–	–
24	0.13	0.14	13.4	11.6	103	83
48	0.15	0.16	14.2	11.1	95	·69
72	0.17	0.15	13.2	9.4	78	62
96	0.17	0.16	12.3	8.6	72	54
120	0.18	0.14	11.6	4.0	64	29
Phosphatidyl glycerol						
12	0.05	0.05	5.1	5.2	–	–
18	–	–	11.2	11.2	–	–
24	0.05	0.05	13.3	13.9	270	280
48	0.06	0.05	7.3	6.2	120	130
72	0.06	0.06	5.7	4.5	95	80
96	0.06	0.06	5.2	2.7	83	45
120	0.07	0.05	4.6	1.1	71	20

†De-embryoed seeds were pulse-labelled with [U-^{14}C] glycerol for 24 h and then incubated for a further 96 h. Incubations were carried out either in the absence or in the presence of 1 μM gibberellic acid (Mirbahar and Laidman, 1982).

This hormone effect was almost entirely due to the metabolism of phosphatidylcholine. The level of phosphatidylcholine did not change during the 18–24 h period, and it is concluded that gibberellic acid causes an increased turnover of the phosphatide. Furthermore, the stimulated turnover coincides with the first appearance of gibberellic acid-induced α-amylase in the tissue (Laidman, 1982). This exciting discovery recalls the earlier claims that gibberellic acid causes an increased phosphatidylcholine turnover during the lag phase leading to α-amylase synthesis in barley aleurone tissue (Evins and Varner, 1971; Koehler and Varner, 1973). It is potentially important to our understanding of the mechanism by which gibberellic acid induces the synthesis and secretion of α-amylase.

Later during the second incubation of the [^{14}C]glycerol-labelled, embryoectomized grains, another feature appeared where the specific radioactivity of phosphatidylinositol declined while its mass remained constant (Table 4.1). In other words, this phosphatide maintained a constant pool size but it was being actively turned over. Moreover, gibberellic acid caused a three-fold increase in the turnover rate. The increase occurred at the time when α-amylase was actively being synthesized and secreted from the aleurone tissue. This finding is especially exciting because similar events have been described in a number of exocytotic animal tissues where enhanced phosphatidylinositol breakdown and turnover occurs in response to external stimuli (Michell, 1979). A common factor in all of these situations in animals is the involvement of an external agonist binding to a receptor site on the plasma membrane, leading eventually to the secretion of enzymes or other active biochemicals from the tissue. The apparently ubiquitous occurrence of stimulated phosphatidylinositol turnover in these systems suggests that the phosphatide plays an important role in agonist-induced secretory mechanisms.

2 *Embryo*

Several groups of investigators have reported increases in phospholipid levels in the embryos of germinating cereals, including rice (Mukherji 1971), wheat (de la Roche *et al.*, 1972; McDonnell *et al.*, 1982), rye (Hallam *et al.*, 1972) and wild oats (Cuming and Osborne, 1978). Figure 4.11a shows our data for wheat embryos as an example of the sort of pattern of increasing phospholipid levels that has been observed. Total phospholipid levels in this species increased slowly up to about 30 h of germination, after which time they increased very quickly. The time of 30 h appears to be particularly relevant, since it is the time when the developing embryo axis begins active cell division (Bray, 1979), and it has been argued that the rapid accumulation of phospholipids after this time reflects the formation of new membranes in the dividing cells (McDonnell *et al.*, 1982).

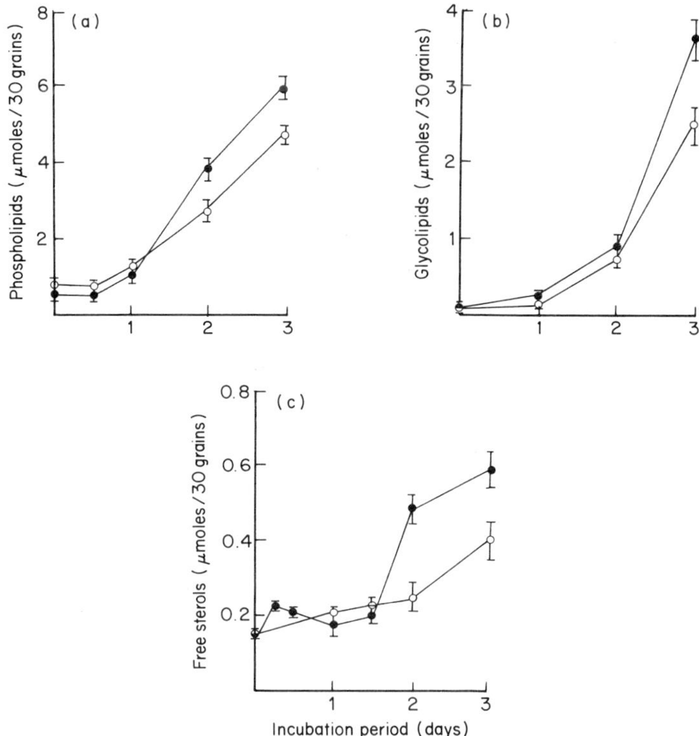

Fig. 4.11 Total phospholipid (a), total glycolipid (b) and total free sterol (c) levels in the embryos of germinated wheat grains. ●, high vigour grains; ○, low vigour grains (from McDonnell *et al.*, 1982).

In their study of wheat, McDonnell *et al.* (1982) studied various commercial lots of grain which had either low- or high-germination vigour. They found that phospholipid accumulation in the low-vigour embryo was delayed compared with that in high-vigour embryos (Fig. 4.11) suggesting a delayed onset of cell division in the former. The sterol/phospholipid ratio was also lower and the turnover of phosphatidylcholine was significantly faster in the low-vigour embryos. These findings lead the authors to speculate that the embryos of germinating low-vigour grains may have characteristically less stable membranes than high-vigour grains. Evidence has also been presented (Mukhtar *et al.*, 1982) that the cellular membranes of low-vigour wheat embryos are less efficient in the process of mineral ion pumping compared with those of high-vigour embryos. Similar differences in the efficiency of nucleic acid and protein synthesis have been reported by Blowers *et al.* (1980). Taken together, these reports provide convincing evidence that vigour in wheat has a biochemical and physiological basis, which must be of considerable importance to the seed producing industry (Stormonth, 1978).

Cuming and Osborne's study (1978) also emphasized the importance of phospholipid turnover in the cereal embryo. They observed a continuous turnover of phospholipids in imbibed, dormant embryos of wild oat and concluded that the turnover was necessary to maintain cellular integrity and hence longevity of the seed during dormancy. Dormancy could be broken by the application of gibberellic acid and only then did net synthesis of phospholipid take place. All phosphatides were accumulated but phosphatidylinositol showed the largest increase. Similar large accumulations of phosphatidylinositol have been noted in germinating hazel seed (Shewry *et al.*, 1973) and in the wheat aleurone tissue (see above). It is tempting to speculate therefore that phosphatidylinositol may play an important role in the newly synthesized membranes of germinating seeds.

D Metabolism of Other Lipids during Germination

Little is known about the quantitatively less important membrane lipids in germinating cereal grains. The lack of interest in these compounds probably reflects the paucity of information about their function in plant cells generally. These compounds include the glycolipids, sterols and tocols. Despite the situation, we have included some data about the levels and the metabolism of these compounds during germination in this account.

1 *Glycolipids*

While phospholipids provide the bulk of the polar lipid in mitochondria and the endomembrane system of plant cells, glycolipids are quantitatively the most important lipid components of plastid membranes. Changes in the glycolipid levels are, therefore, likely to reflect the development of plastids during germination, although the correlation is not necessarily an exclusive one.

The aleurone tissue of cereals is a non-dividing, non-photosynthetic tissue and glycolipid metabolism would not be expected to be important there. This is borne out by analysis of the tissue taken from germinating grains (McDonnell, 1980). Aleurone tissue from ungerminated grains contained about 60 μg lipid sugar/30 grains (expressed as μg galactose). Three classes of glycolipid were present: monoglycosyldiacylglycerol (monoglycosyldiglyceride) accounted for about 20%, and steryl glycoside about 30% of the total; the remaining 50% was mainly accounted for by an unidentified compound, possibly diglycosylmonoacylglycerol which migrated behind diglycosyldiacylglycerol on thin-layer chromatography. Diglycosyldiacylglycerol was not present in detectable quantities. The level of total glycolipid and the relative proportions of the three classes did not change significantly during germination. When radio-labelled glucose was administered to aleurone tissue, it was taken up into the tissue but very little of it was incorporated into the glycolipid fraction. Similarly, incorporation of [^{14}C]glycerol into the fraction was very slow (Mirbahar, 1982). The metabolic inactivity of the glycolipids contrasts with the active metabolism found in the phospholipids in this tissue.

Glycolipid levels in the embryos of germinating wheat follow a similar pattern to that described for the phospholipids (see Section IIIC2). Thus, McDonnell *et al.* (1982) found that total glycolipid levels remained fairly constant during the first 24–30 h of germination and then they increased rapidly. In the quiescent embryo the most abundant glycolipid was steryl glycoside (Table 4.2); smaller quantities of monoglycosyldiacylglycerol and diglycosyldiacylglycerol were found. The increase in total glycolipids during germination was reflected in each of the three classes of glycolipid but the increase in the glycosyldiacylglycerols were proportionately greater than that of steryl glycoside. At 72 h of germination, diglycosyldiacylglycerol was the most abundant glycolipid. The diglycosyldiacylglycerol and the steryl glycoside of the germinated embryo were unusual in having glucose exclusively as their glycosyl moiety; galactose is usually the most common sugar present in plant lipids.

Table 4.2 Glycolipids in the embryos of germinating wheat. Values are presented as percentages of the total glycolipid sugar†

		Germination period (h)			
Vigour rating	Glycolipid	0	24	48	72
Low	Monoglycosyl diacylglycerol	nd	15.0	7.8	10.0
High		nd	10.0	8.0	8.0
Low	Diglycosyl diacylglycerol	13.5	24.4	35.0	55.0
High		12.0	27.0	25.6	60.0
Low	Steryl glycoside	47.5	29.5	20.0	12.0
High		32.7	25.0	22.0	10.0

†Notes:
(1) Each value is the average from two separate experiments.
(2) nd – not detected.
(3) From McDonnell *et al.*, 1982.

When several commerical grain lots with different germination vigours were analysed for their glycolipid contents (McDonnell *et al*., 1982), it was found that the increase in levels in the embryo after 30 h of germination was delayed in the low-vigour lot (Fig. 4.11b). This finding is similar to that reported in the case of the phospholipids, where it was concluded that the delay reflected a delay in the onset of cell division in low-vigour embryos compared with the embryos of high-vigour lots.

2 *Sterols*

Free sterol metabolism in germinating wheat bears similarities to the metabolism of glycolipids described in the previous section. Thus, Pulford (1980) found that the total free sterol content of the aleurone tissues was about 110 μg/30 grains and this level declined to about 70 μg/30 grains after 96 h of germination. The free sterol fraction in the quiescent aleurone tissue was composed of cholesterol (6%), campesterol (20%), stigmasterol (5%), sitosterol (59%) and Δ^5-avenasterol (10%). They all decreased during germination, except sitosterol which remained constant. A similar situation was found by Torres *et al.* (1976) also studying wheat. They demonstrated a fall in total sterol levels over 144 h of germination. At all stages throughout germination, [^{14}C]mevalonate was incorporated into the terpenoid hydrocarbon fraction but never into free sterols or steryl esters (Pulford, 1980). It must be assumed from these results that the aleurone tissue is incapable of synthesizing sterols and that the sterols already present in the quiescent tissue are capable of fulfilling the functions of these compounds during germination.

It is unlikely that sterol synthesis occurs in the starchy endosperm tissue because of the tissue's non-viable nature. Torres *et al.* (1976) reported a sharp fall in sterol levels of wheat endosperm during germination. In the maize endosperm, however, the levels of free sterols and their esters remained constant throughout 14 days of germination (Kemp *et al*., 1967). Assessing their data Torres *et al.* (1976) proposed that the endosperm is not involved in supplying the coleoptile and roots with sterols during germination, but that steryl esters accumulated by the developing embryo serve this purpose. Their data suggested, however, that sterol is transferred from the scutellum to the coleoptile and roots. Kemp *et al.* (1967) had previously suggested that in germinating seeds sterols are transported as their esters.

The total free sterol level in wheat embryos increased slowly during the first 24–36 h of germination from about 60 μg to about 80 μg/30 grains (McDonnell *et al*., 1982). After this time, the level increased more rapidly to reach 150–200 μg/30 grains by 72 h. Marked differences were also found in the rates at which low- and high-vigour embryos accumulated free sterols (Fig. 4.11c). In the quiescent embryo, cholesterol, campesterol, stigmasterol, sitosterol and Δ^5-avenasterol were present in similar proportions to those found in the aleurone tissue. The increase in total sterols during germination was reflected in each of these sterols except cholesterol, the level of which did not change. Stigmasterol levels increased proportionately more than those of the others. Torres *et al.* (1976) reported a similar situation in wheat tissues. They demonstrated a rise in total sterol, mostly free sterols, in both coleoptiles and roots after 72 h of germination. This delay in sterol accumulation compared with that reported by McDonnell *et al.* can easily be explained by the different germination temperatures used by the two groups of investigators. The composition of the free sterol fraction in the quiescent embryo was also very similar to that found by McDonnell *et al.* A sharp fall in total sterol levels was observed in the scutellum during germination. Studying maize, Kemp *et al.* (1967) demonstrated a similar rise in free sterols in the roots and shoots following 4–6 days of germination. Free sterols of the scutellum, however, remained at a constant level throughout the 14 days of germination. The individual sterol composition in the embryo tissues of maize was also very similar to that observed in wheat.

Experiments to measure the incorporation of [^{14}C]mevalonate into sterols have provided interesting results (Pulford, 1980). Embryos from quiescent grains and grains germinated for 12 h were able to incorporate mevalonate into squalene but not into free sterols or steryl esters. After 12 h of germination, incorporation into squalene declined while that into the sterols and steryl esters increased. Similar situations have been reported previously for the embryos of pine (McKean and Nes, 1977) and peas (Capstack *et al.*, 1962). It thus appears that quiescent embryos, including those of wheat, contain the enzymes of the anaerobic pathway from mevalonate to squalene, but that the oxidative biosynthetic pathway from squalene to the sterols appears during germination after a lag period.

Compartition of sterol biosynthesis at the sub-cellular level has been demonstrated in the coleoptiles from germinating maize. Hartmann *et al.* (1978, 1979) isolated two fractions of membraneous vesicles from the tissue. The first fraction was rich in endoplasmic reticulum and contained a high proportion of the enzymes of the sterol biosynthetic pathway. The second fraction was rich in plasma membrane. It contained most of the free sterol and most of the UDP glucose:sterol glycosyl transferase activity. They concluded that free sterol biosynthesis takes place in the endoplasmic reticulum, but the synthesized sterol is transferred to the plasma membrane and glycosylation occurs exclusively in the plasma membrane.

3 *Tocols and Quinones*

To monitor the development of mitochondria and plastids in germinating wheat grains Hall and Laidman (1968) determined levels of ubiquinone

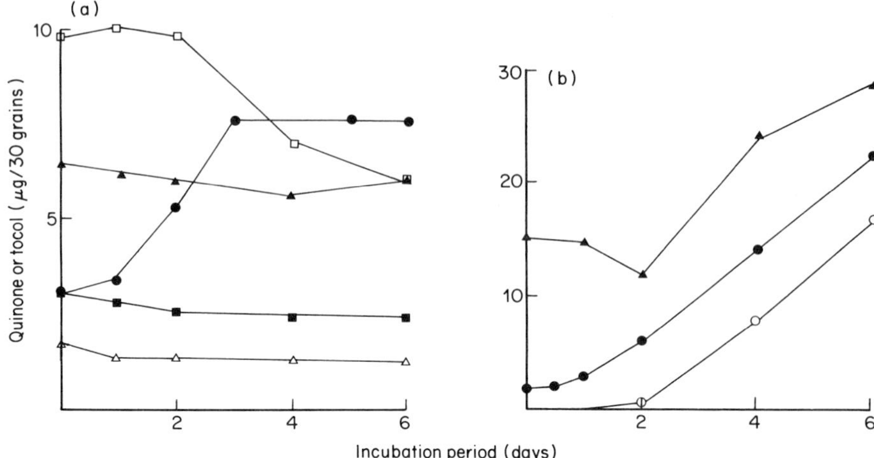

Fig. 4.12 Isoprenoid quinone and tocol levels in the endosperm (starchy endosperm + aleurone layer) (a) and the embryo axis (b) of germinated wheat ●, ubiquinone-9; ○, plastoquinone-9; ▲, α-tocopherol; △, β-tocopherol; ■, α-tocotrienol; □, β-tocotrienol (from Hall and Laidman, 1968).

and plastoquinone respectively. These isoprenoid quinones are components of the electron transport chains located specifically in the respective organelles.

Ubiquinone levels in the aleurone tissue began to increase within 12 h of the commencement of imbibition (Fig. 4.12a). They increased by about two-fold between 12 and 48 h of germination and then remained constant for a further 3 days. A particularly important point about this accumulation of ubiquinone was that it occurred independently of the embryo or of gibberellic acid. This indicates that the development of mitochondria in the aleurone tissue during germination is not controlled by gibberellic acid or by any other factor from the embryo (*cf.* phospholipid metabolism in the endomembrane system; see Section IIIC). Hall and Laidman determined tocol levels in the whole embryoectomized grain rather than in the separated starchy endosperm and aleurone tissue. Nevertheless, their data (see Hall and Laidman, 1968) show that the tocols are probably not synthesized in the aleurone tissue during germination. The levels of α-tocopherol, β-tocopherol and α-tocotrienol remained constant during germination, while the level of β-tocotrienol decreased from about 10 μg/30 grains in the quiescent grain to about 6 μg/30 grains after 6 days. This general picture for the tocols is very similar to that described previously for the glycolipids and sterols.

In the embryo axis, ubiquinone levels also began to increase within 12 h of the start of imbibition and then increased rapidly for the rest of the experimental period of 6 days (Fig. 4.12b). Accumulation of plastoquinone did not commence until some 36 h later than that of ubiquinone, but its level then also increased rapidly. The pattern of α-tocopherol levels in the embryo axis was very similar to that of plastoquinone. It was concluded that mitochondrion development occurs very early during germination, but development of plastids does not occur until the third day of germination. The latter conclusion agrees roughly with that obtained from the glycolipid analyses discussed in Section III.D.1 (allowing for the fact that the two analyses were conducted on different wheat cultivars).

In their study of the germinating wheat embryo, Hall and Laidman (1968) showed that the synthesis of α-tocopherol and ubiquinone is independent of light. This was consistent with earlier reports from maize and barley shoots (Griffiths *et al*., 1967), where levels of α-tocopherol and ubiquinone were shown to be unaffected by light. Griffiths *et al*. also demonstrated that the synthesis of phylloquinone, plastoquinone and α-tocopherolquinone in etiolated tissues were stimulated upon illumination. This correlated with chloroplast thylakoid development. Lichtenthaler (1969) similarly demonstrated that plastoquinone and various terpenoid lipids increased upon illumination of dark-grown barley seedlings. α-Tocopherol levels, however, also increased on illumination which was at variance with the observations of Griffiths *et al*. (1967).

ACKNOWLEDGEMENTS

We express our sincere appreciation to Mrs J. Macrae who typed the manuscript and Miss W. Jones who prepared the figures for this article. We are also indebted to Professor H. H. Mollenhauer, United States Department of Agriculture. College Station, Texas, U.S., who provided us with the electron micrographs in Fig. 4.2.

N.A.C. and M.C.W. are in receipt of postgraduate research studentships from the Science and Engineering Research Council to whom we are indebted.

REFERENCES

Bechtel, D. B. and Pomeranz, Y. (1978). *Am. J. Bot*. **65**, 75–85.
Beevers, H. (1969). *Ann. N. Y. Acad. Sci*. **168**, 313–324.
Ben-Tal, Y. and Varner, J. E. (1974). *Plant Physiol*. **54**, 813–816.

Berjack, P. and Villiers, T. A. (1970). *New Phytol.* **69**, 919–938.
Bewley, J. D. and Black, M. (1978). *Physiology and Biochemistry of Seeds* Vol. 1. Springer-Verlag, Berlin.
Blowers, L. E., Stormonth, D. A. and Bray, C. M. (1980). *Planta (Berl.)* **150**, 19–25.
Bray, C. M. (1979). In *Recent Advances in the Biochemistry and Physiology of Cereals* (D. L. Laidman and R. G. Wyn Jones, eds), pp. 147–173. Academic Press, London and New York.
Breidenbach, R. W. and Beevers, H. (1967). *Biochem. Biophys. Res. Commun.* **27**, 462–469.
Briggs, D. E. (1972). *Planta (Berl.)* **108**, 351–358.
Buckhout, T. J., Gripshover, B. M. and Morré, D. J. (1981). *Plant Physiol.* **68**, 1319–1322.
Buttrose, M. S. (1963). *Aust. J. biol. Sci.* **16**, 768–774.
Buttrose, M. S. (1971). *Planta (Berl.)* **96**, 13–26.
Capstack, E., Baisted, D. J., Newschwander, W. W., Blondin, G., Rosin, N. and Nes, W. R. (1962). *Biochemistry* **1**, 1178–1183.
Chittenden, C. G., Laidman, D. L., Ahmad, N. and Wyn Jones, R. G. (1978). *Phytochem.* **17**, 1209–1216.
Colborne, A. J., Morris, G. and Laidman, D. L. (1976). *J. exp. Bot.* **27**, 759–767.
Cooper, T. G. and Beevers, H. (1969). *J. biol. Chem.* **244**, 3514–3520.
Cuming, A. C. and Osborne, D. J. (1978). *Planta (Berl.)* **139**, 219–226.
De La Roche, I. A., Andrews, C. J., Pomeroy, M. K., Weinberger, P. and Kates, M. (1972). *Can. J. Bot.* **50**, 2401–2409.
Deltour, R. and Bronchart, R. (1971). *Planta (Berl.)* **97**, 197–207.
Doig, R. I., Colborne, A. J., Morris, G. and Laidman, D. L. (1975a). *J. exp. Bot.* **26**, 387–398.
Doig, R. I., Colborne, A. J., Morris, G. and Laidman, D. L. (1975b). *J. Exp. Bot.* **26**, 399–410.
Doig, R. I. and Laidman, D. L. (1972). *Biochem. J.* **128**, 88P.
Evins, W. H. and Varner, J. E. (1971). *Proc. Natl. Acad. Sci. U.S.A.* **68**, 1631–1633.
Firn, R. D. and Kende, H. (1974). *Plant Physiol.* **54**, 911–915.
Frey-Wyssling, A., Grieshaber, E. and Muhlethaler, K. (1963). *J. Ultrastruct. Res.* **8**, 506–516.
Getz, G. S., Bartley, W., Lurie, D. and Notton, B. M. (1968). *Biochim. Biophys. Acta.* **152**, 325.
Griffiths, W. T., Threlfall, D. R. and Goodwin, T. W. (1967). *Biochem. J.* **103**, 589–600.
Hall, G. S. and Laidman, D. L. (1968). *Biochem. J.* **108**, 475–482.
Hallam, N. D., Roberts, B. E. and Osborne, D. J. (1972). *Planta (Berl.)* **105**, 293–309.
Hargin, K. D., Morrison, W. R. and Fulcher, R. G. (1980). *Cereal Chem.* **57**, 320–325.
Hartmann-Bouillon, M. A. and Benveniste, P. (1978). *Phytochem.* **17**, 1037–1042.
Hartmann-Bouillon, M. A. and Benveniste, P. (1979). *Biol. Cell.* **35**, 183–194.
Harwood, J. L. (1979). *Prog. Lipid Res.* **18**, 55–86.
Jacobsen, J. V., Knox, R. B. and Pyliotis, N. A. (1971). *Planta (Berl.)* **101**, 189–209.

Jelsema, C. L., Morré, D. J., Ruddat, M. and Turner, C. (1977). *Bot. Gaz.* **138**(2), 138–149.
Johnson, K. D. and Kende, H. (1971). *Proc. Natl. Acad. Sci. U.S.A.* **68**, 2674–2677.
Jones, R. L. (1969a). *Planta (Berl.)* **85**, 359–375.
Jones, R. L. (1969b). *Planta (Berl.)* **87**, 119–133.
Jones, R. L. (1969c). *Planta (Berl.)* **88**, 73–86.
Jones, R. L. (1972). *Planta (Berl.)* **103**, 95–109.
Jones, R. L. (1980a). *Planta (Berl.)* **150**, 58–69.
Jones, R. L. (1980b). *Planta (Berl.)* **150**, 70–81.
Kemp, R. J., Goad, L. J. and Mercer, E. I. (1967). *Phytochem.* **6**, 1609–1615.
Koehler, D. E. and Varner, J. E. (1973). *Plant Physiol.* **52**, 208–214.
Kyle, D. J. and Styles, E. D. (1977). *Planta (Berl.)* **137**, 185–193.
Laidman, D. L. (1980). In *Gibberellins–Chemistry, Physiology and Use* (J. R. Lenton, ed.), pp. 77–94. British Plant Growth Regulator Group.
Laidman, D. L. (1982). In *The Physiology and Biochemistry of Seed Development, Dormancy and Germination* (A. A. Khan, ed.), pp. 371–405. Elsevier/North Holland.
Lichtenthaler, H. K. (1969). *Biochim. Biophys. Acta.* **184**, 164–172.
Longo, G. P. and Longo, C. P. (1970). *Plant Physiol.* **45**, 249–254.
Lord, J. M. and Roberts, L. M. (1980). *TIBS* **5**, 271–274.
MacLeod, A. M. and Palmer, G. H. (1966). *J. Inst. Brew.* **72**, 580–589.
MacLeod, A. M. and White, H. B. (1962). *J. Inst. Brew.* **68**, 487–495.
McDonnell, E. M. (1980). Ph.D. Thesis, University of Wales.
McDonnell, E. M., Pulford, F. G., Mirbahar, R. B., Tomos, A. D. and Laidman, D. L. (1982). *J. exp. Bot.* **33**, 631–642.
McKean, M. L. and Nes, W. R. (1977). *Lipids* **12**, 382–385.
Michell, R. M. (1979). *TIBS* **4**, 128–131.
Miflin, B. J., Burgess, S. R. and Shewry, P. R. (1981). *J. exp. Bot.* **32**, 199–219.
Mirbahar, R. B. (1982). Ph.D. Thesis, University of Wales.
Mirbahar, R. B. and Laidman, D. L. (1981a). *Biochem. Soc. Trans.* **9**, 440–441.
Mirbahar, R. B. and Laidman, D. L. (1981b). *Biochem. Soc. Trans.* **9**, 441–442.
Mirbahar, R. B. and Laidman, D. L. (1982). *Biochem. J.* **208**, 93–100.
Mollenhauer, H. H., Morré, D. J. and Jelsema, C. L. (1978). *Bot. Gaz.* **139**(1), 1–10.
Moreau, R. A., Liu, K. D. F. and Huang, A. H. C. (1980). *Plant Physiol.* **65**, 1176–80.
Morré, D. J. (1975). *Ann. Rev. Plant Physiol.* **26**, 441–481.
Mukherji, S., Day, B., Paul, A. K. and Sincar, S. M. (1971). *Physiol. Plant* **25**, 94–97.
Mukhtar, N. O. (1981). Ph.D. Thesis, University of Wales.
Nieuwdorp, P. J. (1963). *Acta Bot. Neerl.* **12**, 295–301.
Nieuwdorp, P. J. and Buys, M. C. (1964). *Acta. Bot. Neerl.* **13**, 559–565.
Oaks, A. and Beevers, H. (1964). *Plant Physiol.* **39**, 431–434.
Obata, T. and Suzuki, H. (1976). *Plant Cell Physiol.* **17**, 63–71.
Okamoto, K. and Akazawa, T. (1979). *Plant Physiol.* **63**, 336–340.
Okamoto, K. Murai, T., Eguchi, G., Okamoto, M. and Akazawa, T. (1982) *Plant Physiol.* **70**, 905–911.
Paleg, L. (1960). *Plant Physiol.* **35**, 293–299.
Paleg, L. and Hyde, B. (1964). *Plant Physiol.* **39**, 673–680.

Pulford, F. G. (1980). Ph.D. Thesis, University of Wales.

Schwarzenbach, A. M. (1971). *Cytobiologie* **4**, 145–147.

Sevinate-Pinto, I., Pais, M. S. and Marty, F. (1981). *Ann. Sci. Nat., Bot.* **XIII**, **2** and **3**, 11–25.

Shewry, P. R., Pinfield, N. J. and Stobart, A. K. (1973). *J. exp. Bot.* **24**, 1100–1105.

Sorokin, H. P. (1967). *Amer. J. Bot.* **54**, 1008.

Stormonth, D. A. (1978). In *Developments in the Business and Practice of Cereal Seed Training and Technology* (P. R. Haywood, ed.), pp. 49–76. The Gavin Press, London.

Swift, J. G. and O'Brien, T. P. (1972). *Aust. J. biol. Sci.* **25**, 9–22.

Tavener, R. J. A. and Laidman, D. L. (1972a). *Phytochem.* **11**, 981–987.

Tavener, R. J. A. and Laidman, D. L. (1972b). *Phytochem.* **11**, 989–997.

Tomos, A. D. and Laidmain, D. L. (1979). In *Recent Advances in the Biochemistry and Physiology of Cereals* (D. L. Laidman and R. G. Wyn Jones, eds), pp. 119–146. Academic Press, London and New York.

Torres, J. V., Carbonero, P. and Garcia-Olmedo, F. (1976). *Phytochem.* **15**, 677–680.

Van der Eb and Nieuwdorp, P. J. (1967). *Acta Bot. Neerl.* **15**, 690–699.

Varty, K. and Laidman, D. L. (1976). *J. exp. Bot.* **27**, (99) 748–758.

Vick, B. and Beevers, H. (1978). *Plant Physiol.* **62**, 173–178.

Vigil, E. L. and Ruddat, M. (1973). *Plant Physiol.* **51**, 549–562.

Vredevoogd, C. L. and Ruddat, M. (1974). Gibberellin Control of Phospholipid metabolism in barley aleurone cells. Paper presented at Midwest Section, American Society of Plant Physiologists, Aug. 15–16, Colombus, Ohio.

Wanner, G., Formanek, H. and Theimer, R. R. (1981). *Planta (Berl.)* **151**, 109–123.

Wanner, G. and Theimer, R. R. (1978). *Planta (Berl.)* **140**, 163–169.

Yatsu, L. Y. and Jacks, T. J. (1972). *Plant Physiol.* **49**, 937–943.

Yomo, H. (1960). *Makko Kyokaishi* **18**, 603.

5 Starch Lipids in Barley and Malt

D. J. BAISTED
Oregon State University, Corvallis, Oregon 97331, U.S.A.

I INTRODUCTION

A description of the occurrence, analysis and metabolism of cereal lipids has been given in other chapters of this book. An excellent book by Briggs (1978) has also described a wide variety of studies including the biochemistry and physiology of barley and malt from both the pure and applied viewpoints. This chapter will focus attention on a group of lipids, the starch-included lipids, which may prove to play important physiological roles both in the processes of starch deposition during grain filling and in starch breakdown during germination or malting of the barley grain.

II AMYLOSE–LIPID COMPLEXES

Cereal starches are unique in containing monoacyl lipids which are present as inclusion complexes with the amylose component of the starch (Acker

"Lipids in Cereal Technology"
ISBN 0-12-079020-3

and Becker, 1971; Morrison, 1978). The helical structure of amylose, first proposed by Freudenberg *et al.* (1939) and later confirmed by X-ray diffraction studies of the iodine-amylose complex by Rundle *et al.* (1944), provides the basis for the starch-lipid interaction. The blue complex formed with iodine is produced by the polyiodide chain within the helical segments of amylose. The amylose helical structure generated by the glucose residues possesses a hydrophobic interior (Krog, 1971) which can be occupied by lipophilic substances of suitable dimensions, e.g. aliphatic alcohols, fatty acids and other lipids (Acker and Becker, 1971; Bade, 1974; Krog; 1971). The adducts of amylose formed with butanol (Schoch, 1942) and fatty acids (Schoch and Williams, 1944) display X-ray diffraction patterns (Mikus *et al.*, 1946) largely identical and compatible with the formation of amylose-lipid inclusion complexes. Other evidence consistent with this view is the decreased iodine-binding capacity of lipid-included starch as compared with the iodine-binding capacity of starch extracted with hot water-saturated butan-1-ol (Acker and Becker, 1971, 1972; Bolling and El Baya, 1975). Acker and Becker (1971, 1972) also found that low-amylose, waxy maize starch has very little lipid, and high amylose amylomaize starch has more lipid than normal maize starch. This suggests a correlation between amylose content and lipid content. Finally, lysophosphatidylcholine (LPC) in a mixture with amylose displays characteristic absorption bands which disappear when the same proportion of LPC is present in an inclusion complex with amylose.

The naturally occurring starch lipids, unlike non-starch lipids, are not readily removed with the usual dry organic solvents (Morrison, 1978). This is a reflection of the tight binding of the acyl chain in the hydrophobic amylose helical core. The lipids are removed by repeated extraction with hot water-saturated butanol, a process where the tightly bound lipid is displaced by the butanol.

Barley starch lipids, like the other cereal starch lipids, represent about 1% by weight of the cereal starch. The composition is lysophosphatidylcholine (LPC) 62%; lysophosphatidylethanolamine 6%; lysophosphatidylinositol 3%; and free fatty acid (FFA) 4%, (Acker and Becker, 1971, 1972; Becker and Acker, 1972). Palmitate and linoleate are the major free fatty acids in the starch lipid mixture and also the major fatty acids providing the acyl chain in LPC (Table 5.1) (Acker and Becker, 1971).

The acyl chain of the LPC from wheat starch has been shown to be located exclusively at the C-1 position (Arunga and Morrison, 1971; Becker and Acker, 1974). The former workers showed that phospholipase C treatment of the isolated LPC produced only 1-monoacylglycerol; the latter group demonstrated the resistance of the isolated LPC to a phospholipase A from *Crotalus atrox* venom. The enzyme activity is specific for

Table 5.1 Fatty acid composition of the starch-bound fatty acids and starch-bound LPC of barley†

	Fatty acid (% of total FA)					
Lipid fraction	14 : 0	16 : 0	18 : 0	18 : 1	18 : 2	18 : 3
Fatty acids	2	45	3	9	34	5
LPC	< 1	44	–	4	46	4

†From Acker and Becker, 1971.

hydrolysis of an acyl group adjacent to the phosphate group, even for LPC (van den Bosch and van Deenen, 1965).

The orientation of the lysophospholipids in the amylose helix is as shown in Fig. 5.1 for LPC. It is based on the behaviour of the starch-bound lipid to enzymatic attack and upon the X-ray diffraction data for the V-form of

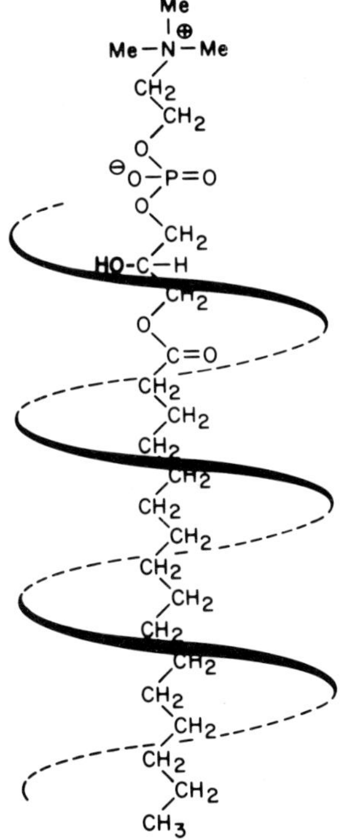

Fig. 5.1 Representation of an inclusion complex of LPC in an amylose helix. The amylose is shown as a left-handed helix with a pitch of 8 Å (from Baisted, 1981).

amylose (Murphy *et al.*, 1975), a form obtained in certain inclusion complexes. The acyl ester group of the polar head of starch-bound LPC is resistant to phospholipase B (Acker and Becker, 1971, 1972) but the choline group is cleaved by phospholipase D (Acker, 1977). Clearly the acyl chain of LPC must be buried inside the amylose helix although the polar head must be sufficiently exposed to allow the phosphorylcholine bond to undergo hydrolysis.

The V-form of the helix has six glucose residues per turn with the pitch of the helix, 8 Å. For an LPC molecule carrying a palmitoyl chain, the lipid chain would be enveloped by 3–4 turns of such a helix. One model for the structure of amylose shows that it exists as an interrupted helix composed of helical segments, 100–120 glucose residues long, joined by short regions of random coil (Holló and Szejtli, 1958; Richter *et al*, 1968). Clearly, each helical segment of amylose could comfortably include two lysophospholipid molecules, each one with its polar head exposed at the ends of the segment (Acker, 1977).

III AMYLOSE-LIPID FORMATION IN DEVELOPING BARLEY

The mechanisms by which the cereal starch lipids come to occupy their compartmented niches in the starch granule are unknown. However, there appears to be a coordination between the activities which produce the included lipid and those which generate the polysaccharide during the grain-filling process. Becker and Acker (1972) have shown that during starch granule synthesis in developing barley, lipid phosphorus and choline accumulation occur in parallel with increasing amylose content. Furthermore, in the early stages of starch deposition, the nonpolar or slightly polar lipids (fatty acids and monoacylglycerols) predominate, only later do the polar lipids, e.g. LPC, increase. The acyl chain composition of the LPC also changes during development. At the earliest stages of grain filling, palmitic and linoleic acid which together constitute 90% of the acyl chains of the LPC, are present in almost equal proportions. As development progresses, the distribution of acyl groups increasingly favours linoleic acid over palmitic acid.

Baisted (1979) found that phosphatidylcholine (PC) and LPC were synthesized from labelled acetate and choline by isolated barley heads at several different stages of development. Maximum incorporation of label into LPC occurred midway through development when the seeds had attained about 60–70% of their maximum fresh weight. A coordination between starch synthesis and starch-bound LPC formation was suggested by the marked reduction in the labelling of LPC when the barley heads were deprived of sucrose, a source of sugar residues for starch synthesis. Although in time-course experiments the labelling patterns for PC and LPC are consistent with PC as the precursor of LPC (Fig. 5.2), there has been no definitive experiment showing this to be true.

In a series of papers Acker and his collaborators (Acker and Müller, 1965; Acker and Gayer, 1968, 1969) have described the isolation of a phospholipase B from barley malt. Prompted by the notion that the

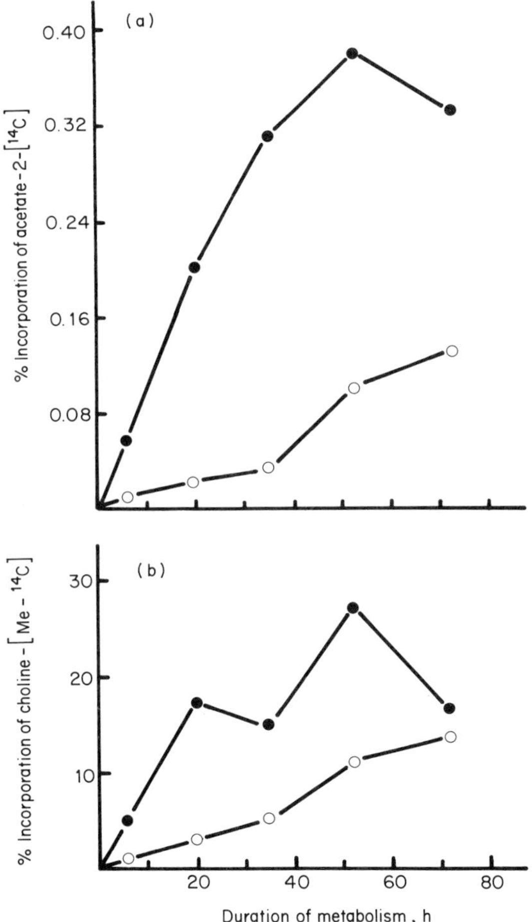

Fig. 5.2 Time course of incorporation of ^{14}C into LPC (○—○) and PC (●—●) from acetate-2-[^{14}C] (a) and choline-Me-[^{14}C] (b) in isolated developing barley heads (from Baisted, 1979).

starch-bound LPC of developing barley may be derived from PC by the action of a phospholipase, Rebmann and Acker (1973) sought such an activity in developing barley. They found that a phospholipase B reached a maximum activity 13 days after flowering; it then declined rapidly to a low, but distinct level. Phospholipase B can act as both a phospholipase A and a lysophospholipase. If the activity described by Rebmann and Acker is indeed the one responsible for the formation of the starch-included LPC, then the lysophospholipase activity of the phospholipase B must be effectively eliminated. This would, of course, be accomplished if the LPC produced by the first hydrolytic step is compartmented inside the amylose helix. However, before the question of LPC synthesis is resolved it will be necessary to show that the enzymatic activity generating the amylose-included LPC is, in fact, located in the starchy endosperm of the developing seed and, in all likelihood, inside the amyloplast in association with the starch synthetase.

What function does the included lipid have in starch synthesis and structure? Experiments have shown that complex formation occurs much more rapidly with saturated monoacyl lipids than with *cis*-unsaturated lipids (Birnbaum, 1971; Krog, 1971; Lagendijk and Pennings, 1970). It is noteworthy then that such a large proportion (approximately 50%) of the acyl chains of LPC is composed of the unsaturated fatty acids. Based on the previous observation of Acker and his colleagues of the parallel syntheses of amylose and the included lipid, Morrison (1978) suggested that the lipids could act as a template for the construction of the helix. The helix is very capable of accommodating the deviations from linear structures created by the presence of the *cis*-unsaturations of the acyl chains.

From the view of the contribution that the included lipids may play a part in starch granule structure it is worth noting that the polar head groups of the included lysophospholipids are the same as those of the phospholipid constituents of cell membranes. In cell membranes these head groups orient themselves to lie parallel to the surface of the membrane bilayer and engage in intermolecular interactions with neighbouring phospholipids (Yeagle *et al.*, 1977; Yeagle, 1978). In the liquid endosperm of the cereal seed during starch deposition it is conceivable that intermolecular interactions of the zwitterionic head groups of the included lysophospholipids may serve in a similar way to provide some structural organization to the amylose segments in the starch granule. The exposed polar head groups which may appear at the surface of the granule may also clearly be contributors to the surface charge of starch granules. They may even act as polar binding sites for the hydrolytic enzymes released into the starchy endosperm during germination.

Downton and Hawker (1975) first suggested that the amylopectin to amylose ratio in starches may be governed by the amylose-LPC interaction. They suggested that the presence of LPC may inhibit the action of the branching enzyme and to support this cited the absence of LPC in low-amylose waxy starches. It is reasonable to suppose that the presence of a polar cap at the non-reducing end of an amylose segment may interfere with either the binding of the branching enzyme or the reaction catalysed by it. It may also be anitcipated that enzymes such as β-amylase and phosphorylase, which attack the non-reducing end of a starch chain, may be similarly inhibited by a lipid-included amylose chain. However, there has been no direct test of this. Vieweg and de Fekete (1976) have shown that amylose in the presence of large amounts of phospholipids is virtually resistant to attack by phosphorylase and only poorly degraded by barley β-amylase; even a bacterial α-amylase is 75% inhibited. Interestingly, starch synthesis from glucose-1-phosphate and phosphorylase proceeded at 60% of the rate of that with lipid-free amylose. A starch synthetase from maize bundle-sheath cells was little affected by the presence of phospholipids, but the branching enzyme of maize endosperm did not produce any amylopectin from the lipid-rich amylose. From these data the authors reach the same conclusion as Downton and Hawker (1975); the presence of lipids can affect the amylopectin to amylose ratio in starch granules. However, the nature of the amylose-lipid association in Vieweg and de Fekete's experiments is not the same as the inclusion complex discussed by Downton and Hawker.

One of the interesting properties given to amylose by the inclusion lipids is the enthalpy of melting observed by differential scanning calorimetry (DSC) (Kugimiya and Donovan, 1981). Iodine-binding is commonly used to determine the amylose content of starches (Bates *et al*., 1943; McCready and Hassid, 1943; Larson *et al.*, 1953; Banks *et al.*, 1970). In recent years, the phase transitions starches undergo on heating in the presence of water have been studied by DSC (Stevens and Elton, 1971; Donovan, 1979; Wooton and Bamunuarachchi, 1979; Donovan and Mapes, 1980). Above the gelatinization temperature, transitions occur in the presence of water which arise from the melting or disordering of starch lipid complexes. Based on these observations Kugimiya *et al.* (1980) hypothesized that the enthalpies of melting of the complexes formed when starches are heated with water and excess LPC were proportional to amylose contents. Accordingly, in laboratories equipped with DSC instruments, this technique now provides a simple alternative to the iodine binding method for the amylose content of starches.

IV AMYLOSE-LIPID COMPLEXES DURING GERMINATION

The breakdown of cereal seed reserves during germination has long been the object of extensive research. The earliest study by Brown and Morris (1890) described the asymmetric modification of the endosperm tissue of barley; the site of synthesis of the diastatic enzyme was the epithelium of the scutellum in the germinating seed. In recent years, the role of the scutellum rather than the aleurone as the site of hydrolase release at the onset of germination has been a point of some controversy (Okamoto *et al.*, 1980; Palmer, 1982). Although there is no doubt that the aleurone layer plays an important role in the gibberellic acid-induced synthesis of several hydrolases (Trewavas, 1976), direct evidence showing the aleurone layer to be the initial site of synthesis of these hydrolases is not available. Indeed, the fact that the scutellum plays a major role in the synthesis, during the initial stages of germination, of the hydrolytic enzymes has been shown through the use of fluorescence techniques (Gibbons, 1979, 1980; Jensen and Heltved, 1982), the substrate film technique (Okamoto *et al.*, 1980), scanning electron microscopy and freezefracture replicas (Gram, 1982 a,b).

The development of lipolytic activity during barley germination has received increasing attention in recent years (McLeod and White, 1962; Narziss and Sekin 1974; Wainwright, 1980; Baisted and Stroud 1982 a,b; Jensen and Heltved, 1982.

The pattern of development of the activity during barley germination has been followed by taking advantage of the hydrolysis of non-fluorescent fluorescein dibutyrate by "lipase" (Guilbault and Kramer, 1964; Guilbault and Hiesenman, 1969). Fluorescein, the product of hydrolysis, is a highly fluorescent material which thereby permits the location of lipase activity in longitudinally cut barley half-seeds to be visualized (Jensen and Heltved, 1982)*.The pattern of development of lipolytic activity parallels that of endosperm cell wall breakdown (Fig. 5.3). The latter was also observed by a fluorescent agent, Calcofluor, which stains barley cell walls (Gibbons, 1980; Wood and Fulcher, 1978). Interestingly, Jensen and Heltved found that the embryo contains a heat-resistant "lipase" which was also present in half-seeds of seeds germinated for six days and which had then been

*A Malt-modification Analyser System is now commercially available. It consists of a system for seed fixation and UV-light box equipped with various filters. Measurement of malt modification and also development of a number of enzymatic activities in both pregermination and preharvest sprouting individual cereal seeds is facilitated. Further information may be obtained from Dr. Svend Jensen, Department of Biotechnology, Carlsberg Research Laboratory, Gamle Carlsberg Vej 10, DK-2500, Copenhagen, Valby, Denmark.

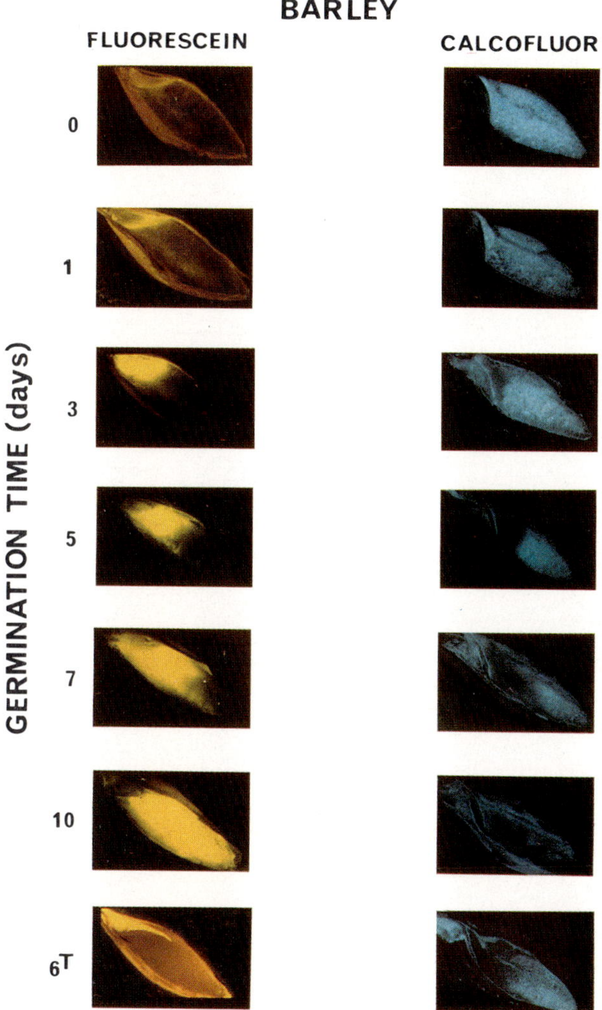

Fig. 5.3 Formation of activity of enzymes catalysing the hydrolysis of fluroescein dibutyrate into fluorscein (left) and progression of cell wall breakdown (right) in germinating barley seeds. Longitudinal sections were treated with fluorescein dibutyrate and Calcofluor, respectively (from Jensen and Heltved, 1982).

exposed to boiling water for 10 min (6^T, Fig. 5.3). Such an activity might well resist the kilning step in the preparation of malt.

In our studies (Baisted 1981; Stroud and Baisted, 1982 a,b) we have examined the fate of the amylose-included LPC during germination of barley. The early events in germination lead to the production of a battery of hydrolases released from the aleurone layer into the starchy endosperm in response to gibberellic acid produced in the embryo (Varner and Ho, 1976). A study (Baisted, 1981) of the potential resistance to attack by degradative enzymes of amylose chains included with lipid was made by measuring amylase activity and the residual starch-bound lipid phosphorus and LPC in seeds germinated up to eight days. The data shown in Fig. 5.4 reveal that the starch-bound lipid phosphorus declines from approximately 0.4 to 0.1 μmols per seed between day 4 and day 7. Amylase becomes active after day 2 and reaches a maximum at day 4. Clearly, 75% of the lipid is lost from the starch during its degradation. Conversely it is also clear that a certain proportion of the starch must be resistant to attack. About 65% of the lipid phosphorus is present as LPC in the dry seed. This percentage remains unchanged during most of the loss of lipid, indicating no selection for amylose chains carrying LPC. Only in the later stages of germination (days 6, 7 and 8) does there appear to be a selection for the degradation of LPC-bearing amylose chains (Fig. 5.4).

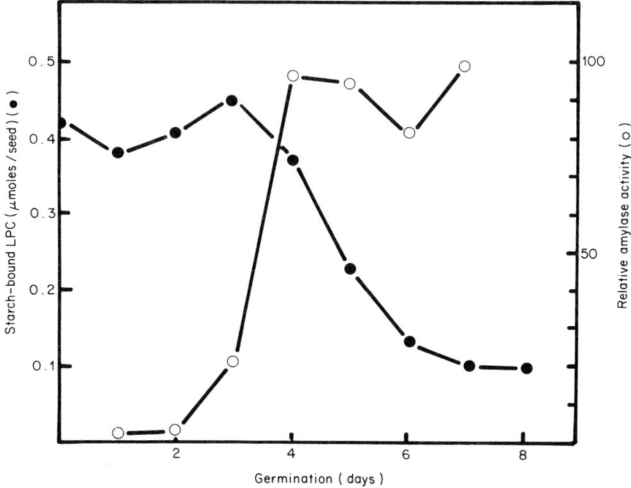

Fig. 5.4 Increase of amylose activity and loss of starch-bound LPC in germinating barley (from Baisted, 1981).

This may reflect a non-uniform distribution of the lysophospholipid within a starch granule or different lysophospholipid distributions between the two sizes of granule common to barley (Banks *et al.*, 1973). A scanning electron microscope study of malting barley showed the small granules (average dia. 5 μm) to be degraded before the large granules (average dia. 15–35 μm) (Palmer, 1972). A correlation of the included lipid content with the amylose content of the specific population of granules is inconclusive as there are conflicting results concerning the amylose and amylopectin composition of the two granule types (Bathgate and Palmer, 1972; Evers *et al.*, 1974; Williams and Duffus, 1977).

Palmitoyl and linoleoyl groups represent about 90% of the acyl chains of the included LPC in the dry seed (Fig. 5.5). Measurement of the acyl composition of the starch-included LPC during germination reveals that during days 4 and 5, when there is the greatest loss of such included lipid (Fig. 5.4) there is proportionally more of the LPC molecules carrying $C_{18:2}$ lost than those carrying $C_{16:0}$. This may reflect the different distributions of the two types of LPC molecule in the starch granule which result during the grain-filling process (Acker and Becker, 1972); there is an increasing proportion of linoleic acid relative to palmitic acid incorporated into the

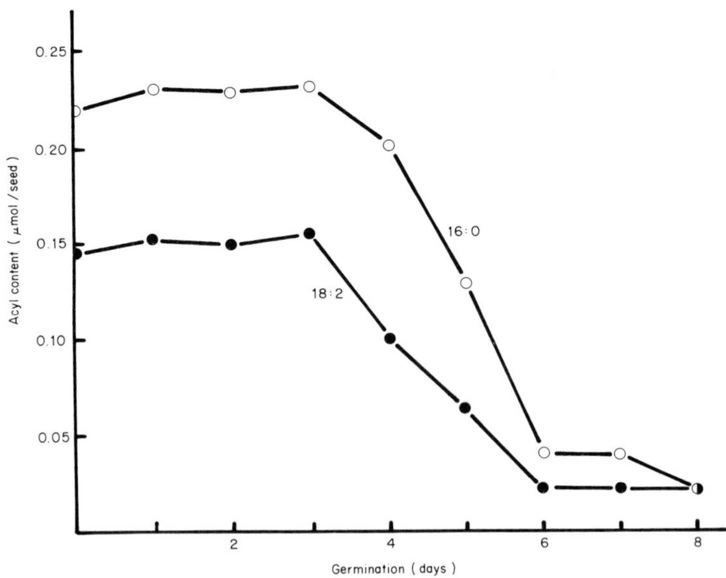

Fig. 5.5 Palmitoyl and linoleoyl content of starch-bound LPC in germinating barley (from Baisted, 1981).

LPC molecule during starch granule synthesis in developing barley. Alternatively, it may reflect the greater susceptibility to amylase attack of a more expanded amylose helix which would be required to accommodate the diunsaturated acyl chain of linoleic acid.

Although 0.3 μmoles/seed of LPC disappears from the starch during germination, the level of free LPC in the seed remains relatively constant between 0.02 and 0.04 μmoles/seed. This indicated to us that an enzyme with lysophospholipase activity must be present at least by day 4 of germi-

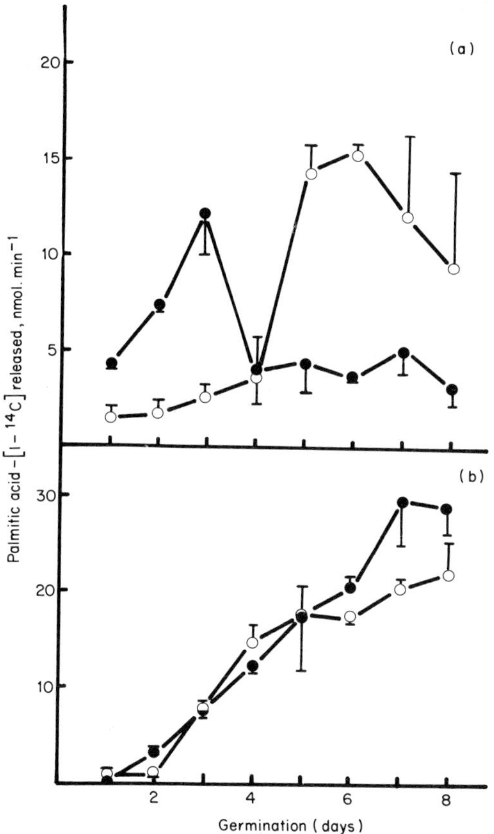

Fig. 5.6 Changes in soluble (●—●) and particulate (○—○) lysophospholipase activities in the starchy endosperm (a) and aleurone (b) of germinating barley. After removal of the embryonic tissue from duplicate groups of 5 seeds each, the starchy endosperm was separated from the remaining tissue leaving the aleurone attached to the testa, pericarp and husk. The enzyme activity was measured as described in the legend to Table 5.2 (from Baisted and Stroud, 1981a).

nation. Such an activity was first described by Contardi and Ercoli (1933). We have identified both a soluble and particulate form of the activity in the starchy endosperm and aleurone layer (Fig. 5.6) and the scutellum (Baisted and Stroud, 1982 a,b) of the germinating seed. The enzyme is active over a broad pH range with a maximum activity occurring at pH 8. Interestingly, the particulate activity in the starchy endosperm becomes bound to the starch fraction after day 4, when the loss of included LPC accelerates (Baisted and Stroud, 1982b; Fujikura and Baisted, 1983).

The steady state level of free LPC we have measured in the germinating seed is 0.02 to 0.04 μmols, which would be minimally equivalent to 0.2 to 0.4 mM. It is important to note that a concentration in the range 3–6 μM of LPC is sufficient to lyse human erythrocytes at room temperature within about 1 min (Collier, 1952; Reman *et al.*, 1969; Weltzien, 1979). The aleurone layer is evidently resistant to this and higher concentrations of LPC. We have found no evidence of a cytoplasmic enzyme, malate dehydrogenase, being released through lysis from aleurones incubated in solutions of LPC in concentrations as high as 0.5 mM (Lundgard and Baisted, unpublished observations). The protection of the aleurone layer is probably given by the particulate lysophospholipase located in the aleurone and by the residual cell wall which is resistant to the cell wall-digesting enzymes (Moll and Jones, 1982).

The barley "lipase" activity revealed by the fluorescent technique of Jensen and Heltved (1982), like the proteinase, RNAase and α-amylase of barley described by Okamoto *et al.* (1980) is initiated in the embryonic tissues. Briggs (1964) had also earlier shown that 6.5% of the endospermic α-amylase of malt was of embryonic origin. A gradient of lysophospholipase activities from the embryo end to the distal end of the germinating seed was shown to be decreased during a two day germination interval (Table 5.2); the ratio of the activities (+/−) declines between days 4 and 6 (Baisted and Stroud, 1982a). This was consistent with the pattern of lipase generation observed by Jensen and Heltved (1982). As the timing of the appearance of lysophospholipase in the starchy endosperm was similar to that of the gibberellic acid (GA)-induced hydrolytic enzymes, the dependence of the activity on GA was investigated both in embryo-free half-seeds and in isolated aleurone layers. In each case, a marked stimulation of enzyme activity was observed (Table 5.3, data for isolated aleurones only).

The nature of the activation and the origin of the starch-bound lysophospholipase activity is presently being studied in this laboratory. Also, as the aleurone becomes an active secretory membrane during germination and lysophospholipids are known to be fusogenic agents, the potential role of lysophospholipids as promoters of the secretory process is being investigated.

Table 5.2 Lysophospholipase activities of fractions from the embryo-containing (+) and embryo-free (−) halves of germinated barley seeds.

Germination period (days)	Amount of palmitate released (nmol min^{-1} total $fraction^{-1}$)											
	Aleurone						Starch endosperm					
	Soluble			Particulate			Soluble			Particulate		
	+	−	+/−	+	−	+/−	+	−	+/−	+	−	+/−
4	15.1	6.0	2.5	5.0	3.4	1.5	6.8	1.8	3.8	4.5	0.8	5.6
6	18.8	12.4	1.5	15.4	11.7	1.3	7.5	3.3	2.3	15.5	8.4	1.8

After 4 and 6 days of germination the roots and shoots were removed from batches of 10 barley seedlings. The remaining tissue was cut transversely to give halves with (+) and without (−) the embryos. Lysophospholipase activities were assayed in soluble and particulate fractions of the aleurone and starch endosperm by measuring the rate of release of palmitic acid-1-^{14}C from 1-[1-^{14}C] palmitoyl LPC. (From Baisted and Stroud, 1982b.)

Table 5.3 Influence of GA on acidic and alkaline lysophospholipase activities of isolated aleurones from embryo-free barley half-seeds

		Amount of palmitate released (nmol min^{-1} total fraction^{-1})		
			Aleurone	
		Incubation medium	Soluble	Particulate
Control	Acidic	< 0.2	4.0	1.6
	Alkaline	7.0	8.3	7.9
1 μM GA_3	Acidic	6.3	53.8	14.0
	Alkaline	15.3	124.8	70.4

Aleurones were separated from surface-sterilized half-seeds which had imbibed water for 3 days from 1% agar plates. Groups of 10 aleurones were incubated in either 1 mM acetate (pH 5) or 1 μM GA in the same buffer. After 24 h at 21° the tissues were washed free of the incubation medium. The acid (pH 5) and alkaline (pH 8) lysophospholipase activities were measured in the aleurone soluble and particulate fractions and in the incubation medium as described in Table 5.2. (From Baisted and Stroud, 1982b.)

V LIPIDS IN MALT

A recent annual report of the U.K. Brewing Research Foundation (1980–81) summarizes the situation: "There is little precise knowledge of which lipids are present in barley and malt and how processing conditions influence the quantities of those which are extracted into malt and which persist in beer." The major problem appears to have been that until recently the reproducibility of the analytical methods has been inadequate.

A review of the lipid analyses of barley and the distribution of the lipid classes among the various seed parts has been given by Morrison (1978). The major fatty acids of barley are linoleic (approximately 53%) and palmitic (approximately 25%). They are predominantly present as glycerolipids: triacylglycerols (72–74%) and the phospho- and glyco-lipids (16–19%). Although the embryo has only 10–15% of the total lipid, it is rich in the fatty acyl lipids.

The value of the lipid analyses of barley and malt has been questioned recently. It is thought that measurement of the enzyme activities that can alter the lipid composition during mashing may be more important (Wainwright, 1980). As described below, however, changes can occur in lipid composition during malting even though specific enzyme activities may be destroyed during kilning. The major changes that occur in the lipids during malting result from the action of lipases, lipoxygenases and hydroperoxide isomerase. The reactions catalyzed by these enzymes in cereals have been described elsewhere in this book and have been previously reviewed (Morrison, 1978). In summary, the work of MacLeod and White (1962) and, more recently, Jensen and Heltved (1982) has revealed the development of lipase activity during the germination phase of malting. This activity is responsible for liberating the unsaturated fat reserves from the acylglycerols (Hernandez *et al.*, 1967). Linoleic acid, the major unsaturated fatty acid produced as a result of lipase action, is oxidized by lipoxygenase to hydroperoxylinoleic acid(s) (Lulai *et al.*, 1981; Baxter, 1982); the latter are the common substrates for hydroperoxide isomerase (Lulai *et al.*, 1981; Gardner, 1970; Zimmerman and Vick, 1970). Little is known of the fate of the oxidation products of these reactions, but it is known that these hydroxyl-, carbonyl and hydroperoxyl-containing compounds are readily extracted from the malt and are implicated in causing staling and off-flavours in beer (Graveland *et al.*, 1972; Meilgaard, 1972; Yabuuchi and Amaha, 1975).

It has been shown that lipoxygenase activity increased during germination (Lulai and Baker, 1975). More recently, two lipoxygenase activities have been observed to increase to differing extents during malting (Baxter, 1982). The amount of activity that develops is clearly dependent upon the barley variety. About 98% of the total activity of green malt is in the embryo, but most of this is destroyed during kilning. As most of the oxidation of linoleic acid by lipoxygenase occurs during the germination phase of malting, the steeping phase of the process presents the best opportunity to control the activity. This may be effected by controlling embryo development by the use of inhibitors (e.g. bromate or acid) or by employing anaerobic conditions during steeping. Such treatments would adversely affect modification and would necessitate GA applications because of the reduced synthesis of gibberellins under such conditions (Gordon, 1980). An alternative approach is to reduce the availability of the lipoxygenase substrate, unesterified linoleic acid, by lowering lipase activity. Selecting appropriate barley varieties with low lipase and lipoxygenase activities would clearly be beneficial in counteracting lipid-derived, off-flavour components produced as a result of these enzyme activities.

The hydroperoxide isomerase activity has been shown to be a particulate enzyme which can be solubilized by Triton X-100 (Lulai *et al.*, 1981). Using this treatment, substantially more of the activity is evidently present in barley than previously determined (Yabuuchi and Amaha, 1976). These investigators previously reported that hydroperoxide isomerase decreased during germination, while Lulai and Baker (1981) have now demonstrated that both lipoxygenase and hydroperoxide isomerase increase during germination. The occurrence of the isomerase in new tissue, especially the acrospire and the rootlets, suggests that the activity is important to plant growth during germination.

Finally, a group of lipids has been suggested as important in the breaking of dormancy in barley. Skarsaune *et al*. (1973) found that the wax hydrocarbons contained about 60% $C_{23,25,27,29}$ alkanes. These decreased to 50% during the five-week period following harvesting when dormancy decreased. The investigators suggested that the lipid in the husk and outer layers of bran and aleurone act as an oxygen barrier. During the five-week post-harvest period, changes in lipid composition result in the outer layer becoming more permeable to oxygen, thereby allowing it to enter the embryo.

REFERENCES

Acker, L. (1977). *Fette Seifen Anstrichmittel* **79**, 1–9.
Acker, L. and Becker, G. (1971). *Staerke* **23**, 419–424.
Acker, L. and Becker, G. (1972). *Gordian* **72**, 275–278.
Acker, L. and Geyer, J. (1968). *Lebensmittel-Unters. u.-Forsch.* **137**, 231–237.
Acker, L. and Geyer, J. (1969). *Lebensmittel-Unters. u.-Forsch.* **140**, 269–275.
Acker, L. and Muller, K. (1965). *Nahrung* **9**, 3–14.
Arunga, R. O. and Morrison, W. R. (1971). *Lipids* **6**, 768–776.
Bade, V. (1974). *Getreide, Mehl Brot.* **28**, 296–299.
Baisted, D. J. (1979). *Phytochemistry* **18**, 1293–1296.
Baisted, D. J. (1981). *Phytochemistry* **20**, 985–988.
Baisted, D. J. and Stroud, F. (1982a). *Phytochemistry* **21**, 29–31.
Baisted, D. J. and Stroud, F. (1982b). *Phytochemistry* **21**, 2619–2623.
Banks, W., Greenwood, C. T. and Muir D. D. (1970). *Staerke* **22**, 105–108.
Banks, W., Greenwood, C. T. and Muir, D. D. (1973). *Staerke* **25**, 153–157.
Bates, F. L., French, D., and Rundle, R. E. (1943). *J. Am. Chem. Soc.* **65**, 142–148.
Bathgate, G. and Palmer, G. H. (1972). *Staerke* **24**, 336–341.
Baxter, E. D. (1982). *J. Inst. Brew.* **88**, 390–396.
Becker, G. and Acker, L. (1972). *Fette Seifen Anstrichmittel* **74**, 324–327.
Becker, G. and Acker, L (1974). *Fette Seifen Anstrichmittel* **76**, 464–466.
Birnbaum, H. (1971). *Baker's Digest* **45**, 22–24, 27, 29.
Bolling, H. and El Baya, A. W. (1975). *Chem. Mikrobiol Technol. Lebensm.* **3**, 161–163.
Brewing Research Foundation Annual Report (1982). *J. Inst. Brew.* **88**, 111–121.
Briggs, D. E. (1964). *J. Inst. Brew.* **70**, 14–24.
Briggs, D. E. (1978). *Barley.* Chapman and Hall, London.
Brown, H. T. and Morris, G. H. (1890). *J. Chem. Soc.* **57**, 458–528.
Collier, H. B. (1952). *J. Gen. Physiol.* **35**, 617–628.
Contardi, A. and Ercoli, A. (1933). *Biochem. Z.* **261**, 275–302.
Donovan, J. W. (1974). *Biopolymers* **18**, 263–275.

Donovan, J. W. and Mapes, C. J. (1980). *Stearke* **32**, 190–193.
Downton, W. J. S. and Hawker, J. S. (1975). *Phytochem.* **14**, 1259–1263.
Evers, A. D., Greenwood, C. T., Muir, D. D. and Venables, C. (1974). *Staerke* **26**, 42–46.
Freudenberg, K., Schaaf, E., Dumpert, G. and Ploetz, Th. (1939). *Naturwiss.* **27**, 850–853.
Fujikura, Y. and Baisted, D. J. (1983). *Phytochemistry* **22**, 865–868.
Gardner, H. W. (1970). *J. Lipid Res.* **11**, 311–321.
Gibbons, G. C. (1979). *Carlsberg Res. Commun.* **44**, 353–366.
Gibbons, G. C. (1980). *Carlsberg Res. Commun.* **45**, 177–184.
Gordon, I. L. (1980). *Cereal Res. Commun.* **8**, 115–129.
Gram, N. H. (1982a). *Carlsberg Res. Commun.* **47**, 143–162.
Gram, N. H. (1982b). *Carlsberg Res. Commun.* **47**, 173–186.
Graveland, A., Pesman, L. and van Erde, P. (1972). *Tech. Quart. Master Brew. Assoc. Amer.* **9**, 98–104.
Goilbault, G. G. and Hiesenman, J. (1969). *Anal. Chem.* **41**, 2006–2009.
Guilbault, G. G. and Kramer, D. N. (1964). *Anal. Chem.* **36**, 409–412.
Hernandez, H. H., Banasik, O. J. and Gilles, K. H. (1967) *Am. Soc. Brew. Chemists Proceedings*, 24–31.
Holló, J. and Szejtli, J. (1958). *Staerke* **10**, 49–52.
Jensen, S. A. and Heltved, R. (1982). Carlsberg Res. Commun. **47**, 297–303.
Krog, N. (1971. *Staerke,* **23**, 206–210.
Kugimiya, M., Donovan, J. W. and Wong, R. Y. (1980). *Staerke* **32**, 265–270.
Kugimiya, M. and Donovan, J. W. (1981). *J. of Food Sci.* **46**, 765–777.
Lagendijk, J. and Pennings, H. J. (1970). *Cereal Sci. Today* **15**, 354–356.
Larson, B. L., Gilles, K. A. and Jenness, R. (1953). *Anal. Chem.* **25**, 802–804.
Lulai, E. C. and Baker, C. W. (1975). *Proc. Am. Soc. Brew. Chem.* **33**, 154–158.
Lulai, E. C., Baker, C. W. and Zimmerman, D. C. (1981). *Plant Physiol.* **68**, 950–955.
MacLeod, A. M. and White, H. B. (1962). *J. Inst. Brew.* **68**, 487–495.
McCready, R. M. and Hassid, W. Z. (1943). *J. Am. Chem. Soc.* **65**, 1154–1157.
Meilgaard, M. (1972). *Brewers Digest* **47**, 48–62.
Mikus, F. F., Hixon, R. M. and Rundle, R. E. (1946). *J. Am. Chem. Soc.* **68**, 1115–1123.
Moll, B. A. and Jones, R. L. (1982). *Plant Physiol.* **70**, 1149–1155.
Morrison, W. R. (1978). *Adv. in Cereal Sci. and Technol.* **2**, 221–348.
Murphy, V. G., Zaslow, B. and French, A. D. (1975). *Biopolymers* **14**, 1487–1501.
Narziss, L. and Sekin, Y. (1974). *Brauwissenschaft* **27**, 311–320.
Okamoto, K., Kitano, H. and Akazawa, T. (1980). *Plant and Cell Physiol.* **21**, 201–204.
Palmer, G. H. (1972). *J. Inst. Brew.* **78**, 326–332.
Palmer, G. H. (1982). *J. Inst. Brew.* **88**, 145–153.
Rebman, H. and Acker, L. (1973). *Fette Seifen Anstrichmittel* **75**, 409–411.
Reman, F. C., Demel, R. A., DeGier, J., van Deenen, L. L. M., Eibel, H. and Westphal, O. (1969). *Chem. Phys. Lipids* **3**, 221–233.
Richter, M., Augustat, S. and Schierbaum, F. (1968). *Ausgewahlte Methoden der Stärkechemie*, Wiss. Verlagsges. mbH, Stuttgart p. 176 (cited from Acker, 1977).
Rundle, R. E., Foster, J. F. and Baldwin, R. R. (1944). *J. Am. Chem. Soc.* **66**, 2116–2120.
Schoch, T. J. (1942). *J. Am. Chem. Soc.* **64**, 2957–2961.
Schoch, T. J. and Williams, C. B. (1944). *J. Am. Chem. Soc.* **66**, 1232–1233.

Skarsaune, S., Banasik, O. J. and Watson, C. A. (1973). *Proc. Am. Soc. Brew. Chem.* 94–97.

Stevens, D. J. and Elton, G. A. H. (1971). *Staerke* **23**, 8–11.

Trewavas, A. J. (1976). In *Molecular Aspects of Gene Expression in Plants* (K. A. Bryant, ed.), 249–298. Academic Press, London and New York.

van den Bosch, H. and van Deenen, L. L. M. (1965). *Biochim. Biophys. Acta* **106**, 326–337.

Varner, J. E. and Ho, D. T. (1976). In *The Molecular Biology of Hormone Action* (J. Papaconstantinou, ed.), 173–194. Academic Press, London and New York.

Vieweg, G. H. and de Fekete, M. A. R. (1976). *Planta* **129**, 155–159.

Wainwright, T. (1980). *Proceedings of European Brewery Convention, Symposium on Malt and Beer*, Helsinki, 118–127. (cited in *J. Inst. Brew.* (1983) 89, 51).

Weltzien, H. U. (1979). *Biochim. Biophys. Acta* **559**, 259–287.

Williams, J. M. and Duffus, C. M. (1977). *Plant Physiol.* **59**, 189–192

Wood, P. J. and Fulcher, R. G. (1978). *Cereal Chem.* **55**, 952–966.

Wooton, M. and Bamunuarachchi, A. (1979). *Staerke* **31**, 201–204.

Yabuuchi, S. and Amaha, M. (1975). *Phytochemistry* **14**, 2569–2572.

Yabuuchi, S. and Amaha, M. (1976). *Phytochemistry* **15**, 387–390.

Yeagle, P. L. (1978). *Acc. of Chem. Res.* **11**, 319–327.

Yeagle, P. L., Hatton, W. C., Huang, C. and Martin, R. B. (1977). *Biochemistry* **16**, 4344–4349.

Zimmerman, D. C. and Vick, B. A. (1970). *Plant Physiol.* **46**, 445–453.

6 Enzymic Degradation of Cereal Lipids

T. GALLIARD

The Lord Rank Research Centre, High Wycombe, Bucks., U.K.

I INTRODUCTION

The major factors in the ubiquitous use of cereal grains are their nutritional value, ease of harvesting and stability during storage. Clean, sound, undamaged grains normally store well but damage and tissue disruption can result in intermixing of cellular components and this can initiate degradative reactions. The degradation of lipids in cereal grains is associated with the loss of functional quality in raw materials or with undesirable changes in the organoleptic properties of products.

Lipid breakdown in cereal grains and in products derived from them is due

"Lipids in Cereal Technology"
ISBN 0-12-079020-3

to the catalytic reactions in which the catalysts may be enzymes or other factors such as metal ions and light. This chapter is concerned specifically with enzymic reactions; the reader who requires information on non-enzymic reactions in food materials should refer to the two recent publications by Simic and Karel (1980) and Allen and Hamilton (1983).

The major pathways of lipid degradation (both enzymic and non-enzymic) are by hydrolytic or oxidative reactions – or by a combination of both. Hydrolytic and oxidative reactions will be discussed separately in this chapter but the interrelationships between the two types of process should be emphasized. For example, hydrolysis may produce free fatty acids (FFA) which can act as substrates for subsequent oxidation reactions. Rancidity may be produced by a hydrolytic reaction (hydrolytic rancidity) or as a result of reactions with oxygen (oxidative rancidity).

In his comprehensive review of cereal lipids, Morrison (1978a) described some of the more important enzymes that catalyse degradation of lipids; a recent review by Gardner (1980) includes a discussion of some enzyme-catalysed lipid degradation reactions in cereals and other food materials; a more general coverage of enzymic degradation of plant lipids has been presented (Galliard, 1980).

II HYDROLYTIC ENZYMES

A Enzymic Formation of Free Fatty Acids (FFA)

Enzymes that catalyse liberation of FFA from acyl lipids, by hydrolytic cleavage of acyl ester bonds, are referred to usually as "lipases". However, among the lipases are enzymes that differ significantly in their specificities and modes of action. For the following discussion, the FFA-liberating enzymes, known to occur in cereal grains, can be divided into two distinct groups.

Triglyceride lipases hydrolyse triglycerides (TG) at an oil-water interface and do not hydrolyse lipids in solution. In contrast, *lipid acyl hydrolases* (LAH) act on monodisperse or micellar forms of lipids and do not hydrolyse water-insoluble TG. LAH enzymes have been variously described as phospholipases A and B, lysophospholipase, galactolipase, monoglyceride lipase, some ill-defined "lipases" as well as lipid (or lipolytic) acyl hydrolases.

1 *Lipases*

The TG lipases that act on water-insoluble TG, should be distinguished

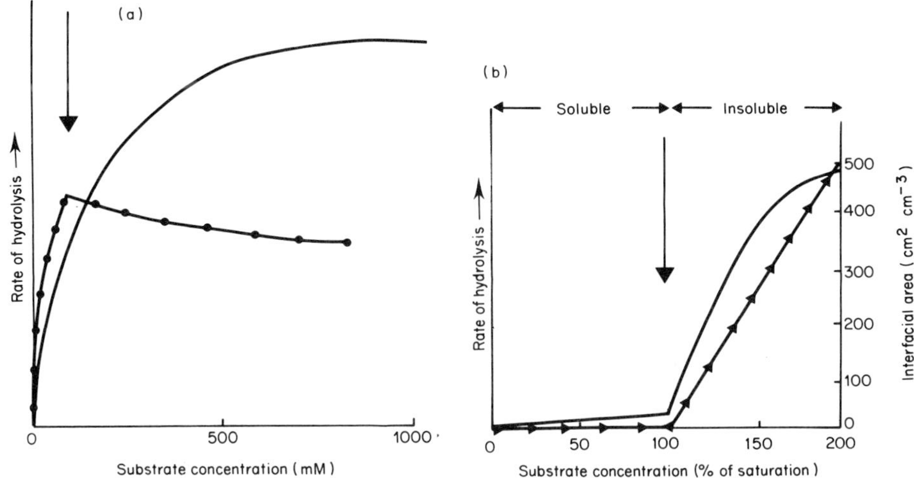

Fig. 6.1 Comparison of substrate concentration curves for (a) an esterase acting on ethyl butyrate (——) and on tributyrin (-●-●-●-●-) and (b) pancreatic lipase acting on methyl butyrate (———). Note that the rate of hydrolysis in the esterase-catalysed reaction does not increase when substrate concentrations exceed that of a saturated solution (arrow in part a indicates saturation point for tributyrin); lipase-catalysed hydrolysis is only effective when substrate concentrations exceed that of a saturated solution (arrow in part b) and the reaction rate rises with increased interfacial area of dispersed substrate (-▲-▲-▲-▲) (adapted from Sarda and Desnuelle, 1958; and from Dixon and Webb, 1964).

from "esterases" that hydrolyse water-soluble substrates, including short-chain TG (e.g. triacetin). The distinction is illustrated in Fig. 6.1 which shows how enzyme activity varies with substrate concentration for a typical esterase (Fig. 6.1a); hydrolysis rates increase with increasing substrate concentrations up to the saturation level, at which point further addition of substrate causes no increase in hydrolysis rate. Lipases (Fig. 6.1b) have little or no activity at substrate concentrations below that at which oil droplets form, and the activity increases as the interfacial area of the oil phase increases. This distinction between lipases and esterases is important in relation to assays for lipase activity (see Appendix 2, p. 404).

Lipases differ markedly from most of the enzymes commonly encountered in foods in having significant activities at very low water activities. This unusual property is illustrated in Fig. 6.2. It can be seen that mixtures of wheat bran and olive oil liberated significant amounts of oleic acid, even when the water activity was below 0.2, and that enzyme activity increased gradually to attain a maximum level at a_w = 0.85 (Drap-

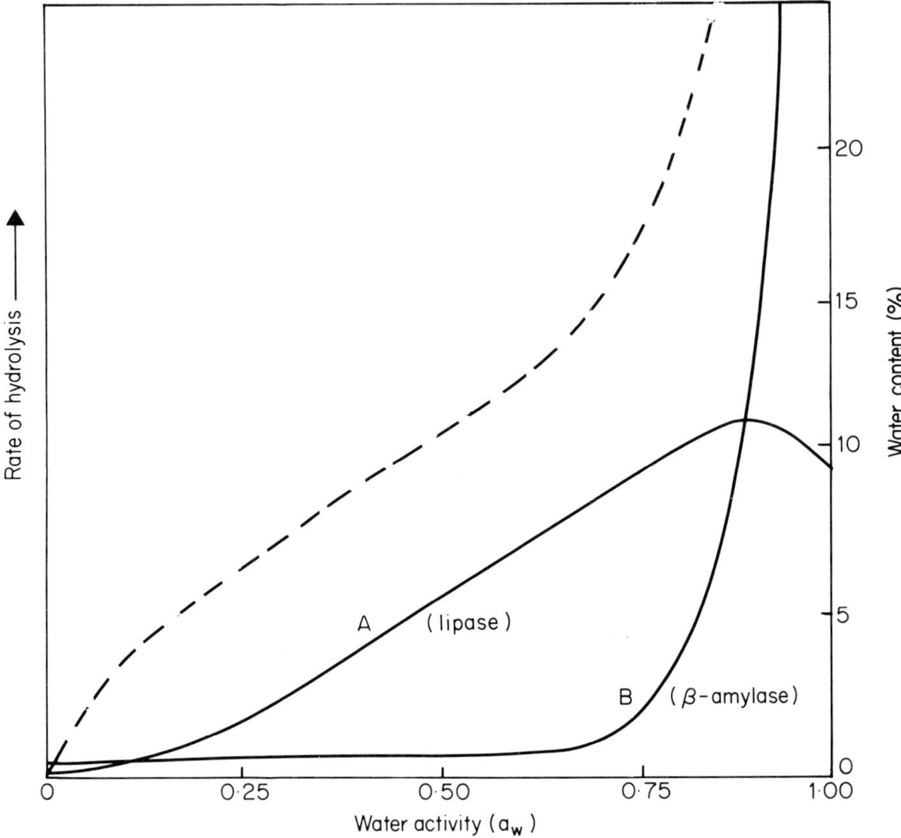

Fig. 6.2 Relationship between water activity (a_w) and enzyme activity for hydrolysis reactions catalysed by (A) wheat lipase acting on a triolein-flour mixture, and (B) β-amylase acting on starch. The broken line represents the sorption isotherm for both systems (adapted from Drapon, 1972, 1983).

ron, 1972). In contrast, β-amylase activity on starch is not detected at moisture contents below the point at which free water is present, as illustrated by the sorption isotherm and β-amylase activity curves in Fig. 6.2B (Drapron and Guillebot, 1962; Drapron, 1983). Other studies with oat lipase (Acker and Beutler, 1965) and wheat germ lipase (Rothe, 1958) confirm this effect. The explanation for the unusual activity of lipases is that the enzymes act at the surface of oil bodies and the rate of reaction depends upon diffusion of liquid oil within the material. Unlike most hydrolytic enzymes, which require water both as solvent and substrate, lipases require only sufficient water to act as substrate in the reaction:

$$RCOOX + H_2O \rightarrow RCOOH + XOH$$

However, for hydrolysis to occur, the lipid must be in a fluid state; Acker and Wiese (1972) have used model systems containing lipid-cellulose mixtures to show that liquid, but not solid fat was hydrolysed by lipase at ambient temperature over a period of 6 weeks at E.R.H. levels as low as 2%. The activity of lipase at low moisture contents indicates the potential for hydrolytic rancidity in stored grain or products and can be used also as the basis for lipase assay.

2 *Lipid Acyl Hydrolases (LAH)*

Lipid (or lipolytic) acyl hydrolases (LAH) are now known to occur widely in the plant kingdom (Galliard, 1980) and are characterized by relatively low substrate specificies; i.e. LAH catalyse deacylation, liberating FFA from a range of lipids, including mono- and diacylphospholipids and galactolipids, mono- and di-glycerides and phenolic esters of long-chain fatty acids (Galliard, 1975). This broad specificity explains the range of names previously given to enzymes of this type.

Three important differences distinguish LAH enzymes from TG lipases. First, LAH do not hydrolyse long-chain TG, although some do hydrolyse the partial deacylation products, i.e. monoglycerides (MG) and diglycerides (DG). Second, LAH probably act on monodisperse forms of lipid substrates rather than at an oil-water interface; substrate concentration curves for LAH follow conventional Michaelis-Menten kinetics (i.e. similar to those shown for esterase in Fig. 6.1a. Third, from the limited evidence available, it appears that LAH enzymes are not active at the low moisture levels at which TG lipases are effective; for example, Acker and Luck (1958) showed that phosphatidylcholine hydrolysis in malted barley did not occur at moisture levels below the free-water inflexion point in sorption isotherms. Thus, LAH characteristics in relation to moisture content probably resemble the β-amylase example (in Fig. 6.2B) rather than the lipase pattern (in Fig. 6.2A).

3 *Measurement of Lipolytic Activity*

TG lipase assays involve the determination of FFA formed from endogenous or added TG. Lipase assays applicable to cereal products are described in Appendix 2. Some methods, described in the literature and using water-soluble substrates, are not recommended for reasons discussed previously, TG lipases act at oil-water interfaces and hydrolysis in the aqueous phase indicates esterase rather than lipase activity. Thus, assays using aqueous solutions may not give useful information on the ability of the material under investigation to catalyse FFA formation from long-chain, naturally-occurring TG. Since lipases act at low a_w levels, assay methods differ from conventional enzyme assays in that little or no addition of water to the test material is necessary or desirable (see Appendix 2).

Assays for LAH generally involve the use of phospholipids or glycolipids as substrate, dispersed in an aqueous buffer with the aid of emulsifying agents to produce a stable micellar dispersion.

4 *Occurrence of Lipolytic Enzymes in Cereal Grains*

(a) *Occurrence in various species.* A recent survey of TG lipase activity in cereal grains by Drapron (1983), using a non-aqueous incubation system (a mixture of olive-oil and ground cereal grain), gave the results shown in Table 6.1. The data from the range of cereals and products confirm the generally observed pattern of high lipase activity in oats and rice, compared with wheat and maize; millet and sorghum appear to have intermediate levels of lipase activity. The data in Table 6.1 refer specifically to lipase activity (olive oil as substrate). Similar comparative studies have not been undertaken for LAH; information on such enzymes has been derived, therefore, from separate studies of individual cereal species (see Table 6.2).

A range of lipase activities can be found within a given species. For

Table 6.1 Lipase activities of cereal grain products†

Sample (Commercial sources)	Lipolytic activity (oleic acid liberated) mg g^{-1} dry weight
Maize	2.5–4
Wheat	2–4.5
Patent wheat flour	1–1.25
Wheat bran	7
Wheat germ	4–4.5
Brown rice	11–13
Milled rice	1.25
Rice bran	20–30
Oat	20
Millet	6–10
Sorghum	6

†Results are expressed in terms of free oleic acid liberated from a mixture containing ground sample (2 g) and olive oil (100 mg) at 30°C for 72 h (a_w = 0.8). (Data from Drapron, 1983, with permission.)

Table 6.2 Lipid acyl hydrolase (LAH) enzymes in cereal grains

Source	Description	Substrates hydrolysed†	Reference
Barley	phospholipase B	PC ≈ lysoPC (not PI)	Rebmann and Acker (1973)
Barley	lysophospholipase (lysoPC acyl hydrolase)	lysoPC (also ρ-nitrophenyl palmitate)	Baisted and Stroud (1982a)
Rice	lipolytic acyl hydrolases:		
	phospholipase	PI > lysoPC > PC ≫ PE	Hirayama and Matsuda (1979)
	galactolipase	DGDG > MGDG > PL	Matsuda and Hirayama (1979)
Wheat	"lipase" (monoglyceride lipase)	MG (not TG)	Stauffer and Glass (1966)

†Relative reaction rates indicated.

example, Frey and Hammond (1975) examined 350 different varieties of oat grown in the same location and found a 21-fold difference between the lowest and highest lipase levels (10–210 μmoles fatty acid h^{-1} g^{-1} of seed). Bell *et al.* (1979) found significant differences between lipase levels in breadmaking flours from different wheat grists (0.85 ± 0.04 and 0.43 ± 0.03 μmoles fatty acid day^{-1} g^{-1} flour respectively), and showed that the measured lipase activities varied with the levels of FFA generated on long-term storage of flours.

Environmental factors also influence lipase levels. Enzyme activities in wheat tend to be higher in wetter seasons (Halton *et al.*, 1959); this is probably due mainly to microbial contamination as discussed below.

Oats The lipase activity in oats has been studied in some detail because of its significance in relation to the quality of oats and oat products. Oats have higher levels of lipid than other cereals (6–10%; see Chapter 16) and interactions of lipase and TG can lead to hydrolytic rancidity in oat products. The enzyme, first described in detail by Martin and Peers (1953) has been further characterized by Berner and Hammond (1970). Oat lipase has a pH optimum around pH 7.4; it is most active in low moisture systems (25–50% of seed weight) and is inactive in excess water. The enzyme removes fatty acids from all three acyl ester groups of TG, preferentially hydrolyzing unsaturated fatty acyl esters.

Rice Rice bran is very susceptible to hydrolytic rancidity, both because of its oil content and the presence of lipolytic enzymes. Funatsu's group in Japan have studied in some detail a lipase in the "bran" fraction from rice milling (Funatsu *et al.*, 1971; Aizono *et al.*, 1976; Aizono and Funatsu, 1978). However, LAH (galactolipase and phospholipase activities) play a major role in causing rancidity in stored rice germ and bran (Matsuda and Hirayama, 1975; Hirayama and Matsuda, 1975). A lipase from commercial rice germ hydrolyses tributyrin and rice bran oil (Aoyagi *et al.*, 1979); this enzyme from rice germ may be the same as that described in commerical rice bran. Lipase and LAH (galactolipase and phospholipase) activities have been identified also in rice endosperm (Matsuda and Hirayama, 1975). It is not clear to what extent the enzyme activities in bran, germ and endosperm fractions are due to cross contamination during the preparation of grain components, or perhaps to contamination with lipolytic micro-organisms (Juliano, 1977).

Wheat Sound, ungerminated wheat has very low levels of lipase activity. Nevertheless, commercial wheat germ and bran are prone to hydrolytic rancidity and the slow accumulation of FFA on the storage of white flours over several months may be due to lipase in the endosperm or in fragments of germ or bran in the flour. The enzyme(s) responsible for lipolysis in wheat have not been fully characterized; the role of the so-called "wheat germ lipase" is uncertain as it does not hydrolyze long-chain TG (e.g. olive oil) in lipase assays and is more accurately described as an esterase. Analysis of stored white flour has showed that the increase of FFA during storage is accompanied by a corresponding decrease in TG levels; this indicates lipase action, although the source of enzyme could be, at least partly, micro-organisms contaminating the wheat. Drapron and his colleagues have studied in some detail the characteristics of lipase activity in wheat (see below). Indirect evidence for the hydrolysis of polar lipids in stored flour was obtained by Clayton and Morrison (1972); this was confirmed by Warwick *et al.* (1979) who demonstrated substantial reductions in PL and GL levels during flour storage.

Barley Little information is available on lipase in barley, despite the practical problems encountered with FFA oxidation and off-flavours in malt and beer (see Chapter 5). MacLeod and White (1961) reported that isolated embryo axes of barley can metabolize their lipid reserves completely within 48 h of germination if exogenous substrates are excluded. This observation indicated the presence of TG lipase and these authors (MacLeod and White, 1962) subsequently showed the presence in ungerminated grain of particulate (i.e. insoluble) lipase activity. Finely-ground grain hydrolyzed triolein or olive oil and also hydrolysed tributyrin. Lipase activity depended upon water activity, showing maximal rates at relatively low moisture contents (40%). Only a small proportion (15%) of the lipase activity could be extracted into aqueous media from the grain sampled before germination or during malting of barley.

Enzymes that catalyse deacylation of PL have been detected in developing and germinating barley grains (Chapter 5). The phospholipase B described by Rebmann and Acker (1973) and the lysophospholipase described by Baisted and Stroud (1982a) may represent the same enzyme, a lipid acyl hydrolase (LAH) similar to the LAH found in other plants (Galliard, 1980) and having a broad specificity for polar lipid substrates.

(b) *Distribution of lipase within grains* The location of lipolytic enzymes in mature, ungerminated cereal grains is uncertain. Little information is available from studies on dissected grain components. The picture is confused further by references to "lipase" for enzymes assayed with inappropriate substrates.

According to Hutchinson *et al*. (1951), 95% of the lipase activity of oats can be removed by scraping away the layers external to the testa. In transversally cut half-seeds of oat, the germ-end had no more activity than the distal end and Hutchinson *et al*. concluded that virtually all of the lipase activity in oats is to be found in the pericarp layer. In wheat, lipase activity is associated with both bran and germ fractions from flour milling (Drapron, 1983). However, lipase activity in dissected germ from sound, ungerminated wheat is low and the "bran" lipase has yet to be characterized. Lipase is found in the germ, bran and endosperm fractions of commercially-milled rice (see above).

Sastry *et al.* (1977) used β-naphthyl laurate as substrate in a histochemical method to locate lipase in rice grains; the enzyme activity that liberated β-naphthol was located mainly in the testa layer, but not in the aleurone layer. In a recent study using fluorescein dibutyrate as a histochemical substrate for lipase, Jensen and Heltved (1982) found no release of fluorescein in ungerminated grains of barley, wheat, rye and sorghum but with each type of grain, the onset of germination was accompanied by the appearance of hydrolytic activity in the scutellum. It is not known whether the β-naphthyl laurate or the fluorescein dibutyrate used in these histochemical methods were detecting enzymes that also would hydrolyze endogenous lipids.

From the limited amount of information available, it seems that lipase activity is associated mainly with the layers external to the starchy endosperm in mature, ungerminated cereal grains. In mature, clean and ungerminated grain, the presence of lipase in the germ of some cereals is questionable. During germination (of wheat) the starchy endosperm and bran fractions account for more of the total grain lipase activity than does the germ (Tavener and Laidman, 1972a). The lipolytic enzymes of germinating grain are discussed in Chapter 4.

The location of lipolytic activity of cereal grains at the sub-cellular level has not been investigated in any detail. In oil seeds, specialized organelles, glyoxysomes, contain the enzymes necessary to convert the TG (stored in the spherosomes), via fatty acids and the β-oxidation pathway, to acetate. Jelsma *et al.* (1977) have shown that the lipid-storing spherosomes in wheat aleurone grains do not contain lipase activity (towards cotton seed oil substrate), but that the aleurone grains do contain lipase activity together with other hydrolytic enzymes; these authors suggest that aleurone grains act, in ungerminated wheat, in a similar role to vacuoles or "lysosomes" in other plant tissues.

(c) *Microbial sources of lipolytic enzymes in cereal grains* The concentration of lipase activity in the outer layers of cereal grains raises the question: "to what extent do microflora contribute towards the enzyme activity?". There is no doubt that contaminated grain stored under conditions allowing fungal growth, yields flours with greatly increased lipolytic activity; this has been shown clearly in wheat flours (Dirks *et al.*, 1955; Halton *et al.*, 1959). For sound wheat grain and flours, the low level of lipase activity may be also due, at least in part, to microbial lipase activity. The saprophytes of cereals include lipolytic fungi and these are known to cause lipolysis of seed lipids in other crops, e.g. safflower (Heaton *et al.*, 1978). Recent studies with oil palm (Tombs and Stubbs, 1982) have shown that the mature fruit contains no endogenous lipase activity and the lipolytic activity commonly associated with oil palm is due solely to contaminating fungi. Juliano (1977) has pointed out that the lipase activity of rice could be due to lipolytic microorganisms; e.g. lipolytic bacteria (*Xanthomonas* spp.) are found in rice. The problem of distinguishing between endogenous and microbial sources of lipases is shown by a simple calculation. In the two samples of stored flours examined by Bell *et al.* (1979), lipase activities were 0.4 and 0.8 μmols fatty acid liberated day^{-1} g^{-1} flour; purified fungal lipases can release 2000 μmols of fatty acid min^{-1} mg^{-1} of enzyme, i.e. about 2×10^9 μmols day^{-1} g^{-1} (see Tombs and Blake, 1982). Thus, contamination of flour by 1 μg kg^{-1} (1 p.p.b.) of fungal lipase could account for the observed lipolytic activity in flour!

B Other Lipolytic Enzymes: Phospholipase D

The discussion of lipolytic enzymes has been limited so far to acyl hydrolases, i.e. enzymes that catalyse hydrolysis of fatty acyl ester bonds to liberate FFA. Cereal grains contain a lipolytic enzyme, phospholipase D, that acts on the phosphate ester group in phospholipids:

$$\begin{array}{c} H_2COOR \\ | \\ R'OOCH \\ | \\ H_2COPO{-}X \\ \| \\ O \end{array} + YOH \longrightarrow \begin{array}{c} H_2COOR \\ | \\ R'OOCH \\ | \\ H_2COPO{-}Y \\ \| \\ O \end{array} + XOH$$

where R and R′ represent fatty acid alkyl residues and X represents choline, ethanolamine, inositol etc. in PL.

In excess water (Y = H), the product of the reaction is phosphatidic acid. The enzyme also will catalyse transferase reactions with alcohols, e.g. methanol ($Y = CH_3$) to produce lipid artefacts, such as phosphatidylmethanol. The physiological role of phospholipase D is not known. However, the enzyme is sufficiently active in cereal grains that it can cause serious problems in lipid analysis if adequate precautions are not taken to inactivate the enzyme. Colborne and Laidman (1975) showed that phosphatidylbutanol was produced when wheat grains were extracted with butan-1-ol. The potential danger of phospholipase D activity in lipid analysis is well illustrated by the work of Roughan *et al.* (1978) who showed that phosphatidylmethanol represented 75% of the total phospholipids of soyabeans that had been soaked in methanol before lipid extraction. To minimize enzyme action, it is essential to inactivate the enzyme rapidly by heating before tissue disruption permits reaction between the enzyme and its substrates.

Phospholipase D activity in cereals has been investigated by Nolte *et al.* (1974) and Nolte and Acker (1975a, 1975b). The enzyme has been identified in wheat, barley, maize, oats and rye. In barley, the enzyme increases in activity during grain development, reaching a maximum at maturity. Morrison (1978a) has tabulated published results for phospholipase D activities in cereals. Typical values (μmols PC hydrolyzed min^{-1} g^{-1} sample dry weight) are: whole wheat (1.0), wheat germ (4.25), wheat flour (0.25), rye (1.6), barley (0.7), maize germ (0.8), oats (0.5).

C Physiological Aspects of Lipolytic Enzymes

1 *Grain Development*

TG lipase is required in the metabolism of stored oil in seeds and LAH enzymes are involved in "turnover" of membrane lipids during active metabolic phases and in defence and repair mechanisms in response to stress (physical damage). Understandably, there is little information on TG lipases in developing cereal grains, since these enzymes are presumed to act primarily during germination. However, if TG lipases are present in mature grain, their biosynthesis would have occurred during grain development.

Some LAH-enzymes, active towards typical membrane lipids (PL, GL), have been investigated in developing grains. Circumstantial evidence for lipolytic activity during grain maturation is provided by the nature of the lipids found in mature grain. Although lipids such as monoacyl polar lipids (lysophospholipids, FFA, for example) do not occur normally in significant levels in healthy, viable plant cells, cereal grains contain substantial proportions of these lipids (Chapters 2 and 5). The only known metabolic routes for formation of such lipids are via hydrolysis of other membrane lipids. Thus, it is reasonable to suggest that the major monoacyl lipids associated with the amylose fraction of cereal starches (i.e. lysoPC and FFA, see Ch. 2) could have been formed by partial hydrolysis of membrane PL during starch granule formation in developing grains; this is discussed further in Chapter 5.

2 *Germination*

Lipase activity increases rapidly in germinating cereal grains. Jensen and Helvted (1982) have developed a sensitive fluorescence method using fluorescein dibutyrate to detect early signs of germination in cereals. Whether this substrate is hydrolyzed by a true lipase, has yet to be ascertained, but the stage at which enzyme activity is detected in wheat (within 12 h at 16 °C) corresponds to that observed by Tavener and Laidman (1972a) who used an assay for TG lipase. Details of TG metabolism and its control in germinating cereal grains are given by Laidman in Chapter 4.

Other lipolytic enzyme activities also increase during germination. Baisted and Stroud (1982a,b) have shown that lysoPC acyl hydrolase (LAH) activity increases during the germination of barley, reaching a maximum at the point when amylase activity is maximal and when the levels of starch-bound lysoPC are disappearing. Thus, a mechanism exists to hydrolyze the lysophospholipids which because of their cytotoxic nature could cause lysis of intracellular membranes. (The prefix *lyso-* was originally given to monoacyl phospholipids because of their ability to cause lysis of erythrocyte membranes.) Acker and Müller (1965) had shown previously that a phospholipid acyl hydrolase activity ("phospholipase B") increased during malting of barley, also reaching maximal activity around 5 days after germination; this may be the same enzyme as the lysoPC acyl hydrolase described above.

III OXIDATIVE ENZYMES

A General Aspects of Lipid Oxidation

Several different pathways exist by which fatty acids are oxidized. However, many of these require highly integrated enzyme complexes and cofactors. Such processes are very efficient in healthy, viable seeds, but do not operate in materials in which metabolic control is dormant or disorganized. Thus β-oxidation is an important process in germinating seeds, but is not considered a major process in mature grain or products formed by tissue disintegration. For practical purposes, only those oxidation processes that can occur in non-viable tissues, will be considered here. This distinction precludes discussion of the oxidation processes occurring in germinating seeds, but the important aspects of this topic are covered by Laidman in Chapter 4.

Generally, the processes that are responsible for lipid oxidation in cereal products are those that do not require biological cofactors (e.g. coenzyme-A, NAD, NADP) or integrated enzyme complexes. Furthermore, the substrates are usually FFA or acyl glycerolipids, rather than the thioester intermediates essential for processes such as β-oxidation. The reactions of primary importance to cereal technology are those involving molecular oxygen. The following discussion will be limited to these peroxidation reactions. It should be stated, however, that the relevance of other pathways in cereal technology has not been explored adequately. A general review of lipid oxidation enzymes in plants has been presented recently (Galliard, 1980).

B Lipid Peroxidation

A major factor in determining the quality and storage life of food products is their susceptibility to oxidative rancidity. This can be defined as the formation of undesirable tastes and flavours by reactions of unsaturated fatty acids with oxygen. An associated problem is the effect of lipid oxidation products on the functionality of food materials.

Although this chapter is concerned mainly with enzymic reactions in cereals, many aspects of lipid peroxidation are common to enzymic and non-enzymic processes. One essential feature of all peroxidation reactions between oxygen and organic compounds, is the requirement for a catalyst that will reduce the activation energy of the reaction:

$$RH + O_2 \rightarrow ROOH \text{ (hydroperoxide)}$$

Direct addition of O_2 in its ground state (triplet) to an organic molecule (singlet) is a "forbidden" reaction and, for the process to occur, either the O_2 or the fatty acid must be "activated" (see Galliard and Chan, 1980). In effect, O_2 must be converted to a reactive form such as its singlet state or to a free radical, or the fatty acid must lose a proton to form a free radical. In practice, such reactions are facilitated by transition metal catalysts (Fe^{++}, Cu^{++}) which can paricipate in one-electron transfer reactions with oxygen or organic compounds. Light, especially uv light, also can initiate oxidation by forming singlet oxygen or by initiating the formation of organic free radicals.

Thus, oxidative rancidity and its prevention are intimately associated with free radical and transition metal reactions. For a more detailed discussion of lipid oxidation, the reactions involved and their relevance to food and other biological materials, please refer to recent comprehensive and authoritative reviews (Simic and Karel, 1980; Chan, 1983). A collection of more practically orientated papers on *Rancidity in Foods* has been compiled recently by Allen and Hamilton (1983).

C Lipoxygenase (LOX)

Lipoxygenase (LOX – formerly known as lipoxidase, carotene oxidase, or "fat oxidase"), is an enzyme that occurs in cereal grains and has long been known to initiate oxidative breakdown of lipids in food materials. The enzyme has been the subject of intensive research for many years and current interest in this important constituent of food materials is demonstrated by the number of recent reviews dedicated to LOX (e.g. Vliegenthart and Veldink, 1982; Nicolas and Drapron, 1981; Galliard and Chan, 1980; Gardner, 1980).

The term "lipoxygenase" covers a group of enzymes that catalyse the peroxidation of polyunsaturated fatty acids containing *cis*, *cis*-1,4-pentadiene moieties ($-CH\underset{cis}{=}CH-CH_2-CH\underset{cis}{=}CH-$) by molecular oxygen. The product is a fatty acid hydroperoxide with conjugated diene group and that has the hydroperoxide group located at one end of the original pentadiene group. The most common substrates for LOX in food materials are linoleic and linolenic acids. For example, with linoleic acid as the substrate, two isomeric hydroperoxydienoic fatty acids are the possible primary products:

$$CH_3-(CH_2)_4-\underset{cis}{CH=CH}-CH_2-\underset{cis}{CH=CH}-(CH_2)_7-COOH$$

(linoleic acid)

lipoxygenase + O_2

$$CH_3-(CH_2)_4-\underset{\substack{|\\OOH}}{CH}-\underset{trans}{CH=CH}-\underset{cis}{CH=CH}-(CH_2)_7-COOH$$

(13-hydroperoxy-octadeca-9-*cis*,11-*trans*-dienoic acid)

$$CH_3-(CH_2)_4-\underset{cis}{CH=CH}-\underset{trans}{CH=CH}-\overset{\substack{OOH\\|}}{CH}-(CH_2)_7-COOH$$

(9-hydroperoxy-octadeca-10-*trans*,12-*cis*-dienoic acid)

As with most enzymes, LOX catalyses sterospecific reactions and the two isomeric hydroperoxy fatty acids are optically active, i.e. the 9-D-(R)- and 13-L-(S)- hydroperoxy fatty acids are formed from linoleic acid (non-enzymic oxidation reactions produce racemic mixtures of products). Although it was thought for many years that LOX had no prosthetic groups, it is now known that for all cases in which the enzyme has been carefully analysed, one atom of Fe is present. The molecular weight of LOX is approximately 10^5 daltons. For reasons previously given and, as reasoned by Chan (1972), the involvement of a transition metal (in this case, Fe) in enzyme-catalysed peroxidation reactions, was expected.

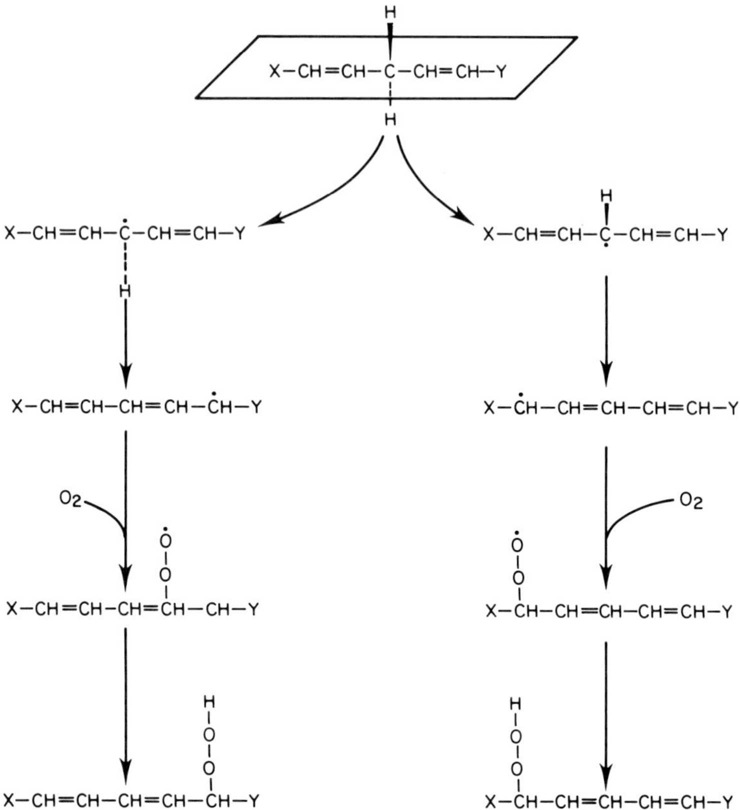

Fig. 6.3 Simplified scheme to show formation of isomeric hydroperoxides via radical intermediates (adapted from Egmond *et al*., 1972).

In general, LOX enzymes show highest activity with FFA substrates, although in some cases MG and TG that contain polyunsaturated fatty acids are oxidized. It should also be noted that although LOX-catalysed reactions are usually studied in aerobic systems, the enzyme is active in the absence of oxygen and catalyses the abstraction of a H• radical from fatty acids. This produces a fatty acid free radical, which may then initiate further radical processes (Veldink *et al*. 1977).

Stereospecific removal of a hydrogen atom from the methylene carbon of the pentadiene group (i.e. C-11 of linoleic acid) is thought to be the initial process of LOX action; subsequently, the diene radical rearranges to the more stable, conjugated *cis*, *trans* form and, under aerobic conditions, O_2 adds to the carbon atom at one end of the diene group to form a hydroperoxy radical which then abstracts a H atom to form the hydroperoxide (Fig. 6.3).

1 *Measurement of LOX Activity*

Several alternative approaches have been used to measure LOX activity. These include

(a) determination of O_2 consumption during linoleic acid oxidation;

(b) spectrophotometric measurement of the conjugated diene system in the hydroperoxide product;

(c) colorimetric reactions of the hydroperoxide and

(d) determination of the bleaching of pigments in co-oxidation reactions. The relative merits of the different approaches are discussed in Appendix 2.

2 *Different Forms of LOX*

Extensive research on LOX from soyabean has revealed the existence of different forms of the enzyme. The "classical" LOX (type I) is now known to be somewhat unusual compared with LOX from other plants; there is also a second, type II, enzyme in soybeans. The type I enzyme has a highly alkaline pH optimum (9–10) and has a relatively low bleaching activity. The type II enzyme from soybean has optimal activity at neutral pH and is more effective in bleaching reactions, i.e. it has a higher co-oxidation activity.

Several forms of LOX have been isolated from cereal grains (see below), but generally they tend to resemble the type II enzyme in pH characteristics. A detailed review of lipoxygenase isoenzymes is not pertinent here; further information is available in the general reviews cited on page 125.

3 *Co-oxidation Reactions of LOX*

An important property of LOX which has major implications in cereal technology is its ability to catalyse the oxidation of other materials concurrently with the fatty acid oxidation. This can be represented as follows:

$$LH + O_2 \rightarrow [LOO\cdot] \rightarrow LOOH$$

$$XH \quad X\cdot \rightarrow \text{co-oxidation products}$$

XH represents an oxidizable co-substrate, such as a carotenoid, chlorophyll, protein-SH or unsaturated fatty acid.

An essential feature of these co-oxidation reactions is that both the LOX substrate fatty acid (LH; e.g. linoleic acid) and the co-substrate must be present initially in the reaction system; the hydroperoxide product (LOOH) does not oxidize the co-substrate. Thus, co-oxidation is part of (and not subsequent to) the LOX- catalysed reaction. The mechanism is though to involve a free radical reaction in which a hydroperoxy radical intermediate (LOO•) interacts with the co-substrate (XH) to produce a radical X•, which subsequently reacts with oxygen or with other molecules, e.g. fatty acids and proteins (Weber and Grosch, 1976; Veldink *et al.* 1977).

Different forms of LOX vary in their activities in the co-oxidation reaction. The "classical" soyabean lipoxygenase (type I) has relatively low co-oxidation activity, whereas the type II enzyme in soyabeans is a potent catalyst of co-oxidation reactions; it is this enzyme that is mainly responsible for the well-known bleaching activity of soya flour in bread doughs. These differences in co-oxidation activities of LOX types explain the observed loss of carotenoid bleaching activity during the purification of the type I enzyme from soyabeans (Kies *et al.*, 1969). LOX enzymes from seeds of other legumes (peas and beans) are potent co-oxidation catalysts (Grosch *et al.* 1976) and horse-bean *(Vicia faba)* flour is used in certain countries as a dough-bleaching agent (see Chapter 10). Wheat LOX enzymes are less active in this respect, although the colour loss in durum pasta production is associated with the action of durum wheat LOX on the carotenoid pigments of the grain (Chapter 13).

Water-soluble compounds are less susceptible to co-oxidation in doughs than are lipophilic substances such as carotenoids and tocopherols. This may be explained by the requirement for co-substrates to penetrate the hydrophobic environment of the fatty acid-LOX enzyme complex where the free radical reactions occur.

4 *Secondary Reactions of Fatty Acid Hydroperoxides (FAHPO)*

Although fatty acid hydroperoxides (FAHPO) are the primary products of LOX-catalysed reactions, they are not present in significant concentrations in plant cells or in products derived from plants. FAHPO are extremely cytotoxic and biological systems usually contain mechanisms for removing hydroperoxides. Moreover, they are chemically unstable and decompose to a number of products. Many of the methods that have been developed for measuring lipid peroxidation, actually determine the secondary products of hydroperoxide breakdown.

Some of the routes by which FAHPO may be transformed by both enzymic and non-enzymic reactions are illustrated in Fig. 6.4. A more thorough and detailed treatment of these and other reactions of FAHPO is given in the recent reviews by Gardner (1980, 1983).

The reactions shown in Fig. 6.4 may be classified as follows: (actual examples in cereal products will be quoted later)

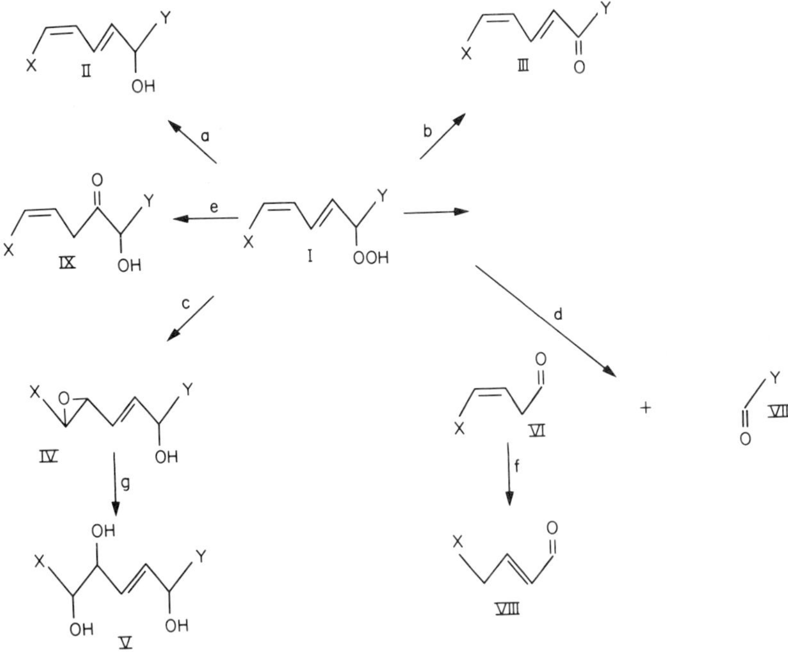

Fig. 6.4 Representation of some of the routes by which FAHPO may be transformed by enzymic and non-enzymic reactions. Only the essential partial structures are shown. For the 9-hydroperoxide of linoleic acid, for example, X and Y would represent $CH_3(CH_2)_4$- and -$(CH_2)_7$ COOH, respectively. H atoms in the fatty acid chain are omitted from the structural formulae.

(1) *Reduction* (reaction a). Enzymic and non-enzymic factors can reduce the hydroperoxide group to the corresponding, more stable, hydroxy acid (II).

(2) *Oxidation* (reaction b). Keto(or oxo)-diene fatty acids (e.g. III) commonly occur in peroxidised lipid systems and are readily detected by their characteristic u.v. absorptions.

(3) *Isomerisation* (reactions c and e). Rearrangement of the hydroperoxydiene structure, leads to epoxy-hydroxy derivatives (e.g. IV) or to keto(oxo)-hydroxy compounds (e.g. IX). Further reactions, by hydrolytic opening of the epoxy group (reaction g) lead to trihydroxy compounds (e.g. V).

(4) *Chain cleavage* (reaction d). In recent years, enzymes that cleave FAHPO have been identified in several different plants (see Galliard and Chan, 1980, Gardner, 1980). The immediate products of cleavage are aldehydes, e.g. from the 9-hydroperoxide of linoleic acid (I), cleavage

produces the volatile, *cis*-3-nonenal (VI) and a non-volatile product carrying the carboxyl group; i.e. 9-oxo-nonanoic acid (VII). *Cis*-3-enals are unstable and isomerize to the *trans*-2-enal forms (VIII). Although the isomerization proceeds without catalysts, enzymes, recently found in some plants, enhance the rate of isomerization (Phillips *et al*., 1979; Schreier and Lorenz, 1981).

(5) *Other reactions*. In addition to the processes summarized in Fig. 6.4, other enzymic and non-enzymic reactions of fatty acid hydroperoxides include polymerizations, addition of nucleophilic groups to the peroxide, further oxygenation of the chain and formation of unsaturated ether derivatives. The recent reviews of Gardner (1980, 1983) describe these processes in some detail.

(6) *Cyclisation of trienoic fatty acids*. Polyunsaturated C_{18}–C_{22} acids containing double bonds at the n-3, 6, 9 positions (e.g. linolenic acid) are converted enzymically, via LOX and a fatty acid cyclase enzyme pathway to prostaglandin-type derivatives. The cyclase enzyme occurs widely in plants (Vick and Zimmerman, 1979) although its role is not yet known.

D LOX and Related Enzymes in Cereal Grains

LOX has been detected in all the cereal species for which analyses have been reported (Table 6.3). Several isoenzymic forms are present in wheat, barley and rye. Product specificities, with linoleic acid as substrate, generally favour peroxidation at the 9-position, under optimal pH conditions (pH 6.0–7.5) (Table 6.3). However, exceptions to this are known; one LOX isoenzyme from wheat produces mainly the 13-hydroperoxide derivative, as does rye LOX; maize germ LOX also predominantly forms 13-hydroperoxide at pH 9.0, but at optimal pH (6.5–7.5) and under conditions more typical for maize products, the 9-position is oxygenated (Table 6.3).

1 *Wheat LOX*

Wheat LOX has received a considerable amount of attention because of its role in several aspects of wheat and flour quality. Recent purifications of the enzyme and its isoenzymic forms have facilitated more detailed studies of its properties (Wallace and Wheeler, 1979; Nicolas *et al.*, 1982). In doughs, wheat LOX catalyses the oxidation of polyunsaturated fatty acids in both FFA and MG (Graveland, 1968; Mann and Morrison, 1974). In this respect, it differs from soya LOX which, when added to doughs as enzyme-active soya flour, catalyses oxidation of the major wheat lipids in dough, i.e. TG, PL, GL, etc. (Morrison and Panpaprai (1975); this general oxidation activity of soya flour may be related to the co-oxidation activity of soya LOX (Type II) described earlier. Nicolas *et al*. (1982) demonstrated that of the 3 isoenzymic forms of LOX that they isolated, one was relatively more active in bleaching carotenoids, although the specific activity was much lower than that of horse-bean LOX; Grosch *et al.* (1976) had shown previously that soya LOX was more potent than wheat LOX in carotenoid bleaching activity.

An unusual property of wheat LOX may be important in understanding its action in wheat products. The enzyme is readily adsorbed on to hydrophobic surfaces (Wheeler and Wallace, 1978). Graveland (1970) had shown previously that wheat LOX is adsorbed on to the glutenin fraction of wheat proteins, but that removal of the lipids from glutenin abolished this adsorption of LOX. Such hydrophobic interactions result in modified kinetics and reaction product profiles compared with those obtained with uncomplexed enzyme. Another unusual aspect of wheat LOX is its capacity for "self-destruction" in flour slurries and doughs (Wheeler and Wallace, 1973). Although LOX from other sources is inhibited by its hydroperoxide products (Smith and Lands, 1972), wheat LOX appears to be especially sensitive; thus, fatty acid oxidation in flour-water slurries may cease before the available substrate or oxygen is depleted. Wheat protein concentrate (Wallace and Wheeler, 1972) and a heat-stable protein factor in flour (Nicolas *et al.* 1981a,b) also cause inactivation of wheat LOX. Conversely, a soluble protein from wheat germ was found to activate both wheat LOX and soya LOX (Wallace and Wheeler, 1979). Purified LOX from wheat is extremely sensitive to inhibition by phenolic compounds, especially 4-nitrocatechol (at 10^{-6}M); however, LOX in crude extracts and doughs is much less sensitive, presumably because the phenolics complex with other proteins (Galpin and Galliard, unpublished observations).

Relatively little is known of the behaviour of LOX during the development or germination of wheat grain. Guss *et al.* (1968) found that LOX activity increased during the early phase of wheat germination (3–5 days) but had fallen again by day 7; a change was noted in the relative amounts of 4 LOX isoenzymes during germination. Although wheat LOX has co-oxidation activity, this is much lower, on a specific activity basis, than that of soya type II (Grosch *et al.*, 1976) or of horse bean lipoxygenases (Nicolas *et al.*, 1982).

2 *Barley LOX*

The enzyme in barley has been studied in some detail in relation to its proposed role in causing off-flavours in beers and malts. LOX activity increases during grain development (Franke and Frehse, 1953) and germination and malting (Lulai and Baker, 1975; Lulai *et al.* 1981; Baxter, 1982). At least two isoenzymes exist, although only one form is present in

Table 6.3 Lipoxygenase: Product specificity and isoenzymic forms described in various cereals and legumes

	No. of isoenzymes	ratio of 9- to 13-HPO†	pH†	Reference
Wheat	4	–	6.9	Wallace and Wheeler (1975)
	3	–	6.0–6.5	Nicolas *et al.* (1982)
	3	(A: 88 : 12		
		(B: 90 : 10	6.9	Galliard (1983b)
		(C: 13 : 87		
	–	85 : 15	5.8	Grosch *et al.* (1976)
		85 : 15	–	Graveland (1973)
Barley	2	(A: (mainly 9-HPO)	6.8	
		(B: (both 9 and 13-HPO)		Baxter (1982)
	–	88 : 12	6.8	Graveland *et al.* (1972)
		96 : 4	6.5	Fuhrling (1975)
		90 : 10	7.0	Heimann and Timm (1977a)
ungerminated	1	96 : 4	7.5	
germinating	2	70 : 30	7.5	Yabuuchi (1976)
	1 (purified)	90 : 10	7.5	

ungerminated	–	70 : 30	6.0	Lulai *et al.* (1981)
germinating	–	40 : 60	6.0	
Maize (germ)	–	91 : 9	7.4	Hamberg (1971
	–	93 : 7	6.5	
	–	83 : 17	6.5	Gardener and Weissleder (1970)
	–	95 : 5	6.6	Egmond *et al.* (1972)
	–	80 : 20	6.6	Veldink *et al.* (1972)
	–	15 : 85	9.0	
Oats	–	90 : 10	7.0	Heimann *et al.* (1973)
Rye	3	20 : 80	7.4	Heimann and Klaiber (1977)
Rice	1	97 : 3	7.0	Yamamoto *et al.* (1980b)
Soyabean		0–30 : 70–100	9.0 (opt)	
mixed isoenzymes	2–3	54 : 46	5.5	Several; see Nicolas and Drapron (1981)
purified, Type I		(mainly 13-HPO)	7.0	
purified, Type II		approx. 50 : 50	7.0	
Vicia faba (horse bean)	–	40 : 60	6.8	Nicolas and Drapron (1977)
	–	41 : 59	–	Vioque and Maza (1971)

†Relative proportions of 9- and 13-HPO (hydroperoxides) from lineolic acid produced with extracts of grain or purified enzymes at the pH values indicated.

ungerminated grain; a second isoenzyme appears during germination and this has predominantly 13-hydroperoxide-forming properties (Yabuuchi, 1976) in contrast to the enzyme in ungerminated barley which forms mainly the 9-hydroperoxide from linoleic acid (Yabuuchi, 1976; Lulai *et al*. 1981).

3 *Rice LOX*

The stale flavour in stored rice has been ascribed to products of LOX action, e.g. hexanal (see Chapter 15). Rice LOX, which is concentrated in the germ has been investigated recently by Yamomoto *et al.* (1980a,b). The enzyme is present in one form only in ungerminated grain and is highly specific for 9-hydroperoxide isomers (see Table 6.3).

4 *Maize LOX*

Franke and Frehse (1953) identified LOX in maize germ and Wagenknecht (1959) ascribed the formation of off-flavours in under-blanched sweetcorn to LOX activity. LOX and linoleic acid hydroperoxide isomerase have been separated and partly purified from maize germ (Gardner, 1970; Gardner and Weisleder, 1970). Maize germ predominantly forms the 9-hydroperoxide isomer (Table 6.3). The major product of the two enzymes acting sequentially in crude extracts of maize germ is an α-ketol derivative of linoleic acid (9-hydroxy-10-oxo-monoenoic acid). Extracts of maize endosperm did not catalyse the LOX-isomerase sequential pathway (Gardner, 1970).

5 *LOX in Other Cereals*

LOX has been studied in less detail in other cereal grains although the product specificities for many have been determined (Table 6.3). The enzyme from rye is unusual in its product specificity, producing predominantly 13-hydroperoxide from linoleic acid (Heimann and Klaiber, 1977a).

6 *Products Formed from Fatty Acid Hydroperoxides in Cereal Extracts*

As previously mentioned, the primary products of LOX action, lipid hydroperoxides, do not accumulate but are converted by secondary reactions to a number of products (Fig. 6.4). In some cases, specific enzymes, catalysing secondary reactions, have been identified in cereals; other reactions may be non-enzymic in nature.

Graveland (1973) examined the products formed from linoleic acid in aqueous suspensions of ground and sieved flours from the grains of several cereal species. (Table 6.4). Although most of the linoleic acid lost can be accounted for by the secondary products indicated in Table 6.4, there are significant differences in the distribution of the products from the different cereal flours. There are many publications on the different products found in various cereal extracts and products; Gardner (1980) has reviewed the extensive literature in this area.

It should be emphasized that measurement of the concentration of oxidation products is of little value to the food technologist unless the flavour thresholds of the individual products are considered.

Reduction to monohydroxydienoic acids is catalysed by "lipoperoxidase" (reaction a, Fig. 6.4); this enzyme activity, as demonstrated in oats (Heimann *et al.* 1973), uses similar electron donors to other peroxidases and may, in fact, be due to a secondary reaction of LOX (Streckert and Stan, 1975). In animals, a glutathione peroxidase enzyme reduces hydroperoxides to the corresponding alcohols. This reaction has been suggested as a possibility in wheat flour (see Chapter 10), and there is some evidence for a reaction of this type in pea seeds (Mapson and Moustafa, 1955).

Lipid hydroperoxide isomerases form ketol derivatives (IX; Fig. 4). Both α-(i.e. 1,2-) and γ-(1,4-) ketols may be formed and such enzymes have been demonstrated in maize germ (Gardner *et al.*, 1975), wheat (Christianson and Gardner, 1975), barley (Yabuuchi and Amaha, 1976), oats (Heimann and Klaiber, 1977b) and rye (Heimann and Klaiber, 1977a).

There is no direct evidence in cereals for enzymic cleavage of fatty acid chains to yield volatile compounds via the hydroperoxide lyase reaction (d in Fig. 6.4). Vick and Zimmerman (1979) failed to find the enzyme in barley and maize, although Lulai *et al.* (1981) found indirect evidence for a

Table 6.4 Products formed from linoleic acid by suspensions of various cereal flours†

		Distribution of reaction products‡				
Product	Structure§	Wheat	Barley	Rye	Oats	Maize
Unreacted linoleic acid	I	60	50	65	60	70
Monohydroxy acids	II	13	8	8	25	5
Ketohydroxy acids	IX	0.1	25	15	–	20
Trihydoxy acids	V	23	10	5	10	0.5

†Linoleic acid (2 mg) was added to suspensions of defatted flour (1 g) in H_2O (10 cm^3). [Data selected from Graveland (1973).]

‡Values expressed as % of added linoleic acid.

§Structures similar, but not necessarily identical, to those shown in Fig. 6.4.

chain-cleavage system yielding volatile "cucumber" odours in rootlets of germinating barley.

Volatile aldehydes, such as hexanal and *trans*-2-hexenal, are formed during LOX-catalysed reactions, probably by a free radical co-oxidation reaction, analogous to that causing bleaching of carotenoids (Grosch *et al*. 1976). Wheat LOX is less active in producing volatile aldehydes than is LOX from beans (*Phaseolus* spp., Grosch *et al*. 1976); a number of volatile products are produced in side reactions of the hydroperoxide isomerase reaction in barley extracts (Heimann and Timm, 1977b). Volatile fragments from hydroperoxides are formed by a variety of non-enzymic processes and result in characteristic off-flavour compounds of oxidative rancidity. Generally, these non-enzymic reactions occur slowly compared with the enzyme-catalysed cleavage reactions. Volatile compounds are also formed in side-reactions of lipoxygenases (for details see Gardner, 1980).

Hydroxyepoxy acids (e.g. IV, Fig. 6.4) are formed in aqueous suspensions of cereal flours (Table 6.4). An enzyme catalysing their formation in oats has been called an isomerase (Heimann and Klaiber, 1977b), but this should not be confused with the isomerase that catalyses the formation of α- and γ- ketols (see above). Formation of epoxyhydroxy- and trihydroxy- (e.g. V, Fig. 6.4) derivatives from fatty acid hydroperoxides has been demonstrated in extracts from several cereal species e.g. oats (Heimann *et al*. 1973), barley (Graveland *et al*., 1972), wheat, maize and rye (Graveland 1973). In most cases, the enzyme(s) responsible for these reactions have not been defined. Heimann and Dresen (1973) have purified from oats an enzyme that forms hydroxyepoxy acids; however, it may be significant that neither this enzyme activity, nor the "lipoperoxidase" activity could be separated from LOX activity by purification. It is possible that these hydroperoxide conversion reactions are caused by a side-reaction of LOX.

7 *Distribution of LOX and Related Enzymes within Cereal Grains*

Available evidence points to a concentration of LOX predominantly, if not exclusively, within the germ of cereal grains. The literature is sometimes confusing because of the use of terms such as "germ" and "bran" to describe both commercial milling fractions and botanical components of grain. For example, the LOX of rice "bran" (Shastry and Rao, 1975) is concentrated in the germ fraction (Yamamoto *et al*., 1980a) that, together with pericarp, seed coat and aleurone layer, constitutes the commercial bran from rice milling.

In ungerminated barley, the LOX activity is virtually all (98%) in the germ, (Lulai *et al*., 1981) as is the second isoenzyme of LOX that appears during germination (Yabuuchi, 1976). The hydroperoxide isomerase of barley is found also mainly in the germ (Yabuuchi, 1976; Lulai *et al.* 1981).

Wheat LOX is concentrated in the germ of the grain and on a fresh weight basis, the scutellum has a higher concentration of the enzyme than has the embryo axis (Galliard, 1983). Vick and Zimmerman (1982) have investigated changes in LOX isomerase and FAHPO cyclase enzymes in germinating maize. All of these enzymes increased in activity (on a per seedling basis) during early stages of germination before falling again after approximately 4 days at 27 °C; the increased activities were found in the growing seedlings, rather than in the residual scutellum and endosperm.

IV EFFECTS OF LIPID-DEGRADING ENZYMES IN CEREAL PRODUCTS

Many of the important aspects of lipid degradation in cereal products are dealt with in appropriate chapters in this book: lipid oxidation in bread-making (Chapter 10); lipid degradation in flour storage (Chapter 12); and lipases in germination (Chapter 4). Lipid degradation in pasta, rice, maize, oats and barley products is discussed also in chapters on the different commodities. The following discussion will concentrate more on aspects that may be relevant to cereal products in general.

A General Aspects of Lipid Degradation

The presence of high FFA levels in raw materials or products is generally recognized as a quality defect and processors or customers will usually include a maximum FFA level in material or product specifications. For example, oats may be considered unfit for processing into food products if the FFA level exceeds 5% of the hexane-extractable lipid. The resulting effects of excess FFA levels can be in organoleptic or functional defects – or both.

An obvious result of lipolysis is hydrolytic rancidity, with the formation of characteristic "soapy" flavours. The objectionable taste is especially acute with fatty acids of intermediate chain length (C_8-C_{12}) which, although not present in high concentration of cereal lipids, may be important constituents of dairy products or, e.g. vegetable oils, with which the cereal may be mixed. A more common organoleptic defect is due to the combined effects of lipolytic enzymes and oxidation processes. Unsaturated fatty acids are more susceptible to enzymic oxidation when unesterified than when esterified in acyl lipids.

Functional defects in foods due to high FFA levels are recognized, but less readily explained. The hydrolysis of unchanged lipids to carboxylic acids will increase [H^+] and produce carboxyl groups that can combine with other compounds or metal ions. Free fatty acids also combine with proteins by hydrophobic interactions and can complex with starch during cooking. FFA can also interfere with the action of surface-active agents, such as polar lipids and emulsifiers.

B Grain Storage

In general, cereal grains are stable if harvested without damage and stored under appropriate conditions of temperature and moisture content to minimize the growth of micro-organisms. The inadequate drying of grain and physical damage, causing disruption of grain structure, can lead to changes in grain lipids and loss of carotenoids and tocopherols. The possible roles of lipolytic enzymes and oxidative enzymes have been reviewed recently by Morrison (1978a) who listed the various manifiestations of lipid degradation in poorly stored grain, including wheat, maize, rice and barley. In well-stored grain, tocopherol content falls slowly; whether this fall is due to a coupled co-oxidation with LOX, or other oxidative processes, is not known.

Although lipolysis is a potential problem with milled oats, the lipids of whole, undamaged grain are not hydrolyzed if the grain is stored at moisture levels below 10% (Youngs, 1978). Welch (1977) found no significant differences between the rate of FFA formation in naked or husked varieties, stored at different moisture levels; this was somewhat unexpected because the naked grain was thought to be more susceptible to handling damage. Even in grains stored at moisture levels >13%, and where lipolysis was evident, no evidence of oxidative degradation of lipids was obtained. It is assumed that the relatively high levels of phenolic and tocopherol antioxidants in oats (Chapter 16) were sufficient to prevent oxidative rancidity.

The lipase activity in oat grains at harvest is determined by both genetic and environmental factors. Frey and Hammond (1975) found a 21-fold range of lipase activity in 445 cultivars. FFA levels (indicative of lipase action) in harvested oats also vary with climate and geographical location. Surprisingly perhaps, FFA levels tend to be higher in dry seasons (Welch, personal communication).

Unpolished brown rice contains a number of lipolytic enzymes and LOX activity that can cause problems of hydrolytic and oxidative rancidity in the stored grain. This area has received a considerable amount of attention because of the commercial importance of rice and is reviewed by Juliano (1977) and in Chapter 15. As with other cereals deterioration is greater at higher temperatures and moisture levels. However, the hydrolytic and oxidative changes during storage are not directly related to changes in milling properties of rice that occur during 3–4 months storage after harvest.

The loss of viability in stored grain has been ascribed to damage to intracellular membrane structures. Seeds stored under N_2 tend to retain germination capacity for longer periods than seeds in air or O_2 (Justin and Bass, 1978). Although it has been assumed that oxidative attack on fatty acids of membrane lipids could explain the loss of viability, this is not proven; the conflicting evidence is discussed by Rudrapal and Basu (1982) and in refs. *loc. cit*. In recent studies with soyabean and safflower seeds (Ohlrogge and Kernan, 1982) no evidence was found for lipid oxidation as the causative factor in the O_2-sensitivity of germination capacity. However, it could be that only minor changes in the structure of a particular type of membrane, could have a major effect on physiological properties and this would be difficult to detect by lipid analysis of grain.

A highly significant correlation between germination capacity and lipid peroxidation was obtained for wheat seeds by Rudrapal and Basu (1982). As a general phenomenon in nature, the deleterious effects of lipid peroxidation on membrane functionality and the protective effects of natural antioxidants (tocopherols, ascorbic acid etc.) are well established (Tappel, 1980).

C Milled Cereal Products

1 *Wheat Flour*

As described by Shearer (Chapter 12), white flours may be stored for several years under appropriate conditions without serious loss of quality. Lipolytic activity in sound flours is low and, although an increase in FFA content is obtained on storage over 60 months at 12 ± 2 °C, there is no direct correlation between lipolytic activity and baking quality of flour; in fact, the functional quality improves over the first 12 months of storage when the FFA content is increasing. Differences in lipolytic activity in flours from different sources have been observed; e.g. the lipase activity of mixed-grist "weak" flours was higher than that of "strong" baking quality flours (Bell *et al*. 1979). However, the actual levels were very low. Only in white flours contaminated with fungi, have relatively high levels of lipolytic activity been detected (see Halton *et al*., 1959; Pomeranz, 1974).

The polyunsaturated FFA levels in stored flours are lower than would be expected from the observed loss of acyl lipids by lipolysis and the difference has been ascribed to oxidative degradation of 18 : 2 and 18 : 3 fatty acids (see Chapter 12). The concurrent formation of hydroxy acids in stored flour can account for part, but not all, of the loss of FFA (Warwick *et al.*, 1979). It is assumed that LOX enzyme activity, together with the related secondary reactions (Fig. 6.4), are responsible for the transformations of FFA. Little recent information is available on the effects of lipid-degrading enzymes in wholemeal and high-extraction flours compared with the extensive research and information on lipid changes in white flour (Chapter 12). The shorter storage life of high-extraction flours is consistent with the fact that the lipid-degrading enzymes are located in the "bran" and germ fractions of wheat. Colas and Charlegue (1974) showed that TG lipase activity was concentrated in the "bran" fractions of wheat milling. LOX activity is also relatively high in low-grade mill streams.

2 *Durum Semolina*

Retention of yellow carotenoid colour in semolina during processing into pasta is important to the quality of pasta products. The relatively mild conditions and long drying period in pasta production permit co-oxidation of carotenoids by LOX. The role of LOX was identified by Irvine and Winkler (1950) and supported by the demonstration of increased bleaching when linoleic acid was added to durum flour slurries (Dahle, 1965) and semolina doughs (Matsuo *et al.*, 1970). McDonald (1979) found a significant correlation between bleaching activity and LOX content of a range of durum semolinas. Plant breeders have produced durum wheats with very low LOX activities to maintain a bright yellow colour in semolina and pasta products (McDonald, 1979). Further aspects of lipid changes in pasta production are discussed by Laignelet (see Chapter 13).

3 *Wheat Germ*

The germ fraction obtained from wheat milling is unstable and, if untreated, develops rancid odours and high FFA levels within a few days (Chapter 19). Lipolysis and LOX-catalysed lipid oxidation are readily demonstrated in wheat germ and stabilization by drying or heating is a necessary step to prevent enzyme activity.

4 *Rice Bran*

The bran fraction obtained from rice milling represents a substantial proportion of the grain weight (5–8%) but the content of oil and lipid-degrading enzymes make the product very unstable. Thus, rice bran and rice bran oil require heating to inactivate lipolytic enzymes, LOX and other related enzymes (see Chapter 15).

5 *Milled Oats*

The relatively high content of both oil and lipase in oats, leads to rapid lipolysis when the lipase (predominantly in the outer layers) and the oil (in spherosomes or oil bodies in germ, aleurone and endosperm tissues) are intimately mixed. A "soapy" off-flavour, characteristic of hydrolytic rancidity, is rapidly generated in untreated oats. Heating the groats before flaking is a necessary step in the production of stable oat products (see Chapter 16).

6 *Maize Germ*

Lipase, LOX and fatty acid hydroperoxide isomerase may all play a role in determining the behaviour of maize germ and maize germ oil with respect to rancidity (Gardner and Inglett, 1971). All three enzymes are readily inactivated by heating; grain drying prior to milling, reduces the enzyme activities. Further moderate heating of maize germ leads to germ oil with low peroxide values. However, heating above 124 °C leads to oils in which peroxide values increase during storage. Gardner and Inglett (1971) proposed that the isomerase may act to remove fatty acid hydroperoxides and prevent generation of oxidative rancidity; heat-inactivation of the isomerase could explain the increase peroxide content of heated germ oils. However, high temperatures would also inactivate endogenous antioxidants in the germ.

7 *Flour–Water Systems*

The importance of lipid-degrading enzymes and the products of their action in wheat flour doughs is emphasized by the detailed coverage of this area in several chapters of this volume.

In breadmaking, wheat lipase activity during the dough mixing and baking processes are considered unimportant because no hydrolysis of flour lipids is observed in doughs (Graveland, 1973; Mann and Morrison, 1974).

Although bread quality is improved by removing non-polar lipids from flour, adding TG lipase to doughs has little effect on baking quality (Nierle and El Baya, 1981); conversely, the addition of LAH (from potato) to bread doughs (Nierle and El Baya, 1981) or cake batters (J. E. Allen and T. Galliard, unpublished) caused drastic failure in baking. Potato LAH hydrolyzes glycolipids and phospholipids (Galliard, 1971), which are known to be important for baking quality of flours (MacRitchie, Chapter 8). Recent results of Kieffer *et al.* (1982), who found no correlation between the content of wheat LOX and the baking quality of wheat, cast doubt upon the role of endogenous wheat LOX in the baking process. Certainly, the LOX added to doughs as enzyme-active soya flour or as horse-bean flour plays a more important role than endogenous wheat LOX in baking.

In wholemeal flour doughs, endogenous lipid-oxidizing enzymes may play a more important role than in white flour doughs. When wheat flour is mixed with water, oxygen is consumed and the amount of O_2 uptake is much greater with low-grade flours (Cosgrove, 1956). Smith and Andrews (1957) showed that the oxygen uptake was due primarily to lipid oxidation. Recent work has confirmed these observations with wholemeal flour suspensions (Galliard, 1983) and has demonstrated that O_2 uptake due to wheat LOX action on endogenous FFA is greater than that due to other components of bread doughs (yeast and soya flour – at conventional levels). Extrapolation of these results to break doughs indicates that the LOX-catalysed oxidation is sufficiently rapid to remove the dissolved O_2 from wholemeal doughs within a few seconds of mixing (T. Galliard and C. Anderson, unpublished). This observation may have implications for the redox system discussed for white flours in Chapter 10. Results on the relative rates of O_2 consumption by aqueous suspensions of white and wholemeal flours and by commercial wheat germ, are reported by Galliard (1983). It should be noted that O_2 is essential for good loaf quality, particularly when ascorbate is used as an improver in mechanical dough development (Chamberlain and Collins, 1979).

In addition to its role as a bread "improver" LOX may play other roles in flour-water systems. Neukom and Markwalder (1978) have shown that LOX plus linoleic acid is one of the peroxidation systems that will catalyse the oxidative gelation of wheat flour pentosans, producing increased viscosities of aqueous suspensions of flour. Zellen and Rasper (1982) have suggested that LOX may play a role in oxidizing the starch fraction in flour to produce effects similar to those obtained by Cl_2 treatment. The bleaching action of LOX has been mentioned previously and it is perhaps significant that all of these proposed roles for LOX could involve a free-radical oxidation process as discussed on page 128.

High FFA levels in wheat gluten are believed to cause a diminution in the functionality of "vital gluten" and the ability of wheat LOX to complex with the glutenin fraction of gluten was mentioned earlier. The participation of LOX in lipid-protein interactions in doughs has been studied by Frazier, Daniels *et al.* and is discussed in Chapter 9.

The internal lipids of wheat starch granules are resistant to oxidative attack (Morrison, 1978b) and to hydrolysis by LAH ("phospholipase B") although the polar group of lysoPC in wheat starch is hydrolysed by phospholipase D (Acker, 1977). These observations have been explained on the basis of the proposed structure for the amylose- monoacyl lipid complex (Chapter 5).

Although oxidative rancidity is considered less important in oat products than in other cereals because of endogenous antioxidant activity, there is evidence for LOX-catalysed reactions in aqueous dispersions of oat flours (Table 6.4). Biermann *et al.* (1980) have shown that the intense bitter taste of aqueous suspensions of unheated oat flour is due to the presence of mono- and tri-hydroxy derivatives of linoleic acid, formed via a LOX-initiated pathway from linoleic acid.

8 *Malt and Beer*

Lipase activity increases during germination of barley (MacLeod and White, 1962), and FFA represent an important component of wort lipids. Despite the importance of FFA as surface-active agents and potential sources of off-flavours, and the fact that malt lipase activity can survive the kilning process (MacLeod and White, 1962), little information is available on the role of lipases in malting and brewing. Further aspects of this area are covered by Baisted in Chapter 5.

9 *Mixed Cereal Products*

For lipid degradation to occur, an active catalyst must be present and able to interact with lipid. Single components of a mixture may be stable in isolation but be subject to, or cause, degradation in mixed products. For example, in defatted wheat germ, the lipase and LOX activities are retained but the material is stabilized by the removal of the lipid; conversely, steamed oat flakes have a relatively high lipid content but are stabilized by enzyme inactivation. Mixtures of the two materials would contain substrates and enzymes with the potential for rancidity development. Reported experiences of problems arising in mixed cereal products include the unacceptable hydrolytic rancidity in wheat flour that had been blended with oat flour (Widhe and Onselius, 1949). Unacceptable flavours have been described in biscuits made with fungi-contaminated wheat and coconut oil and in oat cakes containing palm kernel or coconut oil (Halton *et al.* 1959 and refs. *loc. cit.*); in these cases the problem was traced to lauric acid formed from the oil by lipases in the flour or oatmeal.

Interactions between enzymes and substrates in mixed dry products are very much dependent upon the physical state of the system. Brockmann and Acker, (1977) have shown that in model systems of low a_w containing LOX (from soya) and lipid (FFA or sunflower-seed oil), each impregnated onto an inert cellulose powder, the rate of oxidation was a factor of 10 or more higher in homogenous samples (enzyme and substrate added to the same cellulose sample) than in heterogeneous samples (enzyme and substrate added to separate samples of cellulose and then mixed).

REFERENCES

Acker, L. (1977). *Fette, Seifen, Anstrichmittel*, **79**, 1–9.
Acker, L. and Beutler, H. O. (1965). *Getreide und Mehl.* **15**, 4–6.
Acker, L. and Geyer, J. (1968). *Brauwissenschaft*, **21**, 222–226.
Acker, L. and Luck, E. (1958). *Z. Lebensm. Unters.-Forsch.* **108**, 256–269.
Acker, L. and Müller, K. (1965). *Brauwissenschaft.* **17**, 369–372.
Acker, L. and Wiese, R. (1972). *Lebensm. Unters.-Forsch.* **150**, 205–211.
Aizono, Y. and Funatsu, M. (1978). *Agric. Biol. Chem.* **42**, 757–762.
Aizono, Y., Fumatsu, M., Fujiki, Y. and Watanake, M. (1976). *Agric. Biol. Chem.* **40**, 317–324.
Allen, J. C. and Hamilton, R. J. eds. (1983). *Rancidity in Foods.* Applied Science Publishers., London, in press.
Aoyagi, Y., Yamashita, H., Matsumoto, S. and Obara, T. (1979). *Agric. Biol. Chem.* **43**, 1771–1772.
Baisted, D. J. and Stroud, F. (1982a). *Phytochemistry*, **21**, 29–31.
Baisted, D. J. and Stroud, F. (1982b). *Phytochemistry*, **21**, 2619–2623.
Baxter, E. D. (1982). *J. Inst. Brew.* **88**, 390–396.
Bell, B. M., Chamberlain, N., Collins, T. H., Daniels, D. G. H. and Fisher, N. (1979). *J. Sci. Fd. Agric.* **30**, 1111–1122.
Berner, D. L. and Hammond, E. G. (1970). *Lipids*, **5**, 572–573.
Biermann, U., Wittmann, A. and Grosch, W. (1980). *Fette, Seifen, Anstrichm.* **82**, 236–240.
Brockmann, R. and Acker, L. (1977). *Ann. Technol. Agric.* **26**, 167–174.
Chamberlain, N. and Collins, T. H. (1979). *Bakers Digest*, **53**, 18–24.
Chan, H. W.-S. (1972). *Commun. 11th World Cong. Int. Soc. Fat Res. p. 1978.*
Chan, H. W.-S. ed. (1983). *Autoxidation of Unsaturated Lipids.* Academic Press, London and New York.
Christianson, D. D. and Gardner, H. W. (1975). *Lipids*, **10**, 448–453.
Clayton, T. A. and Morrison, W. R. (1972). *J. Sci. Fd. Agric.* **23**, 721–736.
Colas, A. and Charlegue, A. (1974). *Ann. Technol. agric.*, **23**, 323–334.
Colborne, A. J. and Laidman, D. L. (1975). *Phytochemistry*, **14**, 2639–2645.
Cosgrove, D. J. (1956). *J. Sci. Fd. Agric.* **7**, 668–672.
Dahle, L. (1965). *J. Agric. Food Chem.* , **13**, 12–20.
Dirks, B. M., Boyer, P. D. and Geddes, W. F. (1955). *Cereal Chem.* **32**, 356–373.
Dixon, M. and Webb, E. C. (1964). *Enzymes* (2nd edn.) pp. 90–91, Academic Press, London and New York.
Drapron, R. (1972). *Ann. Technol. Agric.* **21**, 487–499.
Drapron, R. (1983). *Proc. 7th World Cereal and Bread Congress*, Prague, (in press).
Drapron, R. and Guilbot, A. (1962). *Ann. Technol. Agric.* **11**, 175–218; ibid. pp. 275–317.
Drapron, R., Anh, N'G. X., Launay, B. and Guilbot, A. (1969). *Cereal Chem.* **46**, 647–655.
Egmond, M. R., Vliegenthart, J. F. G. and Boldingh, J. (1972). *Biochem. Biophys. Res. Commun.* **48**, 1055–1060.
Franke, W. and Frehse, H. (1953). *Z. Physiol. Chem.* **295**, 333–349.
Frey, K. J. and Hammond, E. G. (1975). *J. Amer. Oil Chem. Soc.* **52**, 358–362.
Führling, D. (1975). Ph.D. Thesis, University of Berlin, W. Germany.

Funatsu, M. Aizono, Y., Hayashi, K., Watanabe, M. and Eto, M. (1971). *Agric. Biol. Chem.* **35**, 734–742.

Galliard, T. (1971). *Biochem. J.* **121**, 371–390.

Galliard, T. (1975). In *Recent Advances in the Chemistry and Biochemistry of Plant Lipids* (T. Galliard and E. I. Mercer, eds), pp. 319–337. Academic Press, New York and London.

Galliard, T. (1980). In *The Biochemistry of Plants* Vol. 4. Lipids: Structure and Function (P. K. Stumpf, ed.), pp. 85–116. Academic Press, New York and London.

Galliard, T. (1983) *7th World Cereal and Bread Congress, Prague*, (in press).

Galliard, T. and Chan, H. W.-S. (1980). In *The Biochemistry of Plants* Vol. 4. Lipids: Structure and Function (P. F. Stumpf, ed.) pp. 131–161. Academic Press, London and New York.

Gardner, H. W. (1970). *J. Lipid Res.* **11**, 311–321.

Gardner, H. W. (1980). In *Autoxidation in Food and Biological Systems* (M. G. Simic and M. Karel, eds), pp. 447–504. Plenum Publishing Corp. New York.

Gardner, H. W. (1983). In *Autoxidation of Unsaturated Lipids* (H. W.-S. Chan, ed.), Academic Press, London and New York (in press).

Gardner, H. W. and Inglett, G. E. (1971). *J. Food Sci.* **36**, 645–648.

Gardner, H. W. and Weisleder, D. (1970). *Lipids*, **5**, 678–683.

Gardner, H. W., Kleiman, R., Christianson, D. D. and Weisleder, D. (1975). *Lipids*, **10**, 602–608.

Graveland, A. (1968). *J. Amer. Oil Chem. Soc.* **45**, 834–840.

Graveland, A. (1970). *Biochem. Biophys. Res. Commun.* **41**, 427–434.

Graveland, A. (1973). *Lipids*, **8**, 606–611.

Graveland, A., Pesman, L. and Van Evde, P. (1972). *Tech. & Master Brew. Assoc. Am.* **9**, 98–104.

Grosch, W., Laskawy, G. and Weber, F. (1976). *J. Agric. Food Chem.* **24**, 456–459.

Grossman, S. and Zakut, R. (1979). In *Methods of Biochemical Analysis* (D. Glick, ed.), Vol. 25, pp. 303–329, J. Wiley & Sons, New York.

Guss, P. L., Macko, V., Richardson, T. and Stahmann, M. A. (1968). *Plant Cell Physiol.* **9**, 415–422.

Halton, P. Knight, R. A., Martin, H. F. and Ottaway, F. J. H. (1959). *J. Sci. Fd. Agric.* **10**, 401–403.

Hamberg, M. (1971). *Anal. Biochem.* **43**, 515–526.

Heaton, T. C., Knowles, P. K., Mikkelsen, D. and Ruckman, J. E. (1978). *J. Amer. Oil Chem. Soc.* **55**, 465–468.

Heimann, W. and Dresen, P. (1973). *Helv. Chim. Acta.* **56**, 463–469.

Heimann, W. and Klaiber, V. (1977a). *Z. Lebensm. Unters. Forsch.* **165**, 131–136.

Heimann, W. and Klaiber, V. (1977b). *Z. Lebensm. Unters. Forsch.* **165**, 144–147.

Heimann, W. and Timm, U. (1977a). *Z. Lebensm. Unters. Forsch.* **165**, 5.

Heimann, W. and Timm, U. (1977b). *Z. Lebensm. Unters. Forsch.* **165**, 12–14.

Heimann, W. Dresen, P. and Klaibor, V. (1973). *Z. Lebensm. Unters. Forsch.* **153**, 1–5.

Hirayama, O. and Matsuda, H. (1975). *J. Ag. Chem. Soc. Japan*, **49**, 569–576.

Hutchinson, J. B., Martin, H. F. and Moran, T. (1951). *Nature* (Lond.) **167**, 758–759.

Irvine, G. N. and Winkler, C. A. (1950). *Cereal Chem.* **27**, 205–218.

Jelsema, C. L., Morré, D. J., Ruddat, M. and Turner, C. (1977). *Bot. Gaz.* **138**, 138–149.

Jensen, S. A. and Heltved, F. (1982). *Carlsberg Res. Commun.* **47**, 297–303.
Juliano, B. O. (1977). *Il Riso.* **26**, 3–21.
Justin, O. L. and Bass, L. N. (1978). *U.S. Dep. Agric. Handbk.* **506**, pp. 57–77.
Kieffer, R., Matheis, G., Belitz, H.-D. and Grosch, W. (1982). *Z. Lebensm. Unters. Forsch.* **175**, 5–7.
Kies, M. W., Haining, J. L., Pistorius, E., Schroeder, D. H. and Axelrod, B. (1969). *Biochim. Biophys. Res. Commun.* **36**, 312–315.
Lulai, E. C. and Baker, C. W. (1975). *Proc. Amer. Soc. Brew. Chem.* **33**, 154–158.
Lulai, E. C., Baker, C. W. and Zimmerman, D. C. (1981). *Plant Physiol.* **68**, 950–955.
MacLeod, A. M. and White, H. B. (1961). *J. Inst. Brew.* **67**, 182–190.
MacLeod, A. M. and White, H. B. (1962). *J. Inst. Brew.* **68**, 487–495.
McDonald, C. E. (1979). *Cereal Chem.* **56**, 84–89.
Mann, D. L. and Morrison, W. R. (1974). *J. Sci. Fd. Agric.* **25**, 1109–1119.
Mapson, L. W. and Moustafa, E. M. (1955). *Biochem. J.* **60**, 71–80.
Martin, H. F. and Peers, F. G. (1953). *Biochem. J.* **55**, 523–529.
Matsuda, H. and Hirayama, O. (1975). *J. Ag. Chem. Soc. Japan*, **49**, 577–583.
Matsuda, H. and Hirayama, O. (1979a). *Agric. Biol. Chem.* **43**, 463–469.
Matsuda, H. and Hirayama, O. (1979b). *Agric. Biol. Chem.* **43**, 697–703.
Matsuo, R. R., Bradley, J. W. and Irvine, G. N. (1970). *Cereal Chem.* **47**, 1–5.
Meilgaard, M. (1972). *Brew. Dig.* **47**, 48–62.
Morrison, W. R. (1978a). In *Advances in Cereal Science and Technology* (Y. Pomeranz, ed.) Vol. 11, pp. 221–348. American Association of Cereal Chemists, St. Paul, Minnesota.
Morrison, W. R. (1978b). *J. Sci. Fd. Agric.* **29**, 365–371.
Morrison, W. R. and Panpaprai (1975). *J. Sci. Fd. Agric.* **26**, 1225–1236.
Neukom, H. and Markwalder, H. U. (1978). *Cereals Food World* **23**, 374–376.
Nicolas, R. and Drapron, R. (1977). *Ann. Techn. Agricol.* **26**, 119–132.
Nicolas, J. and Drapron, R. (1981). *Sciences des Aliments.* **1**, 91–168.
Nicolas, J., Gustafsson, S., Autran, M. and Drapron, R. (1981a). *Sciences des Aliments.* **1**, 203–212.
Nicolas, J., Autran, M. and Drapron, R. (1981b). *Sciences des Aliments* **1**, 213–232.
Nicolas, J., Autran, M. and Drapron, R. (1982). *J. Sci. Fd. Agric.* **33**, 365–372.
Nierle, N. and El Baya, A. W. (1981). *Fette. Seifen. Anstrichmittel*, **83**, 391–395.
Nolte, D. and Acker, L. (1975a). *Z. Lebensm. Unters. Forsch.* **158**, 149–156.
Nolte, D. Acker, L. (1975b). *Z. Lebensm. Unters. Forsch.* **159**, 225–233.
Nolte, D., Rebmann, H. and Acker, L. (1974). *Getreide, Mehl. u. Brot.* **28**, 189–191.
Ohlrogge, J. B. and Kernan, T. P. (1982). *Plant Physiol.* **70**, 791–794.
Phillips, D. R., Matthew, J. A., Reynolds, J. and Fenwick, G. R. (1979). *Phytochemistry*, **18**, 401–404.
Pomeranz, Y. (1974) In *Storage of Cereal Grains and their Products* (C. M. Christensen, ed.) 2nd. edn. pp. 56–114. American Association of Cereal Chemists, St. Paul, Minnesota.
Rebmann, H. and Acker, L. (1973). *Fette. Seifn. Anstrichmittel.* **75**, 409–411.
Rothe, M. (1958). *Ernahrungs forschung*, **3**, 21–51.
Roughan, P. G., Slack, C. R. and Holland, R. (1978). *Lipids*, **13**, 497–503.
Rudrapal, A. B. and Basu, R. N. (1982). *Indian. J. Exp. Biol.* **20**, 465–470.
Sastry, B. S., Ramakrishna, M. and Raghavendra Rao, M. R. (1977). *J. Food Sci. Technol (Mysore)* **14**, 273–274.

Schreier, P. and Lorenz, G. (1981). In *Flavour '81* (P. Schreier, ed.), pp. 495–508. Walter de Gruyter & Co., Berlin.
Shastry, B. S. and Rao, M. R. R. (1975). *Cereal Chem.* **52**, 597–603.
Simic, M. G. and Karel, M. eds. (1980). *Autoxidation in Food and Biological Systems*, Plenum Publishing Corp. New York.
Smith, D. E. and Andrews, J.S. (1957). *Cereal Chem.* **34**, 323–336.
Smith, W. L. and Lands, W. E. M. (1972). *J. Biol. Chem.* **247**, 1038–1047.
Stauffer, C. E. and Glass, R. L. (1966). *Cereal Chem.* **43**, 644–657.
Streckert, G. and Stan, H. J. (1975). *Lipids*, **10**, 847–854.
Tappel, A. L. (1980). In *Free Radicals in Biology* Vol. IV (W. A. Pryor, ed.) pp. 1–47, Academic Press, New York and London.
Tavener, R. J. A. and Laidman, D. L. (1972a). *Phytochemistry*, **11**, 989–997.
Tavener, R. J. A. and Laidman, D. L. (1972b). *Phytochemistry*, **11**, 981–987.
Tombs, M. P. and Blake, G. (1982). *Biochim. Biophys. Res. Commun.* **700**, 81–89.
Tombs, M. P. and Stubbs, J. M. (1982). *J. Sci. Fd. Agric.* **33**, 892–897.
Veldink, G. A. Garssen, G. J. Vliegenthart, J. F. G. and Boldingh, J. (1972). *Biochem. Biophys. Res. Commun.* **147**, 22–26.
Veldink, G. A., Vliegenthart, J. F. G. and Boldingh, J. (1977). *Prog. Chem. Fats Other Lipids*, **15**, 131–166.
Vick, B. A. and Zimmerman, D. C. (1979). *Plant Physiol.* **64**, 203–205.
Vick, B. A. and Zimmerman, D. C. (1982). *Plant Physiol.* **69**, 1103–1108.
Viogue, E. and Maza, M. P. (1971). *Grasas Aceites (Seville)* **22**, 22.
Vliegenthart, J. F. G. and Veldink, G. A. (1982). In *Free Radicals in Biology*, Vol. 5 (W. A. Pryor, ed.), pp. 29–64 Academic Press, London and New York.
Wagenknecht, A. C. (1959). *Food Res.* **24**, 539–547.
Wallace, J. M. and Wheeler, E. L. (1972). *Cereal Chem.* **49**, 92–98.
Wallace, J. M. and Wheeler, E. L. (1975). *J. Ag. Food Chem.* **23**, 146–150.
Wallace, J. M. and Wheeler, E. L. (1979). *Phtochemistry*, **18**, 389–393.
Warwick, M. J., Farrington, W. H. H. and Shearer, G. (1979). *J. Sci. Fd. Agric.* **30**, 1131–1138.
Weber, F. and Grosch, W. (1976). *Z. Lebensm. Unters. Forsch.* **161**, 223–230.
Welch, R. W. (1977). *J. Sci. Fd. Agric.* **28**, 269–274.
Wheeler, E. L. and Wallace, J. M. (1973). *Lebensm. Wiss. Technol.* **6**, 205–208.
Wheeler, E. L. and Wallace, J. M. (1978). *Phytochemistry*, **17**, 42–44.
Widhe, T. and Onselius, T. (1949). *Cereal Chem.* **26**, 393–398.
Yabuuchi, S. and Amaha, M. (1976). *Phytochemistry*, **15**, 387–390.
Yamamato, A., Fujii, Y., Yasumoto, K. and Mitsuda, H. (1980a). *Agric. Biol. Chem.* **44**, 443–445.
Yamamoto, A., Fujii, Y., Yasumoto, K. and Mitsuda, H. (1980b). *Lipids*, **15**, 1–5.
Youngs, V. L. (1978). *Cereal Chem.* **55**, 591–597.
Zellen, W. L. and Rasper, V. F. (1982). *Can. Inst. Food Sci. Techn. J.* **15**, XXIII.

7 Distribution of Wheat Acyl Lipids and Tocols into Flour Millstreams

W. R. MORRISON* and P. J. BARNES†

*University of Strathclyde, Glasgow, U.K. and †The Lord Rank Research Centre, High Wycombe, U.K.

I INTRODUCTION

The lipid content of flour millstreams increases with colour and ash content, hence it is inversely related to flour quality (Kent, 1975; Fortman and Joiner, 1978; Ziegler and Greer, 1978). Since non-starch lipids (Chapter 2) affect the storage stability of flour (Chapter 12) and the performance of flour used for bread, cake, biscuits, cookies (Chapters 8–11) and pasta (Chapter 13), detailed information on the non-starch lipids in various millstreams should be useful.

The composition of the non-starch lipids in mixed-grist flours has been determined many times (Clayton and Morrison, 1972; Lin *et al.*, 1974;

"Lipids in Cereal Technology"
ISBN 0-12-079020-3

Table 7.1 Non-starch lipids in wheat flours (mg 100 g^{-1} dry wt.)

Lipid class	Commercial flours†	Single-variety flours‡
SE	26– 72	11– 48
TG	674– 909	322– 660
DG	60– 86	58– 115
FFA	61– 110	63– 139
MG-ASG	69– 176	62– 96
MGXG	86– 132	64– 138
DGXG	266– 366	218– 340
APE	64– 95	69– 107
ALPE	33– 43	32– 65
DiPL	79– 115	49– 91
LysoPL	41– 66	70– 111
Total	1553–1953	1183–1649

†Five flours, including three grades of soft flour (Morrison *et al.*, 1975, 1978).
‡Sixteen flours, Bühler milled from English-grown single wheat varieties (Morrison, unpublished).

MacMurray and Morrison, 1970; Mann and Morrison, 1974; Morrison, 1978a; Morrison and Hargin, 1981; Morrison *et al.*, 1975). In practice, most of the minor lipids (Chapter 2) are not important in baking technology and some of the others can be grouped together for convenience. Thus, for the purposes of this chapter the non-starch lipids which will be discussed are restricted to sterylesters (SE), triglycerides (TG), diglycerides (DG), free fatty acids (FFA), monoglycerides plus 6-O-acylmonogalactosyldiglyceride plus 6-O-acylsterylglycoside (MG-ASG), monogalactosylglycerides (MGXG), digalactosylglycerides (DGXG), N-acylphosphatidylethanolamine (APE) and its lysoderivative (ALPE), the di-O-acylphosphoglycerides (diPL), and the lysophospholipids (lysoPL). Some representative data for mixed grist and single variety flours are given in Table 7.1.

II FLOUR MILLING

In a modern flour mill many flour streams are produced and blended together to yield a white flour of the appropriate specification. To explain how lipids are distributed into the flour streams it is necessary to give a brief description of the flour milling process.

Flour milling consists of a sequence of breaking, grinding and separating operations (Fig. 7.1). The wheat passes first through fluted break rolls

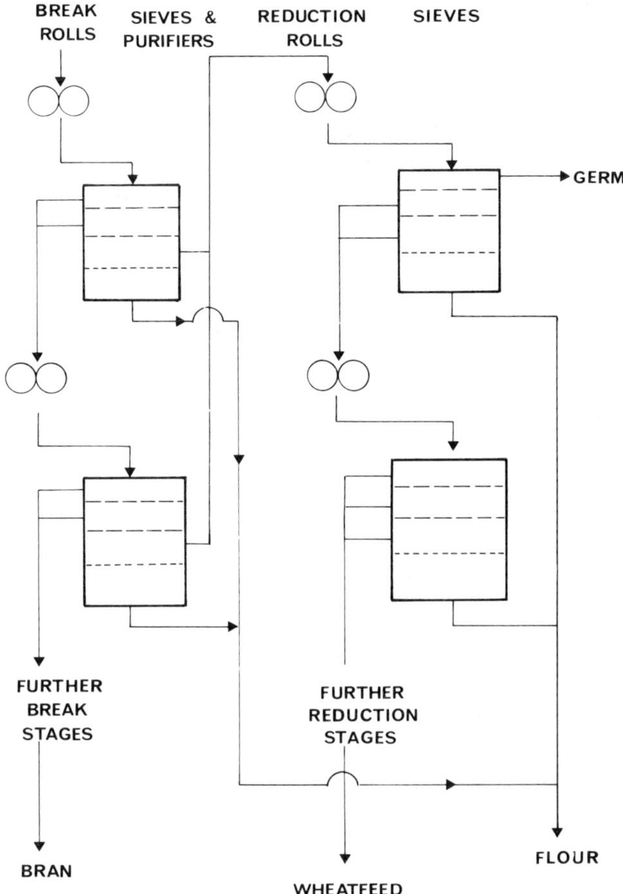

Fig. 7.1 A simplified schematic diagram of the flour milling process.

which split open the grains, releasing some large endosperm fragments (semolina) and a small amount of flour; these fractions are separated by sieving and the remaining bran passes through further break rolls to remove more semolina, finally yielding a relatively clean bran fraction. The semolina fractions from the break rolls are contaminated with free bran flakes which can be removed by sieving and aspiration, but a proportion of the fragments still have bran attached even after these purification stages. If this impure semolina were ground directly into a flour the bran would be pulverized and it would be impossible to effect any further separation. To minimize this possibility, the semolina is ground down gradually on a series of smooth reduction rolls to produce flour of a particle size less than 150 μm which can be sieved out from the unbroken bran particles. This separation becomes increasingly difficult, and the later reduction flours are more contaminated with bran and therefore darker in colour than the early reduction streams. The greatest proportion of flour is released from the early reduction rolls, but even these flour streams contain a small amount of bran.

In a typical U.K. flour mill there may be 6 break rolls and 12 reduction rolls together with further separation on redressers, and a flour stream is produced at each stage. Most of the coarse bran flakes pass along the break system and are collected as a fraction from the final break rolls. Fine bran from the ends of the reduction and break systems and the aspiration steps forms a fraction known as "wheat feed" which is sold to the animal feed compounding industry. Coarse flakes of germ may be separated by sieving on the reduction side of the mill to produce commercial wheat germ as a valuable by-product. If a germ-rich stream is not separated, much of the wheat germ will enter the wheat feed fraction. A proportion of the germ becomes finely pulverized during milling and it is inevitable that the flour streams will contain germ to varying extents.

III THE REDISTRIBUTION OF WHEAT GRAIN LIPIDS DURING FLOUR MILLING

Wheat flour will be expected to contain the original endosperm lipids together with lipids transferred from other parts of the grain. These transferred lipids may be present in fragments of the original germ and bran tissue or as oil released when the tissues are disrupted during milling. The darker coloured millstreams contain the highest proportions of germ and bran particles and it is usually assumed that the relatively high concentration of oil in these streams is at least partly due to the presence of the contaminating particles.

A Redistribution of Lipid-Rich Tissues

Early studies gave an indication of the distribution of embryo axis and scutellum among millstreams following observations by Hinton (1944) that thiamine is concentrated in the scutellum and can be used as a marker for this tissue. Hinton (1944) also showed that both embryo axis and scutellum are rich in petroleum-extractable material (15% and 30% dry weight respectively in wheat) and therefore contamination of flour with either of these tissues might lead to an increase in lipid content. Using thiamine determination and particle counting it was shown (Kent *et al.*, 1944; Kent *et al.*, 1949) that a large proportion of the embryo axis is released intact, or only coarsely-divided, on the early break rolls and passes with the semolina to the early reduction rolls. While passing through these rolls it is compressed into large flakes and then separated to yield commercial wheat germ. Unlike embryo axis, the scutellum is less easily detached from the grain and it is released on the later break rolls. The scutellum is more friable than the embryo axis and tends to be pulverized rather than flaked so that it passes with the finer stocks to the later reduction rolls where it contaminates the flour streams. Jones and Moran (1946) measured the ratio of thiamine content to fat content for a set of commerical millstreams and showed that this ratio was relatively high for the later reduction flours; this suggested that these flours were rich in scutellum compared with the embryo axis.

B Redistribution of Germ Oil

The oil content of purified commercial wheat germ was known to be less than that of dissected embryo axis and this was assumed to be due to transfer of oil from compressed flakes to the endosperm fragments during passage through the mill rolls. Stevens (1959) used an ingenious technique to demonstrate that about 30% of the petroleum-extractable oil in laboratory-milled flour was germ oil. Populations of Confused Flour Beetles (*Tribolium confusum*) were used selectively to remove the germ from wheat grain without bran or endosperm loss. A comparison was then made of the petroleum-extractable oil in flour milled from this wheat with that in flour from the corresponding intact wheat. It was estimated that half of the germ oil in the laboratory milled flour from intact wheat would be accounted for by losses from the germ flakes during milling by comparing the oil content of the dissected embryo axis and purified commercial germ. However, the laboratory milling procedure was rather different from commercial practice and the flour streams were combined to give four groups for examination, thus limiting the amount of information gained from the results. Furthermore, this work was concerned only with germ oil and did not consider the possibility that transferred lipids in the endosperm flour could also arise from contamination with aleurone tissue or aleurone oil. Typical aleurone components such as phytin and niacin are known to appear in greater quantities in lower grade millstreams (Clegg and Hinton, 1958; Kent, 1975; MacMasters *et al.*, 1978).

C Redistribution of Individual Lipid Classes

1 *Laboratory Milling*

Hargin and his colleagues quantified the acyl lipids in the principal tissues of wheat (Hargin and Morrison, 1980; Hargin *et al.*, 1980) and then tried

Table 7.2 Composition of non-starch lipids in endosperm and flours milled from whole and degermed Atou wheat (mg $100g^{-1}$ dry wt)†

Lipid class	Aleurone-free endosperm	Flour	
		whole	degermed
SE	15	14	7
TG	222	407	243
DG	46	52	49
FFA	62	83	73
MG-ASG	71	61	58
MGXG	44	47	47
DGXG	188	213	218
APE	71	76	76
ALPE	50	55	55
DiPL	30	35	35

†From Morrison and Hargin, 1981.

to determine the contributions of aleurone and germ lipids to flour lipids (Morrison and Hargin, 1981).

The dissected aleurone-free endosperm of Atou wheat was used as the basis for comparisons. Whole wheat and manually degermed wheat were milled on a two-break roll micromill to give straight-run flour, coarse offal, fine offal, bran and bran flour from each wheat sample. The composition of the non-starch lipids in the endosperm and straight-run flours is shown in Table 7.2.

These results show clearly that the lipids in the degermed flour were quantitatively identical to those in the endosperm, indicating that no aleurone lipids had been transferred to the flour. The lipids in flour from the whole wheat were likewise identical except for significantly more TG (and possibly SE) so it appeared that the only lipids to be transferred to the flour were those derived from the internal region of germ spherosomes (see Chapter 4). The micromill had only two break rolls so the flour yield was low and the bran contamination was much less than for a commerical flour, perhaps explaining the absence of aleurone lipids. Recent studies (described in Section IIIC2) show that aleurone lipids are also transferred in commerical-scale milling.

2 *Commercial-scale Milling*

Morrison and Hargin (1981) quantified the non-starch acyl lipids in 11 of the millstreams from an all-English soft-wheat grist. Milling degermed wheat was impractical on this scale, and, since aleurone and germ acyl lipids are almost identical (Hargin and Morrison, 1980), no distinction could be made between these two sources of lipid when transferred to the endosperm flour.

This study showed that there were approximately the same levels of MG-ASG, MGXG, DGXG, APE and ALPE in endosperm and all the mill-streams, but progressively more SE, TG, DG, FFA, diPL and lysoPL with increasing values for ash and flour colour grade in the millstreams. These results are entirely consistent with the concept of germ (and possibly aleurone) spherosome lipids being transferred to endosperm in greater quantities in the lower grade millstreams.

It was then discovered that the distribution of tocols in the germ, bran and starchy endosperm of wheat is much more specific than had been reported previously (Chapter 3), and the tocols could be used as markers for germ and aleurone lipid in millstreams (Morrison *et al.*, 1982). α- and β-Tocopherols are almost exclusively in the germ, and are mar-

Table 7.3 Ash, protein and colour grade of flour millstreams†

Millstream	Abbreviation	Ash‡	Protein‡	FCG§
I Break	I Bk	0.66	14.7	5.5
II Break	II Bk	1.03	15.4	3.7
III Break Coarse	III BkC	1.41	17.4	5.2
III Break Fine	III BkF	1.53	18.5	6.5
IV Break Coarse	IV BkC	1.30	18.7	8.6
IV Break Fine	IV BkF	1.29	16.5	8.4
IV Break Redresser	IV BkRe	1.54	17.9	8.6
Break Middlings Redresser	BMR	0.94	16.9	6.3
A Reduction	A	0.52	10.6	1.0
B Reduction	B	0.44	11.8	0.0
A and B Redresser 1	ABRe 1	0.41	12.9	0.3
A and B Redresser 2	ABRe 2	0.41	13.2	−0.4
C Reduction 1	C 1	0.49	12.5	0.6
C Reduction 2	C 2	0.49	12.3	−0.1
B_2 Reduction	B_2	0.73	12.8	1.4
D Reduction	D	0.65	12.9	1.7
E Reduction	E	0.89	13.4	8.1
F Reduction	F	0.70	12.6	2.8
G Reduction	G	1.15	13.6	7.8
H Reduction	H	2.07	15.7	18.5
J Reduction 1	J 1	1.94	15.0	14.3
J Reduction 2	J 2	2.21	14.7	13.5
J Reduction 3	J 3	2.42	14.9	16.0

†Morrison *et al.*, 1982.
‡Percentage of flour dry weight.
§Flour colour grade units.

kers for germ lipid. α-Tocotrienol is mainly in the bran (almost certainly only in the aleurone) and is a secondary marker for aleurone while β-tocotrienol is found in the triploid aleurone and endosperm tissues.

Twenty-three millstreams from a bread wheat mixed grist were analysed for ash, protein, flour colour grade (Table 7.3), acyl lipids (Table 7.4) and tocols (Table 7.5). The data confirmed that lower grade millstreams contained more ash, protein, colour, SE, TG, DG, FFA, diPL, lysoPL, and tocols, and showed for the first time that aleurone lipids were being transferred (α-T-3 highly correlated with TG and other transferred lipids). The results for α-T indicate the extent of germ lipid contamination of the millstreams regardless of whether the lipid is present in germ tissues or transferred as oil to endosperm particles.

High correlations between all transferred components permitted the calculation of the ash, protein, flour colour grade, acyl lipid and tocol contents of the average, or basic, endosperm and of the material transferred to it from germ and aleurone from regression analyses of the data (Table 7.6).

The composition of the basic endosperm flour was typical of dissected aleurone-free endosperm (Hargin and Morrison, 1980; Morrison and Hargin, 1981). Flours ABRe, B and C were nearly pure endosperm material, as may be expected since these millstreams are derived from the cleanest semolina fractions and, together with the A stream, account for the bulk of the high-grade flour produced on this mill. Although the A flour is also high-grade, it contains significantly more TG and α-T. Coarse germ from the break rolls passes with the coarse semolina through the A roll where it is flaked and then separated. The expression of germ oil during flaking results in transfer to endosperm particles and thus explains the higher TG and α-T values for the flour. The very high content of germ lipid in the IV BkF flour is because this set of rolls receives germ-rich fine material derived from the germ separator.

It is claimed that the high protein content of break and later reduction flours is largely because these flours have a greater proportion of the protein-rich outer layers of starchy endosperm (MacMasters *et al*., 1971). More recently it has been shown that gradients of lipid concentration exist across the wheat endosperm (Hargin *et al*., 1980) but the results in Table 7.4 suggest that the gradients of endosperm lipids are well dispersed during milling even though the protein gradient *appears* to survive (relatively high protein content in the break flours). Jones and Moran (1946) used the high lipid content of outer endosperm to explain the relatively high lipid levels in the later break flours; they assumed that these flours contained only low levels of germ because the thiamine content was low. However, thiamine is a marker for scutellum only – not embryo axis – and can be used only to detect oil transferred in particles of scutellum tissue. A comparison of the

Table 7.4 Composition of acyl lipids in millstreams (mg 100 g^{-1} dry wt.)†

Millstream	SE	TG	DG, FFA	MG-ASG	MGXG	DGXG	APE	ALPE	DiPL	LysoPL
I Bk	44	516	139	72	102	301	101	57	84	79
II Bk	45	776	148	67	93	301	115	66	93	93
III BkC	53	1161	146	62	99	275	107	60	99	85
III BkF	57	1572	215	72	86	303	88	44	92	85
IV BkC	78	1700	180	61	81	232	97	57	122	85
IV BkF	88	2308	223	69	75	254	90	45	120	89
IV Re	71	1917	200	61	86	212	96	51	124	94
BMR	50	1159	166	69	98	290	101	54	113	80
A	48	602	107	82	96	275	90	45	76	92
B	42	438	105	73	97	289	94	47	62	86
ABRe1	44	463	110	73	98	305	87	44	74	70
ABRe2	42	426	112	73	97	259	107	54	82	89
C1	40	472	113	75	96	275	91	44	72	63
C2	40	365	94	70	94	262	94	43	78	57
B2	58	981	118	68	98	248	85	44	73	70
D	46	663	138	66	92	291	90	44	103	77
E	58	1231	141	74	84	253	86	44	136	90
F	55	1222	140	69	89	297	102	54	114	103
G	60	1587	155	67	85	265	89	46	197	125
H	83	2393	250	80	86	270	78	40	340	138
J1	86	2654	223	82	85	248	96	43	398	177
J2	95	2750	245	79	90	279	92	45	395	185
J3	92	2151	229	88	93	270	101	48	331	162

†Morrison *et al.*, 1982.

Table 7.5 Composition of tocopherols and tocotrienols in millstreams ($mg\ kg^{-1}$ dry wt.)†

Millstream	α-T	β-T	α-T-3	β-T-3
I Bk	2.2	0.8	2.1	20.2
II Bk	6.5	3.1	1.5	21.5
III BkC	11.2	4.7	2.6	22.6
III BkF	14.6	7.8	1.9	19.5
IV BkC	15.1	6.2	3.6	21.9
IV BkF	30.8	12.9	3.3	19.9
IV BkRe	15.8	8.1	2.6	19.3
BMR	10.2	5.3	1.9	21.3
A	5.2	2.4	1.1	17.9
B	2.8	1.9	1.0	13.5
ABRe 1	4.1	1.7	1.5	21.9
ABRe 2	2.7	1.4	1.3	19.4
C 1	3.3	1.9	1.4	21.7
C 2	2.6	1.4	1.3	20.0
B_2	9.9	5.2	1.3	20.5
D	6.9	3.1	1.5	17.7
E	14.8	6.6	2.4	20.4
F	14.2	6.0	2.1	18.7
G	23.2	11.0	2.8	23.6
H	35.4	16.5	4.5	31.6
J 1	41.3	20.7	3.7	25.8
J 2	38.8	17.7	3.7	21.8
J 3	25.0	13.5	2.9	24.5

†Morrison *et al.*, 1982.

TG and α-T results in Tables 7.3 and 7.4 shows that the high oil content of the later break flours is accounted for by the presence of germ lipid or germ tissue.

In recent studies of lipids in millstreams (Morrison and Hargin, 1980; Morrison *et al.*, 1982) good correlations were obtained between the percentages of palmitate and linoleate in the steryl esters and the total content of steryl esters in the flours (Table 7.7). These results show that contrary to the observations of Berry *et al.*, (1968) the steryl palmitate content rises slightly as millstream quality deteriorates – this is consistent with the model of a basic endosperm flour containing palmitate-rich steryl ester which has linoleate-rich steryl ester added to it in lipid transferred from aleurone and germ (Table 7.7).

LysoPL in the wheat endosperm is found mainly tightly bound within the starch granules (Chapter 2), but in flours and milled dissected endosperm a small proportion is more readily extracted and is presumably unbound. The concentration of lysoPL in dissected germ and aleurone tissue is not

Table 7.6 Calculated composition of basic endosperm and lipid-rich material transferred to it from aleurone and germ†

Component	Basic endosperm	Transferred material
	mg 100 g^{-1} endosperm	g 100 g^{-1} lipid
SE	35	2.0
TG	300	83.8
DG,FFA	105	5.0
MG,ASG	72	–
MGXG	91	–
DGXG	272	–
APE	96	–
ALPE	49	–
DiPL	65	4.8
LysoPL	54	4.4
α-T	–	0.1257
β-T	–	0.0603
α-T-3	0.09	0.0096
β-T-3	1.94	0.0128
ash	0.38 g 100 g^{-1}	58.2
protein	11.8 g 100 g^{-1}	129.9
FCG	−0.5 units	587 units

†Morrison *et al.*, 1982.

Table 7.7 Relationship between fatty acid composition (%16 : 0 or %18 : 2) and quantity of steryl ester (mg SE 100 g^{-1} flour) in flour millstreams†

First study†

16 : 0 = 66.76 − 0.97SE (r = −0.961, n = 30)
18 : 2 = 16.93 + 0.89SE (r = +0.890, n = 30)

FA in SE of basic endosperm (17 mg 100 g^{-1}) contains 50.3% 16 : 0 and 32.1% 18 : 2

FA in SE of transferred oil contains 3.7% 16 : 0 and 74.8% 18 : 2

Second Study‡

(i) SE < 57 mg 100 g^{-1}

16 : 0 = 125.47 − 1.44SE (r = −0.824, n = 29)
18 : 2 = −15.46 + 1.10SE (r = +0.781, n = 29)

(ii) SE > 57 mg 100 g^{-1}

16 : 0 = 66.39 − 0.41SE (r = −0.712, n = 27)
18 : 2 = 25.59 + 0.38SE (r = +0.772, n = 27)

FA in SE of basic endosperm (35 mg 100 g^{-1}) contains 75.1% 16 : 0 and 23.0% 18 : 2

FA in SE of transferred oil contains negligible 16 : 0 and 86.5% 18 : 2 overall

†Morrison and Hargin, 1981.
‡Morrison *et al.*, 1982.

high (Hargin and Morrison, 1980) and the contamination with these tissues cannot account for the relatively high lysoPL content in the low-grade reduction millstreams (Table 7.4). The more intensive milling during the later reduction stages leads to greater mechanical damage to starch granules (Ziegler and Greer, 1978) and it is possible that the high lysoPL values for these flours reflect the greater accessibility to extraction solvent of the starch lipid in the damaged granules.

The composition of the transferred acyl lipid (Table 7.6) is remarkably similar to that of aleurone and germ lipid (Hargin and Morrison, 1980) except that there are less phospholipids in most millstreams. The lowest grade millstreams (flours H, J_1, J_2 and J_3) contain about twice the phospholipid content predicted from the regression analyses.

The results are interpreted as evidence that material was transferred from aleurone and germ to all millstreams. The composition of the transferred material indicates that in most millstreams spherosome lipid and protein bodies were present, but in the lowest grade millstreams there were likely to be greater proportions of whole particles of germ, aleurone, testa and pericarp tissue to account for the particularly high values for diPL, ash and colour.

This transferred material is of no benefit technologically because it increases the proportion of non-polar lipids relative to polar lipids, adds proteins with no gluten-forming potential and adds phytin (source of much of the ash) and colour to the flour – all contrary to the objectives in milling white flour. There is no nutritional benefit from the tocols (vitamin E) because these will be largely destroyed in bleaching and breadmaking (Drapon *et al.*, 1971, 1974; Frazer and Lines, 1967; Slover, 1971), although the storage stability of unbleached flours may be greater when high levels of tocols are present.

The "free' hexane-extractable lipids in flour have a considerable influence on its baking quality (Chapters 8–11). In the study by Morrison *et al.* (1982) previously described, the lipids consisted of all the SE, TG, DG and FFA, just over half of the MGXG and APE, about half of the diPL and ALPE and rather less than half of the lysoPL regardless of whether the lipids were from the basic endosperm or from the transferred material. Thus, the free lipid content of a flour may be predicted, and to some extent controlled, from a knowledge of the total acyl lipids in the millstreams.

D The Effect of Grain Moisture Content on Redistribution of Germ Lipid

Another factor affecting the lipid composition of flour is the moisture content of the wheat entering the first break rolls. This is known to influence the extent of bran contamination of the millstreams (Lockwood, 1960) and the proportion of germ released on the early break rolls (Kent *et al.*, 1949). Recent experiments using a Bühler experimental mill have shown that moisture content has a significant effect on the α-T content, and thus germ lipid content, of the millstreams. This is illustrated by the histogram in Fig. 7.2 which shows the distribution of α-T from 10 kg of wheat grain.

Contamination of the flour streams with germ lipid was reduced by milling the wheat at the higher moisture content. At low moisture, germ lipid passes with the semolina from the break to the reduction side of the mill where it contaminates the reduction flours and finally enters the wheatfeed. In contrast, at high moisture most of the germ lipid is carried through the break system with the coarse bran and results in a very high concentration of α-T in the bran stream. The germ cannot be recovered as a separate stream because the experimental mill is much simpler than a commercial mill, and it must, therefore, enter either the bran or wheatfeed (millfeed). In commerical milling practice a proportion of the germ can be recovered as a relatively pure fraction but some germ always contaminates the lower grade millstreams and enters the wheatfeed. The range of moisture content used in this experiment was more extreme than encountered

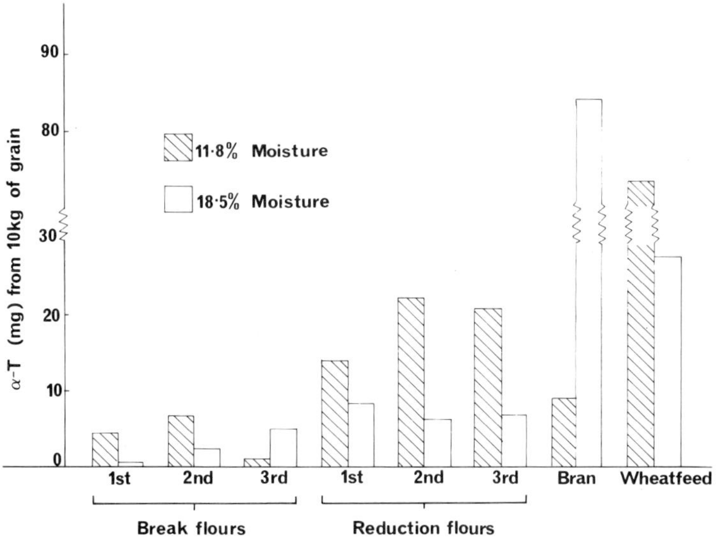

Fig. 7.2 The effect of wheat moisture content on the distribution of α-T into millstreams on a Bühler experimental mill (P. J. Barnes, unpublished results).

in practice, but the results confirm that moisture content has an important influence over the distribution of germ lipid among commercial flour millstreams.

IV SUMMARY

Most flour millstreams can be described in terms of their ash, protein, colour, acyl lipid and tocopherol contents by a simple model of flour derived from average starchy endosperm to which variable quantities of material are transferred from aleurone and germ. Deviations from this model occur in millstreams abnormally enriched with high levels of particulate bran and germ tissue.

REFERENCES

Berry, C. P., Youngs, V. L. and Gilles, K. A. (1968). *Cereal Chem.* **45**, 616–626.
Clayton, T. A. and Morrison, W. R. (1972). *J. Sci. Food Agric.* **23**, 721–736.
Clegg, K. M. and Hinton, J. J. C. (1958). *J. Sci. Food Agric.* **9**, 717–731.
Drapron, R., Beaux, Y., Cormier, M. and Geffroy, J. (1971). *C. R. Acad. Agric. Fr.* **57**, 245–251.
Fortman, K. L. and Joiner, R. R. (1978). In *Wheat: Chemistry and Technology* (Y. Pomeranz, ed.), 2nd edn., pp. 493–522 American Association of Cereal Chemistry, St. Paul, Minnesota.
Frazer, A. C. and Lines, J. G. (1967). *J. Sci. Food. Agric.* **18**, 203–207.
Hargin, K. D. and Morrison, W. R. (1980). *J. Sci. Food Agric.* **31**, 877–888.
Hargin, K. D., Morrison, W. R. and Fulcher, R. G. (1980). *Cereal Chem.* **57**, 320–325.
Hinton, J. J. C. (1944). *Biochem. J.* **38**, 214–217.
Jones, C. R. and Moran, T. (1946). *Cereal Chem.* **23**, 248–265.
Kent, N. L. (1975). *Technology of Cereals*, 2nd edn., pp. 62–67, 180–183. Pergamon, Oxford.
Kent, N. L., Simpson, A. G., Jones, C. R. and Moran, T. (1944). *Milling*, **103**, 294–300.
Kent, N. L., Thomlinson, J. and Jones, C. R. (1949). *Milling*, **113**, 46–54.
Lin, M. J. Y., Youngs, V. L. and D'Appolonia, B. L. (1974). *Cereal Chem.* **51**, 17–33.
Lockwood, J. (1960). *Flour Milling*, pp. 168–175. Northern Publishing Co. Ltd., Liverpool.
MacMasters, M. M., Hinton, J. J. C. and Bradbury, D. (1978). In *Wheat: Chemistry and Technology* (Y. Pomeranz, ed.), 2nd edn., pp. 51–113. American Association of Cereal Chemists, St. Paul, Minnesota.
MacMurray, T. A. and Morrison, W. R. (1970). *J. Sci. Food Agric.* **21**, 520–528.
Mann, D. L. and Morrison, W. R. (1974). *J. Sci. Food Agric.* **25**, 1109–1119.

Morrison, W. R. (1978a). *J. Sci. Food Agric.* **29**, 365–371.
Morrison, W. R. (1978b). *Adv. Cereal Sci.* Technol. **2**, 221–348.
Morrison, W. R., Coventry, A. M. and Barnes, P. J. (1982). *J. Sci. Food. Agric.* **33**, 925–933.
Morrison, W. R. and Hargin, K. D. (1981). *J. Sci. Food Agric.* **32**, 579–587.
Morrison, W. R., Mann, D. L., Wong, S. and Coventry, A. M. (1975). *J. Sci. Food Agric.* **26**, 507–521.
Slover, H. T. (1971). *Lipids*, **6**, 291–296.
Stevens, D. J. (1959). *Cereal Chem.* **36**, 452–461.
Ziegler, E. and Greer, E. N. (1978). In *Wheat: Chemistry and Technology* (Y. Pomeranz, ed.), 2nd edn., pp. 115–199. American Association of Cereal Chemists, St. Paul, Minnesota.

8 Role of Lipids in Baking

F. MacRITCHIE

CSIRO, Wheat Research Unit, North Ryde, N.S.W., Australia

"Lipids in Cereal Technology"
ISBN 0-12-079020-3

I INTRODUCTION

Baking cereal products involves the natural lipids that occur in the grain as well as diverse lipid materials that may be included in the various baking formulas. Wheat has traditionally been the grain used for baked goods and most of the discussion will be centred on the use of wheat flour. Rye ranks next in importance while other cereals assume a role in specialized products and in the preparation of composite flours. The amounts of lipid occurring in different cereal grains and the composition and mode of distribution within the grain are discussed in detail in other chapters. In general, lipid is a minor component comprising, for example, 2–4% by weight of the whole wheat grain (Morrison, 1978). Despite this, small amounts of lipid can exert important effects in baking and this has stimulated a good deal of research.

The role that lipids play has proved to be a challenging problem, causing some earlier reviewers to comment on the confusing, and at times contradictory, results that have been reported. Some of the reasons for these anomalies are now better understood. Certain lipid solvents are known to change the functional properties of flours. A very unusual relationship exists between baking parameters (e.g. loaf volume and texture) and the amount of the natural lipid that has been extracted from a flour (see Section II). This explains at least some of the apparently conflicting earlier results because different lipid solvents extract varying proportions of the total lipid from grain or flour. The lipid of cereal grains is a chemically complex system; for example, more than 20 distinct chemical species can be separated from wheat flour lipid extract. When we consider that the baking process is also very complex with regard to the physical and chemical changes which are undergone it is not surprising that the role of lipid proves difficult to resolve. Many baking formulas contain added materials such as shortenings or surfactants (emulsifiers). Frequently there are complex interactions between these and the native flour lipids which makes it difficult to separate the contributions of each.

Research into the role of lipids requires separation, fractionation and reconstitution methods that preserve the functional properties of both the lipid and other cereal components so a detailed discussion of the relevant techniques is given in Appendix 3. The effects of the natural flour lipids in baking bread on the one hand and cakes and biscuits on the other are described in Sections II and III. Although these baked products differ widely in their nature, there appears to be much similarity in the observed effects of lipids. The effects of lipid additives are discussed in Section IV and the part that lipids play at different stages of the baking process is considered in Section V. A number of studies have been concerned with exploring possible relations between lipid content and composition of grain and flour and baking performance. The results are summarized in Section VI. Theories for explaining the role of lipids have been a controversial area, using both biochemical and physical approaches. Different theories and mechanisms for lipid action are treated in Section VII. Finally, Section VIII summarizes the general features of lipid contributions to baking and highlights some of the areas where future work may be usefully directed.

II THE EFFECTS OF FLOUR LIPIDS ON DOUGH PROPERTIES AND BREADMAKING PERFORMANCE

A The Effects of Lipids on Physical Properties of Flour and Dough

Rheological properties of doughs are principally determined by the gluten protein and lipids do not appear to make an important contribution. Some of the changes in dough properties that have been attributed to lipid extraction of flours have subsequently been shown to be caused by effects on the gluten protein caused by the lipid solvents. Fortunately, a number of common lipid solvents have been found to have minimal effects on the peak mixogram development times of doughs. Removing increasing amounts of lipid from a flour causes a progressive strengthening of the dough as shown by instruments such as the extensograph or alveograph. After mixing, the dough of defatted flour usually has a less smooth texture but during fermentation the texture becomes smoother. The effects on dough properties are reversible: i.e. the original dough properties are recovered on re-addition of the extracted lipid. Although the lipid does not appear to form a continuous structure in dough (as the gluten protein does), it evidently modifies interactions between proteins causing the observed changes in properties.

Dry defatted flour is a finer powder than the untreated flour. Apparently the lipid causes an increased degree of stickiness between flour particles. Foaming properties are considerably stronger in aqueous suspensions of defatted flour and this needs to be considered when carrying out such measurements as the falling number test. The extraction of lipid also increases the whiteness of flours.

B The Effects of Flour Lipids on Volume and Characteristics of Baked Loaves

A very general, although unusual, relationship is found between the content of the natural flour lipid and the baked loaf volume measured in an

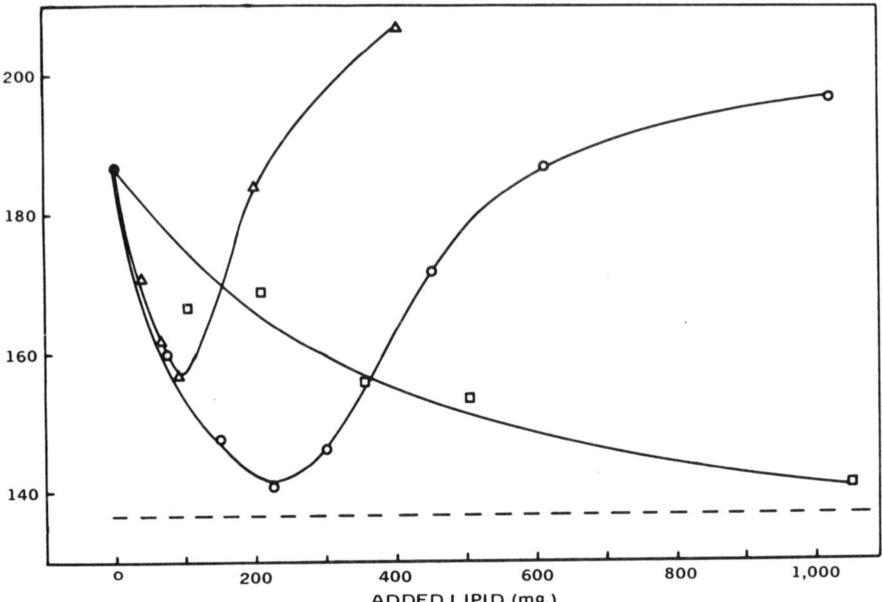

Fig. 8.1 Loaf volume in cc as a function of lipid content, for polar plus non-polar lipid (○), polar lipid (△) and non-polar lipid (□). Dashed line represents volume at end of proof. Additions were made to a dry flour weight of 30.2 g (MacRitchie and Gras, 1973).

optimized baking test. The measurements are usually made by adding progressive increments of lipid to the defatted flour rather than by extracting increasing amounts from the original flour. The behaviour, which is illustrated in Fig. 8.1, shows that loaf volume is high for a flour in which the non-starch lipid has been removed. As this lipid is re-added, volume decreases to a minimum at a lipid content intermediate between the defatted and untreated flours, after which it rises, approaching a constant value at lipid contents higher than the natural value. This pattern has been found for all flours studied, using both short-time and long fermentation baking procedures. The changes in volume are paralleled by corresponding changes in texture. Defatted flour gives bread which is notable for its white colour and very fine crumb grain. With increasing addition of lipid, whiteness diminishes and the grain deteriorates up to the minimum in loaf volume, thereafter improving, but not usually attaining the fineness observed with defatted flour.

C The Effects of Specific Lipid Fractions

1 *Polar and Non-polar Fractions*

Several studies have been made of the separate effects of the polar and non-polar flour lipid fractions on baking performance. All have agreed in showing polar lipids to be favourable and the non-polar fraction to be detrimental. Daftary *et al.* (1968) measured volumes of loaves baked from petroleum-extracted flours as a function of non-polar lipid at several fixed levels of polar lipid. In the absence of polar lipid, loaf volume decreased sharply with increasing addition of non-polar lipid, reaching a constant value at about 0.5% by weight of lipid. As the fixed polar lipid content was increased from 0 to 0.5%, increasingly high constant values for the loaf volume were obtained, showing that the deleterious effects of the non-polar lipid were being counteracted. Ponte and De Stefanis (1969) assessed the effects of the polar and non-polar fractions on loaf volume, texture (grain) score and loaf compressibility of an untreated control flour. The polar lipid fraction produced small texture score improvements and substantial loaf volume increases, especially when lard was absent from the formulation. By contrast, the non-polar fraction markedly depressed loaf volume and increased crumb firmness; however, the crumb texture appeared to be finer and more uniform. MacRitchie and Gras (1973) measured the effects on loaf volume of increments of polar and non-polar lipid fractions on a flour that had been defatted with chloroform. The non-polar fraction produced a progressive decline in loaf volume while the polar fraction depressed loaf volume initially (up to 0.3% by flour weight); it then increased loaf volume to values above that of both the defatted and untreated flours (Fig. 8.1).

2 *Individual Lipid Components*

Daftary *et al.* (1968) used silicic acid chromatography to separate flour lipid into individual components. Their results showed that mono-, di- and triglycerides had no significant effect on the loaf volume of bread baked from petroleum-extracted flour without added shortening. Glycolipids were most effective in improving loaf volume both in the presence and absence of shortening. Phospholipid had no effect in the absence of shortening but appreciably increased loaf volume in the presence of shortening.

In a study aimed at identifying the components of the non-polar fraction responsible for the deleterious effects in baking, De Stefanis and Ponte (1976) separated steryl esters, diglycerides, triglycerides and free fatty acids (FFA) from the mixture and measured their effects on loaf volume. Fractions were added at a level 50% higher than their natural level in the

Table 8.1 Effects of non-polar flour lipid fractions and components on loaf volume†

		Loaf volume (in cc)		
Lipid system	Addition (x level in flour)	Defatted flour (0 % lard)	Intact flour (0 % lard)	Intact flour (3 % lard)
Control	X1	2434	2606	2721
Non-polar fraction (intact)	X1	2684	2581	2647
Steryl esters	X1	2639	2590	2712
Triglycerides	X1	2681	2639	2741
Diglycerides	X1	2561	2610	2704
Free fatty acids	X1	2376	2569	2712
Intact flour	X1.5			2704
Intact flour + NL‡	X1.5			2663
Intact flour + reconstituted NL classes	X1.5			2557
Control		2602	2639	2733
Palmitic acid	X1	2512	2647	2733
Palmitic acid	X3	2569	2598	2725
Linoleic acid	X1	2458	2577	2782
Linoleic acid	X3	2196	2278	2655

†De Stefanis and Ponte (1976).
‡Non-polar lipids.

flour to enhance effects. Three dough systems were compared: flour that had been defatted with petroleum and no lard added, untreated flour with no lard added and untreated flour with 3% lard. The results, which are summarized in Table 8.1, clearly show that the FFA were the components responsible for the depression of loaf volume, although the presence of lard tended to annul the effects. Comparing the effects of palmitic and linoleic acid indicated that it is the unsaturated fatty acids (e.g. linoleic, which constitutes more than half the FFA present in wheat flour) that contribute to loaf volume depression.

MacRitchie (1977) used the batch fractionation methods introduced by De Stefanis and Ponte (1969) to separate flour lipid into five fractions and tested the effects of these fractions by adding them to the defatted flour. Figure 8.2 shows the Tlc patterns of the five fractions and Fig. 8.3 shows the corresponding curves for loaf volume as a function of lipid content. Fractions 1 and 2 are mainly non-polar in nature and fractions 3, 4 and 5 are predominantly polar. Fraction 2, which contains the highest proportion of FFA, gives the steepest decrease in loaf volume per percent of lipid.

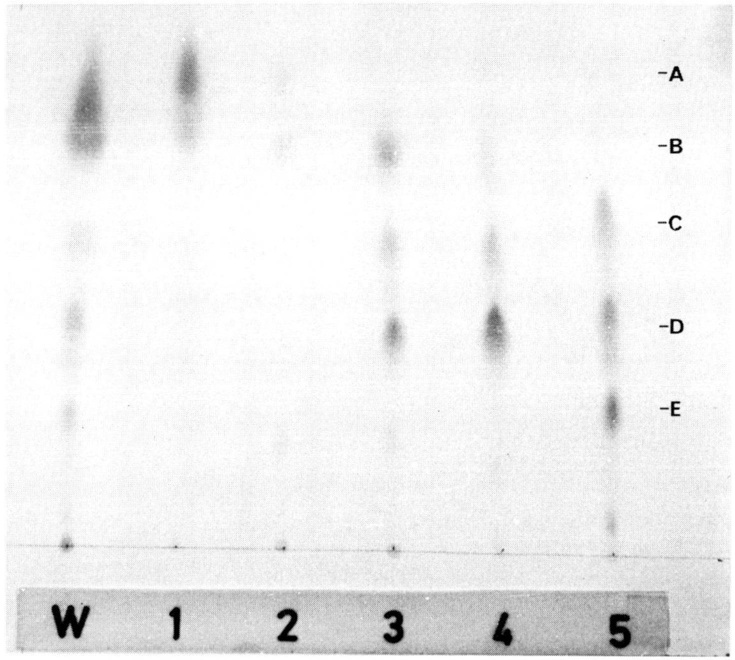

Fig. 8.2 Tlc patterns for whole lipid and five sub-fractions obtained by selective elution from silica gel. Assignments of bands are A: neutral components including triglycerides; B: free fatty acids; C: glycolipids; D: digalactosyl diglycerides; E: phospholipids. Band B was incorrectly assigned in original reference (MacRitchie, 1977).

Fraction 3 contains a relatively strong band for FFA and small additions of this fraction cause depression of loaf volume. These results suggest that the initial depression in loaf volume observed for additions of polar lipid fractions (Fig. 8.1) is due to the presence of some FFA in these fractions.

D Lipid-Protein Interaction

Loaf volume increases linearly with increases in protein content of flours (Finney and Barmore, 1948) in contrast to the unusual relationship found between loaf volume and lipid content (Fig. 8.1). Thus, the baking performance of a flour depends mainly on contributions from the protein and lipid components. MacRitchie (1978) studied the loaf volume-lipid content relations for pairs of flours, each pair having similar characteristics (e.g. protein content and starch damage level) except that one member of the pair performed well and the other poorly in baking. A baking formulation was used which omitted any lipid additives. Several kinds of typical

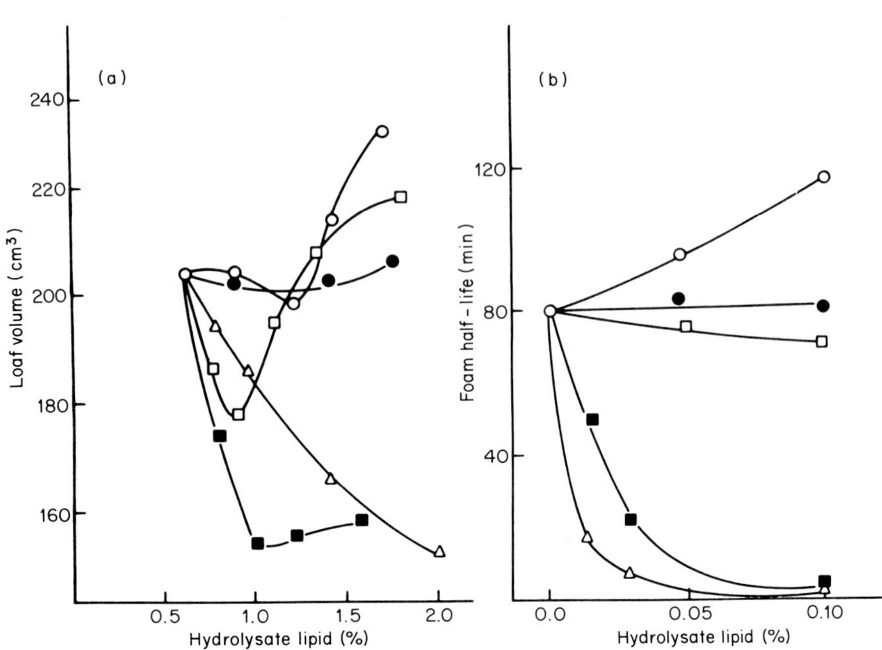

Fig. 8.3 Loaf volume (left) and foam half-life (right) as a function of hydrolysate lipid content of flour (left) and solution (right), for five lipid fractions separated from the flour. Fractions are those for which tlc patterns are shown in Fig. 2: △, 1; ■, 2; □, 3; ●, 4; ○, 5. (MacRitchie, 1977).

behaviour patterns were found; these are illustrated in Fig. 8.4. Loaf volumes of the poorer flours were below those of their paired samples over ranges of lipid content which varied from a small region near the natural lipid content of the flours to the complete range. By interchanging components between the flour pairs it was shown that the variations in the curves were caused by differences in the protein and not by differences in the lipid. It is evident that there is an interaction between the lipid and protein leading to differences in the curves from one flour to another. This does not necessarily mean that there is a chemical association between the two components. It appears possible that lipid and protein make individual contributions to the loaf-volume potential of flours; i.e. loaf volume varies with lipid content according to the general form of the curves of Fig. 8.4 and the effect of gluten protein quality is superimposed at each lipid level. At certain lipid levels (particularly in the vicinity of the natural level found in flour) baking performance appears to be more vulnerable to failure so that differences in protein quality show up more than at other lipid levels.

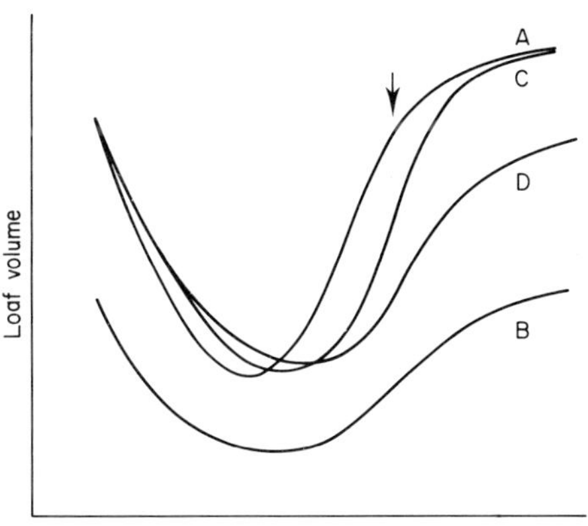

Fig. 8.4 Loaf volume-flour lipid content curves, which illustrate several types of general behaviour found for wheat flours of differing baking performances. Arrow represents the natural flour lipid content (MacRitchie, 1980).

It can be seen from the curves in Fig. 8.4 that loaf volume of a flour, in the absence of lipid additives, is determined by the shape of the loaf volume–lipid content curve. A displacement of the minimum in this curve to higher lipid contents generally signifies a lower loaf volume potential since the natural lipid content then occurs at a lower point on the rising part of the curve. High loaf volume may be achieved in some cases by adding extra lipid or shortening but not always (e.g. curve D of Fig. 8.4). Although only the baking performance of wheat flours has been studied systematically, it has been reported that lipid extraction also improves loaf volume potential of rye flour (MacRitchie, 1977). This suggests that this cereal behaves in a similar manner with respect to the effects of lipid.

III THE EFFECTS OF WHEAT FLOUR LIPIDS IN CAKES AND COOKIES

A Cookies (Biscuits)

Cookie quality, like bread quality, is very sensitive to the content and composition of the natural flour lipid. The similarity of the effects are striking even though the manufacturing procedures and quality assessment of the two products are quite different. Cookie quality is measured by cookie spread, top grain score and internal structure. Extracting the flour lipid causes cookie spread and top grain score to be greatly reduced. These effects have been found by Cole *et al.* (1960) using water-saturated butan-1-ol (WSB) and Kissell *et al.* (1971) using petroleum as the extracting solvents. In the study by Kissell *et al.* (1971), effects on cookie quality of adding flour lipid to the defatted flour were measured. Cookie spread and top grain score increased as lipid additions were increased up to four times the natural level. Improvements were obtained with half the natural level and, with the natural level or double, cookies were equal to or better than the controls (untreated flours). Baking performance was only partly restored by adding either the polar or non-polar fractions of flour lipid, although top grain was significantly better with the polar fraction. By combining the two fractions, cookies identical to the original were obtained, confirming that the fractionation and reconstitution techniques were satisfactory. The interchange of lipid between four flours of different quality showed that there were no varietal differences in lipid composition. The variations in quality, therefore, were in the base flour.

Similar results were reported by Cole *et al.* (1960). However, they showed that complete recovery of cookie properties could be obtained by adding a phosphorous-containing fraction of the lipid to the defatted flour. A phosphorous-free lipid fraction, by contrast, gave no improvement. These results agreed with a recent detailed study by Clements and Donelson (1981). To determine the source of functionality the free flour lipids were separated quantitatively by tlc into 10 fractions. Two of these fractions, which corresponded to digalactosyl diglyceride (DGDG) plus phosphatidyl choline (PC) and monogalactosyl diglyceride (MGDG), gave the highest degree of restoration of quality when added to the defatted flour. Pure commercial samples of DGDG added alone at 0.10% by flour weight and PC at 0.05% gave essentially complete recovery of original properties. However, MGDG gave little response up to levels of 0.15%.

A parallel can be seen between these results and the effects of lipid in bread baking, described previously. The initial removal of lipid from the flour leads to a deterioration in quality. However, unlike bread, there is no suggestion that cookie quality improves again with a further removal of lipid (Cole *et al.*, 1960). Of course, it must be remembered that we are comparing two different baked products and that the cookie formulation contains an appreciable quantity of shortening, while the results which were discussed for bread were obtained in the absence of shortening. Another similarity is that the polar glycolipids and phospholipids are found to be the functionally beneficial lipids in cookies as well as bread. Furthermore, it is becoming increasingly clear that the importance of lipids lies in the role they play in forming and stabilizing the gas cell structure of baked products (see Section V).

B Cakes

Cakes differ from bread and cookies in that they enter the baking phase as batters. Many different formulations exist for cake making. Some recent studies using full formula layer cakes will be referred to here since they are thought to illustrate typical behaviour and can be generally applied. Cakes are assessed, as in bread, by their volume, texture and retention of freshness. In place of dough properties, however, parameters such as whipping quality and batter structure are more pertinent as these govern the volume, appearance and crumb structure (Seibel *et al.*, 1980).

Soft wheats find more general use for cake manufacture and it is customary to treat such flours, usually by chlorine bleaching, to enhance their baking potential particularly for use in high-ratio cakes. Guy and Pithawala (1981) have recently carried out rheological measurements of high-ratio cake batters to elucidate the mechanism of improvement of flours by chlorination or heat treatment. The effects of these treatments appear to be complex at a molecular level but effects on lipids have been implicated by some workers.

In a study where lipids were interchanged between flours that had been chlorinated at different levels, it was shown that chlorination functionally modifies both the extracted lipid and the base flour (Kissell *et al.*, 1979). This contrasts somewhat with the conclusions reached by Spies and Kirleis (1978) and Johnson *et al.* (1979). Another method which has been successfully used for enhancing cake-baking potential is to expose the flour to

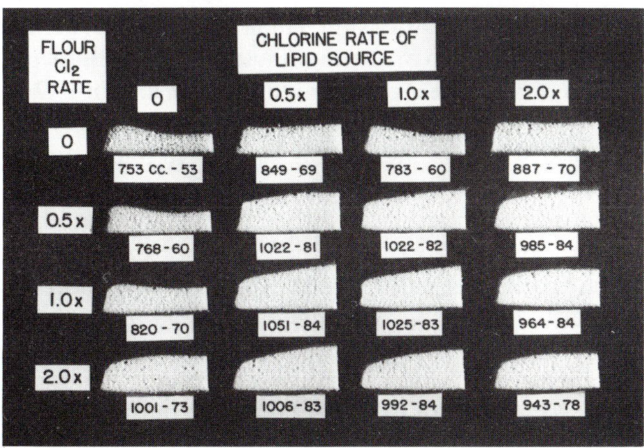

Fig. 8.5 Cross-sections of cakes prepared from flours that had been sequentially chlorinated, defatted and reconstituted with interchange of lipids and base flours in all possible combinations. Numbers below are volumes in cc (left) and grain score (Kissell *et al.* 1979).

moving air, referred to as ageing (Clements and Donelson, 1982). The degree of oven expansion was found to be a function of the flour as well as the length of exposure to the air. Since the final volume of a cake does not always provide an indication of oven expansion because of the variable degree of collapse at the end of baking, Clements and Donelson (1982) devised a method for measuring oven expansion of cakes. This technique illustrated, using interchange experiments, that the degree of oven expansion for both chlorinated and aged flours is a function of the lipid component. Conversely, the degree to which oven expansion can be translated into cake volume depends on the non-lipid components (flour base). The expansion of an unchlorinated base frequently results in a cake structure that is unable to support the potential volume (Kissell *et al*., 1979), while expansion generated on a bleached base is largely retained. This behaviour is illustrated in Fig. 8.5.

IV THE EFFECTS OF NON-FLOUR LIPIDS AND SHORTENINGS IN BAKING

Non-flour lipid additives are commonly included in baking formulations. The two main groups are the saturated fats or shortening and the surfactants or emulsifiers. Both groups are important in improving the baking potential of flours. Shortening is specially recommended in short-time bread processes, such as the Chorleywood Bread Process, to give adequate loaf volume. It is also important in enchancing slicing properties and retaining freshness. The relevance of the melting range of the fats in baking and the mechanism by which their presence delays the loss of CO_2 from loaves during baking have been studied in detail by the Chorleywood workers (Fisher *et al*., 1966; Daniels and Fisher, 1976). An interesting approach to evaluating the effects of shortening by vacuum expansion of developed doughs at proof temperature has been described by Bell *et al*. (1981). This method permits separation of the effects of pure expansion on a dough piece from other complicating processes which occur during oven baking.

Surfactants have become increasingly used in baking formulations in recent times and a relatively large industry has formed around the production of these specialized additives. In general, surfactants improve loaf volume potential and texture in bread and also contribute to extension of shelf-life. Surfactants such as monoglycerides and sodium stearyl lactylate (NaSL) are utilized in composite flours to improve performance. The types of compound that are in use and their modes of action have been summarized by Krog (1981). The characteristic property of surfactants is that their molecules contain hydrophilic and lipophilic parts. A measure of the ratio of the hydrophilic character of the molecule to the hydrophobic or lipophilic is given by the hydrophile-lipophile (HLB) value. In practice, it is found that there is an optimum range for the HLB value for surfactants to be most effective as improvers in breadmaking (Pomeranz, 1968). We are concerned here mainly with the role of lipids in baking so we will consider some recent results which throw light on the function of surfactants in breadmaking.

Junge *et al.* (1981) made a detailed study of the origin of fine grain imparted to dough mixed to peak development by certain surfactants using measurements of dough density and scanning electron microscopy (SEM) of cryofractured dough. These measurements were then compared with the performance of corresponding baked loaves. The different effects on loaf volume and crumb structure for a range of surfactants are summarized in Table 8.2. Measurements showed that the dough mass starts with a density of 1.20 g cm^{-3} and occludes little air after the density reaches 1.10 g cm^{-3}. The optimum development time for the flour-water dough was 3.5 min at which the density was 1.16 g cm^{-3}. Thus, air was still occluded rapidly during the first phase of overmixing. Junge *et al.* (1981) found that rheological properties can change when certain surfactants are added and that the time when the dough starts to occlude air changes as a result. For example,

Table 8.2 Effects of additions† of surfactants‡ on baked loaf characteristics of a flour (from Junge *et al.*, 1981)

Dough treatment	Proof height (in cc)	Loaf volume (in cc)	Crumb grain
Control	7.6	950	medium
No shortening	7.5	855	open
+ NaSL	7.8	955	fine
+ EMG	7.7	970	open
+ Poly 60	7.7	960	medium
+ PGME	7.6	790	very fine
+ F108	7.9	990	open
+ DATEM	7.8	945	slightly open
+ MONO	7.5	860	open
+ (PGME = EMG)	7.6	925	fine
+ CORN OIL (3 %)	7.6	920	open

†Addition of 0.5 % unless otherwise stated.

‡NaSL (sodium stearyl lactylate), EMG (ethoxylated monoglycerides), Poly 60 (polyoxyethylene sorbitan monostearate), PGME (polypropylene glycol monoesters), F108 (pluronic polyol). DATEM (diacetyltartaric acid esters of monoglycerides), MONO (distilled monoglycerides).

NaSL delayed mixogram optimum development from 3.75 to 7 min and mixing stability increased, although adding 1.5% sodium chloride shortened mixing time to 4.75 min. The important conclusion reached in this work was that the surfactants did not change the amount of air occluded but did change the crumb texture. For example, NaSL addition produced dough with more numerous and smaller air cells. Dough containing ethoxylated monoglycerides (EMG) had larger air cells while that containing polypropylene glycol monoesters (PGME) or a mixture of PGME and EMG had larger numbers of smaller cells. These results correlate with the findings on baked bread (Table 8.2). EMG improved loaf volume but gave an undesirable open grain. PGME gave fine grain although it depressed loaf volume. Adding both EMG and PGME improved both volume and grain. SEM was used to assess the effects of punching (moulding) operations on crumb texture. With NaSL, to a greater degree than other surfactants, formation of more air cells was observed during mixing followed by further sub-division of cells during punching. The greater number and smaller size of cells which were maintained throughout punching, and thus present in the final product, are responsible for the fine texture in baked loaves.

V THE EFFECTS OF LIPIDS AT DIFFERENT STAGES IN THE BAKING PROCESS

A Mixing Stage

The pioneering work of Baker (1941) and Baker and Mize (1941) showed that air is beaten into doughs during the latter part of the development stage and that subsequently no new cells are introduced, although the gas cell structure can be considerably modified by the sub-division of existing cells during punching and moulding operations. Lipid materials play an important role in these operations as shown by Junge *et al.* (1981), described in Section IV. Incorporating air into dough involves a process similar to the production of a foam. By analogy, the surface-active compounds which adsorb at the interface between the air and the aqueous phase largely determine the manner in which gas bubbles form, their number, size distribution and stability. The surface-active constituents of dough are the proteins and lipids. The relative proportions of proteins and lipids as well as the lipid composition are important in determining the nature of the gas cell structure. An obvious example of this is seen in the very fine texture of bread baked from defatted flour, resulting from an initially very fine distribution of gas cells in dough, stabilized by protein only. Of course the involvement of lipids in foam formation and stabilization of batters, processes basic to the development of cake structure, has become well recognized (Clements and Donelson, 1982).

B Proof Stage

In addition to affecting the formation of gas cells during mixing by contributing to the properties of the stabilizing adsorbed film surrounding the cells, lipids appear to play a significant role in subsequent punching and moulding steps (*cf.* Section IV). This is suggested by the results depicted in Fig. 8.6. Loaf volume is shown as a function of flour lipid content for two baking tests which differed only in that one included an intermediate moulding step after an initial short fermentation (20 min) while in the other, no manipulations were performed on the dough after initial mixing and moulding. Although differences are small for defatted flour and flours with low lipid levels, there is considerable divergence at higher lipid levels, confirming a strong interaction between lipid content and moulding treatment. An examination of the gas cell structure during proof gives an insight into the reasons for differences in loaf volume behaviour between the two treatments. Fig. 8.7 shows some typical results for cross-sections of dough pieces which have been taken at the end of the proof stage, frozen, freeze-dried and photographed. A direct comparison may be made with the

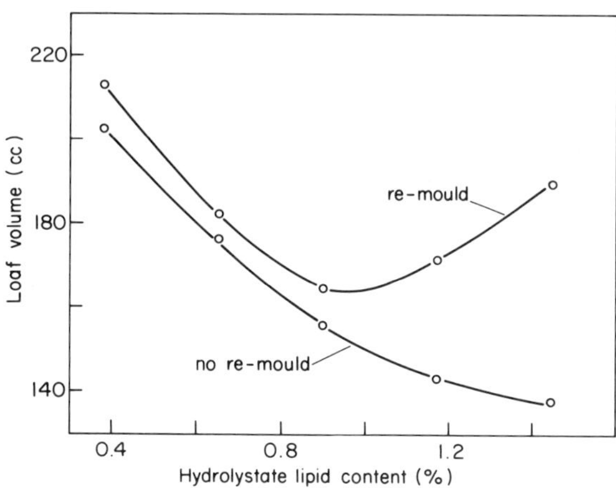

Fig. 8.6 Loaf volume as a function of hydrolysate lipid content of a flour in a baking test using 30.2 g dry flour. Upper curve is for a test in which an intermediate moulding step after 20 min fermentation was included. Lower curve is for a test in which the moulding step was omitted (MacRitchie, 1976).

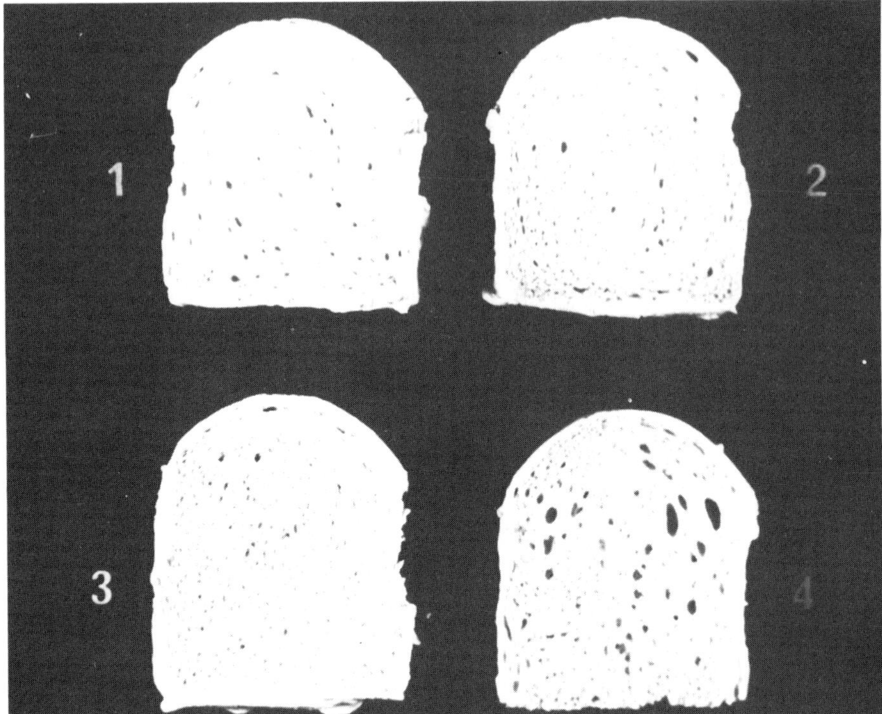

Fig. 8.7 Effects of different treatments on gas cell structure of dough pieces at the end of proof. Dough pieces were frozen, sliced and freeze-dried before photographing. 1: defatted flour; 2: flour of lipid content corresponding to minimum in loaf volume-lipid content curve (Fig. 1); 3: whole flour, remoulding step (after 20 min) included; 4: whole flour, remoulding step omitted. Cross sections may be compared with loaf volumes of Fig. 6 (MacRitchie, 1981).

results of Fig. 8.6. In the absence of the intermediate moulding step, gas cells are heterogeneous in size, leading to their instability. Passage through rollers, as occurs in moulding, appears to produce a larger number of more uniformly sized cells, thus forming a structure with good gas retention properties. Interestingly, the photographs show no obvious differences in gas cell distribution which could explain the nature of the upper curve in Fig. 8.6. However, it is evident from the final loaf volumes that there is a variation in the stability of the gas cells as the lipid content changes.

C Baking Stage

In general, no differences in the volume of dough pieces in bread baking tests, resulting from different lipid treatments, are noted up to the end of the proofing stage. It is during the early stages of oven baking that these differences are manifested in varying degrees of expansion. At this time, the greatest stress is placed on the dough structure and any inherent weaknesses are exposed. The failure of dough pieces in the oven may occur by different mechanisms and these may be identified by an experienced test baker. For example, a heterogeneous gas cell structure may develop in which the larger bubbles grow and quickly collapse. Alternatively, a uniform gas cell structure may suddenly become unstable and simultaneous rupture of most of the cells may occur at a given time.

These different mechanisms give rise to very different types of crumb grain in the baked bread. Examining the crumb texture under magnification and paying attention to the size and shape of the gas cells and gas cell wall thickness can yield information about the origin of instability. However, very few quantitative studies have been made along these lines and the reasons for the premature collapse of dough structure during baking are not well understood. Carlson and Bohlin (1978) have made preliminary measurement of gas cell radii by photomicrography of thin slices of dough. Further quantitative measurements along these lines would be helpful to understand reasons for variations in stability of dough structures. Daniels and Fisher (1976) measured the release of CO_2 from dough during the baking stage and showed the importance of delaying this release to obtain good loaf volumes. However, the relative importance of cell rupture and gas permeability of the walls of gas cells has not been clearly identified.

VI SURVEYS OF LIPID CONTENT AND COMPOSITION OF WHEAT VARIETIES AND FLOURS OF VARYING BAKING QUALITY

Because of the proven influence of lipids in baking, some studies have been directed to trying to ascertain whether differences in baking quality can be related to flour lipid content and composition. The most complete work has been done by Fisher *et al.* (1966) in which lipid content and composition was measured in flours from a range of wheat varieties over a number of crop years. Early results in this study (Fisher *et al.*, 1964) showed that wheat flours varied in lipid content and composition and that certain tentative correlations appeared between these variations and quality. However, in the final analysis of this detailed study it was concluded that although significant variations did exist between varieties and crop years, these variations could not be linked definitively to quality. The few reported experiments in which lipid samples have been interchanged between flours have shown that differences in baking performance could not be accounted for by the lipid fraction (MacRitchie, 1978).

Table 8.3 Lipid composition and content together with loaf volume data for a series of flours. (From P. E. Marston and F. MacRitchie, unpublished results)

			Lipid composition			
Flour	Protein (in %)	Lipid by chloroform extn. (%)	Polar (in %)	FFA (in %)	Triglyceride (in %)	Loaf volume (30 g flour) (in cc)
A	13.1	1.56	36	11	39	171
B	12.9	1.44	38	11	40	183
C	12.5	1.42	39	11	39	171
D	12.5	1.32	29	10	41	182
F	10.1	1.41	35	13	31	162
G	10.7	1.58	37	8	42	176
H	10.5	1.39	36	7	46	201
J	13.1	1.65	33	15	39	177
K	11.1	1.50	42	8	40	204
L	12.3	1.35	21	11	47	174

Some data from a recent study of Marston and MacRitchie (unpublished results) are shown in Table 8.3. Lipid content and composition for a range of flour types were compared with baking performance. Here again, no systematic relation between lipid and quality is evident although some results appear interesting. For instance, flour K, which performed best, is characterized by the highest polar lipid content and a relatively low FFA content. Flour J, which had the highest FFA content, performed poorly considering its protein content. Our present knowledge suggests that the "quality" factors in flour lipid (see Section II) are a high non-starch lipid content, a high ratio of polar to non-polar lipid and a low FFA content.

The problem in trying to correlate lipid parameters with quality is that the quality contribution by the lipid is often swamped by that of the protein. Since protein quality varies from one flour to another, it may be expected that consistent results from lipid contributions would therefore not be observed. Nevertheless, it is clear that differences in lipid content and composition exist between flours and that from the experimental work described in Section II such differences can affect quality. In view of this, lipid content and composition should not be ignored when considering parameters to be used as a basis for selection in plant breeding-programs even though protein composition appears to be the major factor determining flour quality.

Although correlations between wheat flour lipid and baking quality have not been clearly established, the interchange of wheat flour lipid with lipid from other cereals has shown major effects (MacRitchie, 1977). When the lipid from wheat flour was replaced by lipid from rye flour, a similar loaf volume–lipid curve was obtained but loaf volumes were appreciably lower. Maize lipid added to defatted wheat flour gave loaf volumes which progressively decreased with increasing lipid content. These effects could be related to the decreasing ratio of polar to non-polar lipid in these cereal flours. It therefore appears that not only is the protein from wheat flour superior to that of other cereals for breadmaking purposes, but the lipid is as well.

VII MECHANISM AND THEORIES OF ACTION OF LIPIDS

Some of the mechanisms by which lipids influence baking performance are discussed in other chapters including the interactions which occur between proteins and lipids during dough development, the effect of changes in lipid with flour storage, the action of lipoxygenase and the relation between the physical state of lipids and their effects in baking (Chapters 9–12). Therefore this discussion will be restricted to aspects not treated in other chapters.

A Protein-lipid Interactions

The way in which protein and lipid interact to produce variations in the loaf volume–lipid content relations for different flours has been discussed in Section II (see Fig. 8.4). It was shown that these variations resulted from inherent differences in protein quality. However, the mechanism responsible for the effects is not clear. A good deal of attention has been paid to the phenomenon of "lipid-binding" (Wootton, 1966) as this could conceivably affect baking performance. Non-starch lipid may be extracted from flour with organic solvents. However, if the flour be wetted with water, the lipid becomes "bound" and only a small fraction can then be extracted even after the flour has been dried. Some of the factors that govern the extent of binding, such as the degree of dough development, have been studied in detail (Daniels *et al.*, 1966; Daniels, 1974). (See also Chapter 9.) Based on the nature of the loaf volume–lipid content curve (Fig. 8.1) and the effects of specific lipid fractions (Fig. 8.3) it is tempting to predict that the differences observed in Fig. 8.4 may result from varying degrees of lipid binding. For example, it is known that polar lipids are preferentially bound to protein after wetting. If protein from different flours would bind the polar lipids to varying degrees, this might then alter the effective ratio of polar to non-polar lipids, causing the displacements observed in the curves of Fig. 8.4.

Differences in lipid binding by flours of varying baking quality have been observed. No correlation was found between the two parameters by MacRitchie (1978) using chloroform extraction of freeze-dried dough to measure the bound fraction. However, Chung *et al.* (1979a) have reported measurements on the composition of the free lipid (i.e. lipid extracted by petroleum) from 21 samples of hard red winter (HRW) wheat flours of varying breadmaking quality. Significant linear correlations were found between the loaf volume and polar lipid, polar (PL)/non-polar (NL) ratio and lipid galactose content. No correlation was observed between loaf volume and total extractable lipid. When the loaf volume–NL/PL plots were compared for five HRW wheats using six solvent systems for lipid extraction, petroleum gave the steepest slope and therefore differentiated best between the flours (Chung *et al.*, 1980). The slope was zero for WSB, indicating that there were no qualitative differences between the total non-starch lipid extracts. The results of Chung *et al.* (1980) may be interpreted on the basis that more of the polar lipids are extracted from the superior flours, suggesting that these lipids are bound less strongly than in flours of inferior baking performance. Further measurements along the lines of Chung *et al.* (1980) seem justified to establish whether this trend can be confirmed for a wider range of flours.

Protein–lipid interactions in wheat flour have been discussed by Pomeranz (1973) and MacRitchie (1980). The binding of lipids following the addition of water to flour evidently results from purely physical association of proteins into structures of low free energy. Re-orientation of molecules occur to try to maximize contact of polar and non-polar portions of the lipid with polar and non-polar groups of the protein respectively. The factors governing these associations are similar to those involved in the formation of micelles in aqueous media (see Chapter 11).

B Foam Model for Action of Lipids

Many results point to the role of lipids as stabilizers or de-stabilizers of the foam structure in batters, doughs and bread. Some of these have been mentioned in previous sections. It appears significant that very small increments of lipid fractions (less than 0.1% by flour weight) can cause appreciable changes in bread loaf volume (see Fig. 8.1). This is a result which is difficult to reconcile with alterations in bulk properties of doughs and, in fact, rheological measurements confirm that no appreciable changes in these properties take place with changes in lipid levels. On the other hand, very small amounts of lipid suffice to cover an interface with a monolayer and therefore to exert effects at the air-aqueous interface of gas bubbles in dough.

Flour lipid contains a mixture of compounds of varying surface activity. When flour lipid is fractionated, the effects of the different fractions on loaf volume parallel their effects on the foam stability of aqueous extracts from doughs (see Fig. 8.3). Thus, glycolipids and phospholipids, in association with protein, act as foam stabilizers while the non-polar lipid fractions act as foam breakers. Among the non-polar components, the presence of FFA appears to correlate with low loaf volume and foam instability (Fig. 8.3). In the absence of flour lipid, both loaf volume of bread and foam stability (stabilized by protein only) are high. In addition, bread crumb and foam structure are characterized by fineness and uniformity of gas bubble size.

The role of lipids in foam production and stability of cake batters is possibly more obvious, since batters correspond more closely to classical foams. Nevertheless, bread doughs contain a relatively large quantity of "free" liquid and there is evidence that this liquid is distributed throughout the dough as a lamellar network (MacRitchie, 1976). This resembles classical foams except that a solid dough phase is incorporated in the structure. The role that lipids play in the internal gas cell structure of cookies has also been stressed in recent work. Yamazaki and Donelson (1976) have drawn attention to the importance of the internal structure and have introduced a system of scoring for internal appearance. Clements (1980) has devised embedding and staining procedures which enable objective demonstrations of the deleterious effects of lipid removal on the internal appearance of cookies. Cross sections showed how gas cell walls break down during oven expansion, causing coalescence and, in the worst cases, formation of large pockets enclosed by a thin shell. The structure solidifies at an early stage in the baking process, causing limited spread and almost total absence of top-grain. It appears, therefore, that similar processes occur for cookies, cake and bread and that causes of failure as a result of different lipid treatments may be attributed to the inherent instability of the gas cell structure for all these baked products.

C Staling

In addition to their effects in influencing volume and texture of baked products, lipids play a role in staling mechanisms. Generally, the presence of lipid enhances shelf-life although the mechanism has not been clearly resolved. It has been suggested that in bread, lipids retard retrogradation of starch molecules. Generally, however, breadcrumb containing added lipid is initially softer than controls, suggesting that it is not simply a kinetic effect. It seems possible that lipids may concentrate around the surface of starch granules so that, during oven baking, contacts between partly gelatinized granules may be reduced. Since the rigidity of the retrograding starch network is mainly responsible for crumb firmness, the effect of lipid would be to reduce this rigidity. Staling is also discussed in Section VII of Chapter 11.

VIII CONCLUSIONS AND AREAS FOR FUTURE WORK

The role that lipids play in baking has proved difficult to unravel. One reason is that most baking processes employ lipid additives such as shortenings and surfactants, which interact in a complex manner with the natural flour lipids. When lipid additives are omitted from bread formulations, a very general relation is found between loaf volume and the natural flour lipid content. This relation is characterized by a Morse-type curve with a minimum corresponding to a lipid content intermediate between the defatted and untreated flour. Volume changes are accompanied by parallel changes in crumb texture. The reasons for this unusual behaviour have not been explained. Greater understanding may come from studies on model systems, e.g. foams with proteins and lipids as stabilizers.

Although defatted flour is very sensitive to the type of lipid added (e.g. polar lipids enhance and non-polar lipids depress loaf volume), untreated flours are much less sensitive. Loaf volumes of untreated flours are often found to respond in a similar manner to additions of natural lipid or shortening. The fat response of a flour is therefore dependent on the point of the loaf volume–lipid content curve on which a given flour falls. If it falls on the steeply rising portion, a larger fat response may usually be predicted than when it falls on a point where the curve is approaching a constant loaf volume value.

Because the addition of fat or other lipids tends to reduce differences in performances between flours, baking tests which omit these additives are often preferable for research studies on baking quality. When loaf volume–lipid content curves are compared for wheat flours of different baking quality, the curves frequently show displacements which clearly indicate the reasons for the quality differences. In general, a shift of the minimum in the curve to higher lipid contents is synonymous with a worsening baking performance. These shifts indicate a protein–lipid interaction although, in the few cases where pairs of flours have been compared, the shifts were shown to result from differences in protein quality. Despite this, surveys have shown that variations in both lipid content and composition occur in lipid samples extracted from different flours. No systematic relations have been found between these differences and baking quality. It does appear, however, that it would be difficult to establish systematic correlations because of the superimposed and often relatively large effects caused by differences in protein quality.

One way in which to eliminate the protein variable is to study the effects of whole lipid extracts from a range of flours by adding them to one defatted flour, which acts as a control, and measuring the baked loaf characteristics. This could then be extended to several control flours of different baking quality. It would also be interesting to study the effects of varying the protein content using a control lipid; i.e. measure loaf volume–lipid content relations for a given reconstituted flour with different levels of its own protein, using the same lipid sample. Such experiments should, of course, employ a baking test in which lipid additives are absent.

The contribution of the lipid component to flour quality should not be ignored when considering parameters to be used as a basis for selection in plant breeding programmes even though protein composition appears to be the major factor which determines flour quality. The "quality factors" in flour lipid, as presently known, are a high non-starch lipid content, a high ratio of polar to non-polar lipid and a low FFA content. Deficiencies in loaf volume or crumb texture in particular may be corrected by addition of shortening in short-time bread baking processes. If, however, such deficiencies could be eliminated, cost savings in lipid additives for bakery processes might be possible.

It has become increasingly evident from recent studies of baking, not only bread but cakes and cookies, that the main functional role of lipids is to influence the gas cell structure and stability. Lipids apparently exert their effects on the thin films surrounding the gas cells of doughs or batters during mixing, expansion and baking. The factors governing gas retention are not well understood. Methods introduced by Clements (1980) have contributed to knowledge of how failure occurs in cookie structure. It would appear that a more intensive study of the gas cell structure of dough at different stages of bread processing is justified to better understand the reasons why variations in bread volume and texture arise. In this regard, measurements of gas cell size and distribution as a function of different lipid treatments, as well as microscopic studies of gas cell wall structure and the manner in which gas is lost, would be particularly useful.

REFERENCES

Baker, J. C. (1941). *Cereal Chem.* **18**, 34–41.
Baker, J. C., and Mize, M. D. (1941). *Cereal Chem.* **18**, 19–34.
Bell, B. M., Daniels, D. G. H., and Fisher, N. (1981). *Cereal Chem.* **58**, 182–186.
Carlson, T., and Bohlin, L. (1978). *Cereal Chem.* **55**, 539–544.
Chung, O. K., Pomeranz, Y., Finney, K. F., and Shogren, M. D. (1978). *Cereal Chem.* **55**, 31–43.
Chung, O. K., Pomeranz, Y., and Finney, K. F. (1979a). *Cereal Foods World*, **24**, 452 (abstract).

Chung, O. K., Pomeranz, Y., Hwang, E. C., and Dikeman, E. (1979b). *Cereal Chem.* **56**, 220–226.

Chung, O. K., Pomeranz, Y., Jacobs, R. M., and Howard, B. G. (1980). *J. Food Sci.* **45**, 1168–1174.

Clements, R. L. (1980). *Cereal Chem.* **57**, 445–446.

Clements, R. L., and Donelson, J. R. (1981). *Cereal Chem.* **58**, 153–154.

Clements, R. L., and Donelson, J. R. (1982). *Cereal Chem.* **59**, 121–124.

Cole, E. W., Mecham, D. K., and Pence, J. W. (1960). *Cereal Chem.* **37**, 109–121.

Daftary, R. D., Pomeranz, Y., Shogren, M., and Finney, K. F. (1968). *Food Technol.* **22**, 79–82.

Daniels, D. G. H., and Fisher, N. (1976). *J. Sci. Food Agr.* **27**, 351–357.

Daniels, N. W. R. (1974). In *Water Relations of Food* (R. B. Duckworth, ed.), pp. 573–586. Academic Press, London and New York.

Daniels, N. W. R., Richmond, J. W., Russell Eggitt, P. W., and Coppock, J. B. M. (1966). *J. Sci. Food Agri.* **17**, 20–29.

De Stefanis, V. A., and Ponte, J. G. (1969). *Biochem. Biophys. Acta* **176**, 198–201.

De Stefanis, V. A., and Ponte, J. G. (1976). *Cereal Chem.* **53**, 636–642.

Finney, K. F., and Barmore, M. A. (1948). *Cereal Chem.* **25**, 291–312.

Fisher, N., Broughton, M. E., Peel, D. J., and Bennett, R. (1964). *J. Sci. Food Agr.* **15**, 325–341.

Fisher, N., Bell, B. M., Rawlings, E. R., and Bennett, R. (1966). *J. Sci. Food Agr.* **17**, 370–382.

Guy, R. C. E., and Pithawala, R. H. (1981). *J. Food Technol.* **16**, 153–166.

Johnson, A. C., Hoseney, R. C., and Varriano-Marston, E. (1979). *Cereal Chem.* **56**, 333–325.

Junge, R. C., Hoseney, R. C., and Varriano-Marston, E. (1981). *Cereal Chem.* **58**, 338–342.

Kissell, L. T., Donelson, J. R., and Clements, R. L. (1979). *Cereal Chem.* **56**, 11–14.

Kissell, L. T., Pomeranz, Y., and Yamazaki, W. T. (1971). *Cereal Chem.* **48**, 655–662.

Krog, N. (1981). *Cereal Chem.* **58**, 158–164.

MacRitchie, F. (1976). *Cereal Chem.* **53**, 318–326.

MacRitchie, F. (1977). *J. Sci. Food Agr.* **28**, 53–58.

MacRitchie, F. (1978). *J. Food Technol.* **13**, 187–194.

MacRitchie, F. (1980). *Bakers Dig.* **54**, 10–13.

MacRitchie, F. (1981). *Cereal Chem.* **58**, 156–158.

MacRitchie, F., and Gras, P. W. (1973). *Cereal Chem.* **50**, 292–302.

Morrison, W. R. (1978). *Advan. Cereal Sci. Technol.* Vol. II, 221–348.

Pomeranz, Y. (1968). *Advan. Food Research*, **16**, 335–455.

Pomeranz, Y. (1973). *Advan. Food Research*, **20**, 153–188.

Ponte, J. G. and De Stefanis, V. A. (1969). *Cereal Chem.* **46**, 325–329.

Seibel, W., Ludewig, H. G., and Bretschneider, F. (1980). *Getreide Mehl Brot* **34**, 298.

Spies, R. D., and Kirleis, A. W. (1978). *Cereal Chem.* **55**, 699–704.

Wootton, M. (1966). *J. Sci. Food Agr.* **17**, 297–301.

Yamazaki, W. T., and Donelson, R. J. (1976). *Cereal Chem.* **53**, 998–1005.

9 Lipid-Protein Interactions During Dough Development

PETER J. FRAZIER

Dalgety Spillers Ltd., Group Research Laboratory, Cambridge, U.K.

I INTRODUCTION

When flour and water are mixed together a progression of physical and chemical changes takes place leading ultimately to a material which exhibits both the viscous flow characteristics of a liquid and the elastic resilience associated with a rubber-like solid. The functional properties of such a viscoelastic dough, and hence its suitability for baking into bread,

"Lipids in Cereal Technology"
ISBN 0-12-079020-3

biscuits or pastry are dependent on the proportions of water added, the "strength" of the flour used (itself dependent on the quantity and quality of the protein present), the presence of various minor constituents and additives and the time-scale and vigorousness of the mixing employed.

Modern developments in breadmaking have involved greater control over wheat varieties and flour strength together with the use of more sophisticated additives, particularly oxidant mixtures and emulsifiers. However, the greatest single change has been the use of very short-time, high-energy mechanical dough development processes to replace the long-bulk fermentation stage needed to mature conventional bread doughs. While enabling good bread to be made from weaker flours, these processes have highlighted the importance of one of the minor constituents of all flours – the lipid fraction. This chapter will outline some of the early studies of changes in lipid distribution during dough mixing and then examine in more detail the location of triglyceride lipid in developed doughs.

II RHEOLOGICAL AND BAKING EFFECTS OF LIPID BINDING

A Lipid Extraction: Free and Bound Lipid

Approximately 1% of oil can readily be extracted from fresh, commercially milled wheat flour using a non-polar solvent such as light petroleum (b.p. 40–60 °C). Increasing the polarity of the solvent gives a higher yield of extractable lipid, up to about 2% with water-saturated butan-1-ol (see Chapter 7); this is close to the total lipid content of the flour by acid hydrolysis. In 1947 Olcott and Mecham found that the petrol-extractable or "free" lipid fraction decreased substantially as flour was wetted and mixed into a dough. The consequent increase in the level of residual "bound" lipid was attributed to the formation of lipoprotein complexes in the dough (Mecham and Pence, 1957; Pomeranz, 1967).

An alternative solvent for total lipid was introduced by Tsen *et al.* (1962) involving a two-phase system of chloroform, methanol and water. This system was adopted by Daniels *et al.* (1966) and has been used extensively since to extract residual bound lipid from freeze-dried doughs after prior removal of free lipid with light petroleum.

Although little direct evidence for lipoprotein complexes in dough had been obtained, in 1961 Grosskreutz proposed a lipoprotein model of wheat gluten structure in which lipid bilayers provided a "slip-plane" to explain viscous flow behaviour in an elastic protein network. Other dough models have involved interactions between glycolipids, proteins and starch, and a specific complex between gliadin, glycolipid and glutenin (Pomeranz, 1971). Nevertheless, until the very recent work of Frazier *et al.* (1981), the only flour lipoprotein that had been isolated and studied was lipopurothionin. This is present in light petroleum extracts of wheat flour and yields a crystalline globulin-like protein after disruption of the lipoprotein complex (Balls *et al.*, 1942; Fisher *et al.*, 1968). Phosphatidylcholine, glycolipids and a steryl ester have been identified as associated lipids, but mono-, di- and tri-glycerides appeared to be absent (Redman and Fisher, 1968). An analogue of purothionin, hordothionin, has been isolated from barley flour (Redman and Fisher, 1969), leading to the conclusion that these lipoproteins are unlikely to be significantly involved in the unique properties of wheat dough structures. Purothionin has not been implicated in lipid binding during dough mixing, since it is preformed in flour.

B Mechanical Dough Development

The mechanical processes involved and the rheological changes undergone by doughs in modern breadmaking processes have been reviewed by Frazier *et al.* (1975), and will only be described briefly here. Eliminating the need for bulk fermentation can be achieved by intensive mixing of the dough, either continuously (e.g. Do-maker and Amflow processes) or batchwise (e.g. the Chorleywood Bread Process).The amount of mechanical energy expended on the dough is of the order of 40 kJ kg^{-1} dough (i.e. 5 Wh per lb dough or 11 Wh kg^{-1}) over a period of time not exceeding 5 min (i.e. an average mixing power of 8–10 kJ kg^{-1} min^{-1} or 130–160 W for each kg dough capacity) (Axford *et al.*, 1963). Coupled with the presence of a suitable level of oxidizing improver in the dough (Frazier, 1974, 1979; Frazier *et al.*, 1979), this results in optimum dough rheological properties for baking after a final proof period of 40–45 min.

Conventional dough mixing processes would take 20–25 min instead of 4–5 min to achieve the 40 kJ kg^{-1} level of work input. Furthermore, because the rate of work input is then much slower, such mixes would never achieve the same breadmaking quality even if mixing times were prolonged (see Frazier *et al.*, 1975). In practice, conventional doughs are mixed to much lower work levels and rely on a bulk fermentation of 3 h or longer to modify, or "mature", the dough proteins adequately. Alternatively, reducing agents such as cysteine together with slower acting oxidizing agents can be used to speed up, or "activate", the dough development process without using intensive mixing methods.

Since differences in mixing intensity (work rate) affect the dough protein structure, it is likely that interactions such as lipid binding would be similarly affected. Higher levels of bound lipid were indeed found in continuously mixed doughs by Baldwin *et al.* (1963) and these were confirmed by Daniels *et al.* (1966) for Chorleywood Bread Process (CBP) doughs.

C Lipid Binding during Dough Development

A drop in the proportion of readily-extractable free lipid is an immediate consequence of adding water to flour (Olcott and Mecham, 1943). This was shown to be true even if mechnical work was rigorously excluded by blending flour and powdered ice together under a blanket of liquid nitrogen at -190 °C and then allowing the mixture to warm up to 30 °C without disturbance (Davies *et al.*, 1969). By varying the moisture content, it was found that the onset of lipid binding coincided with the appearance of free water in the unworked dough systems (Davies *et al.*, 1969; Daniels, 1975), and the critical moisture content was found to be 24.8% by differential scanning calorimetry (Davies and Webb, 1969).

When work was introduced very gently (Farinograph bowl on Brabender Do-Corder at 15 rev min^{-1}) into work-free moistened doughs, a somewhat higher moisture content (29%) was needed before the flour particles began to cohere and register an increasing mixer torque (Wood *et al.*, 1972). This gentle work had little further effect on bound lipid between 24.5 and 29% moisture. However, above 29% moisture, when sufficient free water was evidently present for interprotein interactions to occur and the gluten complex to develop, levels of free lipid decreased rapidly during the first 2–3 kJ kg^{-1} of work input (Wood *et al.*, 1972).

Investigations of mixing at moisture contents typical of bread doughs (45% moisture, or approximately 57% water absorption on a 14% moisture flour) showed that the level of free lipid continued to decrease rapidly, and conversely that of the bound lipid to rise, as mechanical work input increased through the CBP work level of 40 kJ kg^{-1} and well beyond, levelling out only as work input approached 400 kJ kg^{-1} (10 × CBP work level). Such dramatic changes in lipid distribution were found to occur only when oxygen was largely excluded from the mixing bowl (e.g. by mixing under vacuum or in nitrogen) (Daniels *et al.*, 1967, 1969, 1970). Lipid binding was also dependent on the rate of work input and was sensitive to the presence of oxygen in the mixer atmosphere.

The overall effects on lipid distribution of wetting and working dough are summarized in Fig. 9.1 (Daniels, 1975). It can be seen that free lipid fell by about 10% as flour moisture was raised from 14 to 45%, even though mixing was rigorously excluded. In the absence of oxygen, mechanical dough development then rapidly reduced free lipid content by a further 40%. However, when oxygen was present, the fall in free lipid was checked and some recovery occurred. This release of bound lipid was found to be

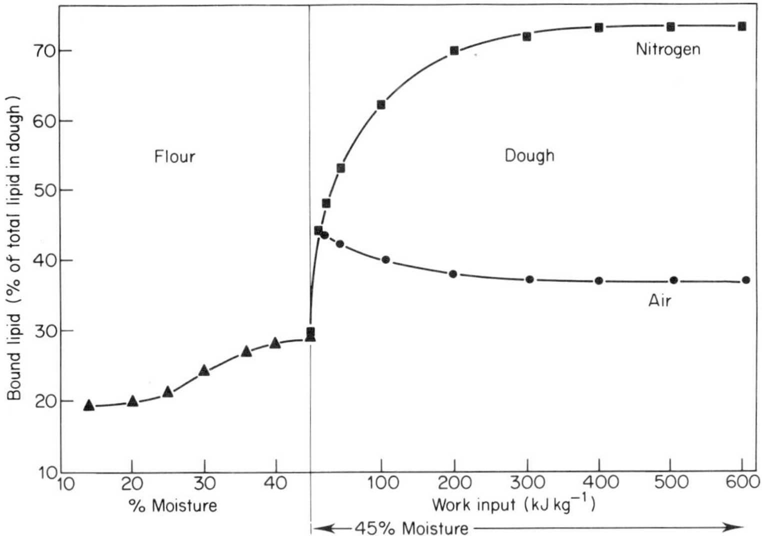

Fig. 9.1 Lipid binding as a function of work-free wetting in flour and of mechanical development in a commercial dough at 45% moisture in air and oxygen-free nitrogen (from Daniels, 1975).

most significant in commercial doughs containing a small proportion of lipoxygenase-active soya flour (Daniels *et al.*, 1970).

D Lipoxygenase Action during Dough Development

The release of bound lipid during dough development in air occurs before an increase in the level of lipid peroxides in dough can be observed. Daniels *et al.* (1970) also showed that addition of peroxidized lipid did not prevent lipid binding in nitrogen-mixed doughs. Daniels therefore suggested that lipid was released as a result of structural changes in dough proteins following oxidation of thiol groups at hydrophobic binding sites. He proposed that the oxidation pathway was coupled with the lipoxygenase-catalysed oxidation of polyunsaturated fats in a manner analogous to the mechanism of pigment bleaching (Sumner and Sumner, 1940; Sumner, 1942) (see also Chapter 10, Section IIID). Thus, the protein sites would act as secondary lipoxygenase substrates, becoming oxidized and releasing bound lipid without the intervention of lipid peroxides, the latter being formed in quantity in dough only after such coupled oxidation had neared completion. (See Chapter 10, Fig. 10.2.)

One consequence of this mechanism was that oxidative improvement of the dough should be observable whenever lipid release was occurring.

Early evidence for this was obtained using the Brabender Extensograph on doughs mixed in air and nitrogen with and without enzyme-active soya flour (Daniels *et al.*, 1971). Later, Frazier *et al.* (1973) made rheological measurements of dough stress relaxation time and determined free and bound lipid over a range of work levels. A marked rheological improvement of the dough was found in air at work levels above 150 kJ kg^{-1} (Fig. 9.2), by which time the release of bound lipid was largely complete (Fig. 9.3). Baking tests (Fig. 9.4) confirmed this effect (Frazier *et al.*, 1974) and, in a later series of experiments, Frazier *et al.* (1977) showed conclusively, by lipid extraction and replacement, that in addition to the lipoxygenase enzyme and atmospheric oxygen an oxidizible, free-lipid fraction had to be present in the dough system for the rheological effect to be observed. Furthermore, addition of nordihydroguaiaretic acid (NDGA), known to be an efficient lipoxygenase inhibitor (Oliveto, 1972; Siddiqi and Tappel, 1957; Blain and Shearer, 1965; Yasumoto *et al.*, 1970), greatly reduced the formation of peroxides in doughs but did not prevent the rheological improvement.

This evidence strongly supported a coupled oxidation mechanism since

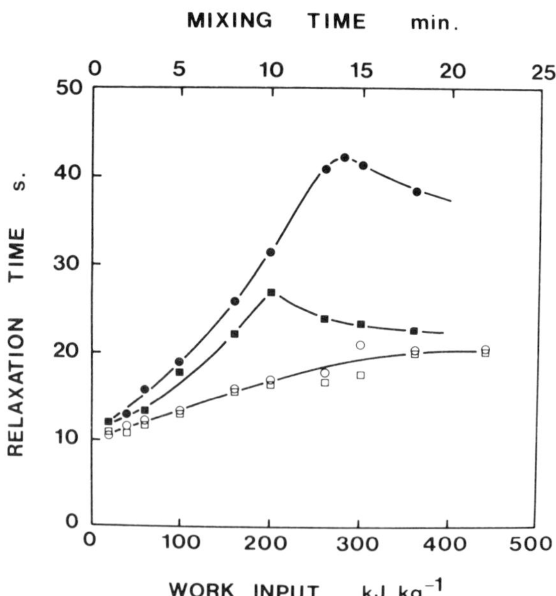

Fig. 9.2 Effect of lipoxygenase-active soya flour on the rheological properties of doughs mechanically developed in air (closed symbols) and nitrogen (open symbols). ■ □ : dough mixed from flour, salt and water only; ● ○ : dough plus enzyme-active soya flour (from Frazier *et al.*, 1973).

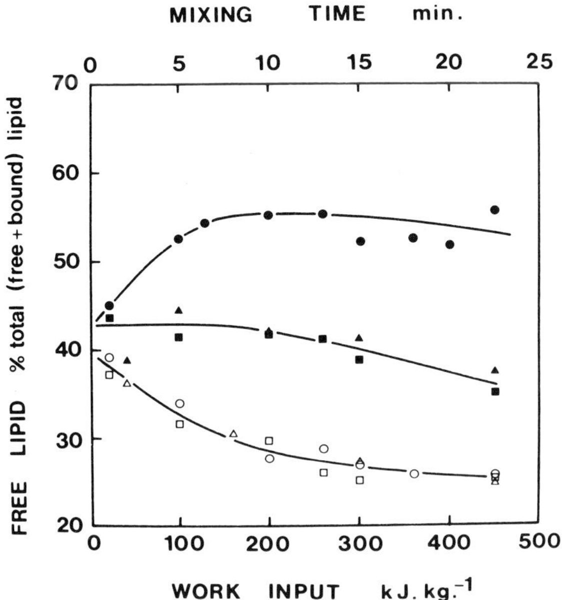

Fig. 9.3 Effect of lipoxygenase-active soya flour on the proportion of free lipid in doughs mechanically developed in air (closed symbols) and nitrogen (open symbols). ■ □ : dough mixed from flour, salt and water only; ● ○ : dough plus enzyme-active soya flour; ▲ △ : dough plus heat-denatured soya flour (from Frazier *et al.*, 1973).

lipoxygenase-catalysed oxidation *actively occurring* in the dough during mixing was shown to be the fundamental requirement for the rheological improvement, irrespective of whether the lipid oxidation process led to lipid peroxides or oxidized NDGA. It is still not known whether this mechanism depends on simultaneous turnover of lipid and protein at the active site of the enzyme (which seems unlikely for steric reasons, although reversible adsorption of lipoxygenase on glutenin has been reported [Graveland, 1970]), or on the formation of a lipid oxidation intermediate (possibly a free radical). Schaich and Karel (1975) have demonstrated a transfer of free radicals from peroxidizing methyl linoleate to a protein (lysozyme) forming protein free radicals and resulting in extensive protein cross-linking at water activities between 0.4 and 0.75. Such a mechanism could well be operative in dough systems when localized water activities may be low in hydrophobic areas of the developing gluten network. Radical transfer, possibly involving oxygen exchange via the hydroperoxy radical (Chan *et al.*, 1979) and radical recombination would be particularly enhanced by the intensive mixing of dough promoting contact between

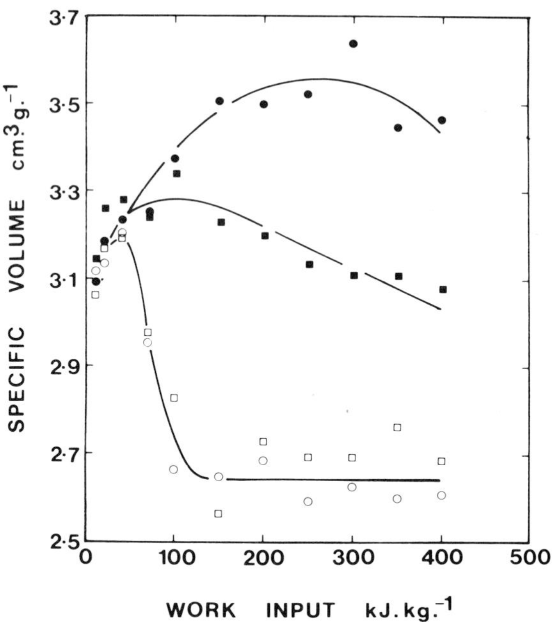

Fig. 9.4 Effect of lipoxygenase-active soya flour on bread specific volume from doughs mechanically developed in air (closed symbols) and nitrogen (open symbols). ■ □ : dough mixed from flour, salt, water, shortening fat and yeast; ● ○ : dough plus enzyme-active soya flour (from Frazier *et al.*, 1974).

lipid and protein. The improving effect of lipoxygenase action in doughs has been shown to be greatest at high rates of work input (Frazier *et al.*, 1973). A free radical mechanism has also been proposed to explain the extended mixing tolerance of doughs containing active lipoxygenase (Hoseney *et al.*, 1980). This effect apparently is different from rheological improvement since it also occurs in the absence of atmospheric oxygen, providing fast-acting oxidants (e.g. potassium iodate) are present in the dough.

When mixing under nitrogen so that the lipoxygenase oxidation mechanism is inactive, the rheological improving effect is absent and the development of the dough is impeded compared with mixing in air (Frazier *et al.*, 1973) (see Fig. 9.2). Thus, development proceeds at a slower rate and the dough reaches a plateau of lower relaxation time than the peak development reached in air. Throughout such mixing, lipid binding increases (Fig. 9.3) and the ultimate effect on loaf volume is most dramatic (Frazier *et al.*, 1974). As shown in Fig. 9.4, after an initial small increase loaf volume fell very rapidly with increasing work input indicating that high levels of bound lipid drastically modify the gluten network and reduce its capacity to retain gas.

III DISTRIBUTION OF TRYGLYCERIDE LIPIDS IN DOUGH FRACTIONS

A Fractionation of Whole Dough

To establish precisely which components of bread doughs were involved in lipid binding and in the consequent reduction in loaf volume, Frazier *et al.* (1981) used radiotracers to follow the fate of triglycerides throughout mixing and subsequent fractionation of doughs. A mixture of glycerol tri-(1-^{14}C) oleate (abbreviated to ^{14}C-GTO) in olive oil was added to flour at a level of 0.72% to give an activity of 5 μCi per dough mix (80 g). Doughs from unbleached, untreated strong bakers' flour were developed to a range of work levels from 10 to 300 kJ kg^{-1} in atmospheres of both nitrogen and air. After mixing, the doughs were freeze-dried, soxhlet-extracted with light petroleum to remove free lipid and fractionated into water-solubles, acetic acid-solubles and starch residue by the procedure summarized in Fig. 9.5. The aim was to maximize protein solubility and provide quantitative recovery of both labelled lipid and Kjeldahl protein.

Typical results for doughs mixed under nitrogen are shown in Table 9.1. Protein recoveries averaged just over 90% and little change occurred across work levels. Approximately 20% of the protein was recovered in the water-soluble fraction, 75% in the acetic acid-soluble fraction and 5% in the starchy residue. Thus, an efficient system of protein dispersion at each stage had avoided any proportionality problems resulting from an increase or decrease in protein solubility with mixing (Tsen, 1967; Patey *et al.*, 1977).

Overall labelled lipid recoveries (also in Table 9.1) averaged 97%. The majority of the labelled lipid appeared as "free" lipid in the petrol extract, decreasing from 55 to 40 μCi kg^{-1} dough (89% to 60% of recovered lipid) as work input increased from 10 to 300 kJ kg^{-1}. Of the "bound" lipid, most occurred in the acetic acid-soluble fraction, increasing three-fold from 6 μCi kg^{-1} dough at 10 kJ kg^{-1} to 18 μCi kg^{-1} dough at 300 kJ kg^{-1} (*i.e.* from 9.5% to 30% of recovered lipid). Lipid recovered from the water-soluble protein was an order of magnitude lower, averaging 0.4 μCi kg^{-1} dough (0.6% of recovered lipid) and not changing with dough work input. Residual solids, mainly starch and accounting for almost 70% of the dough dry mass, attracted the lowest proportion of labelled lipid (under

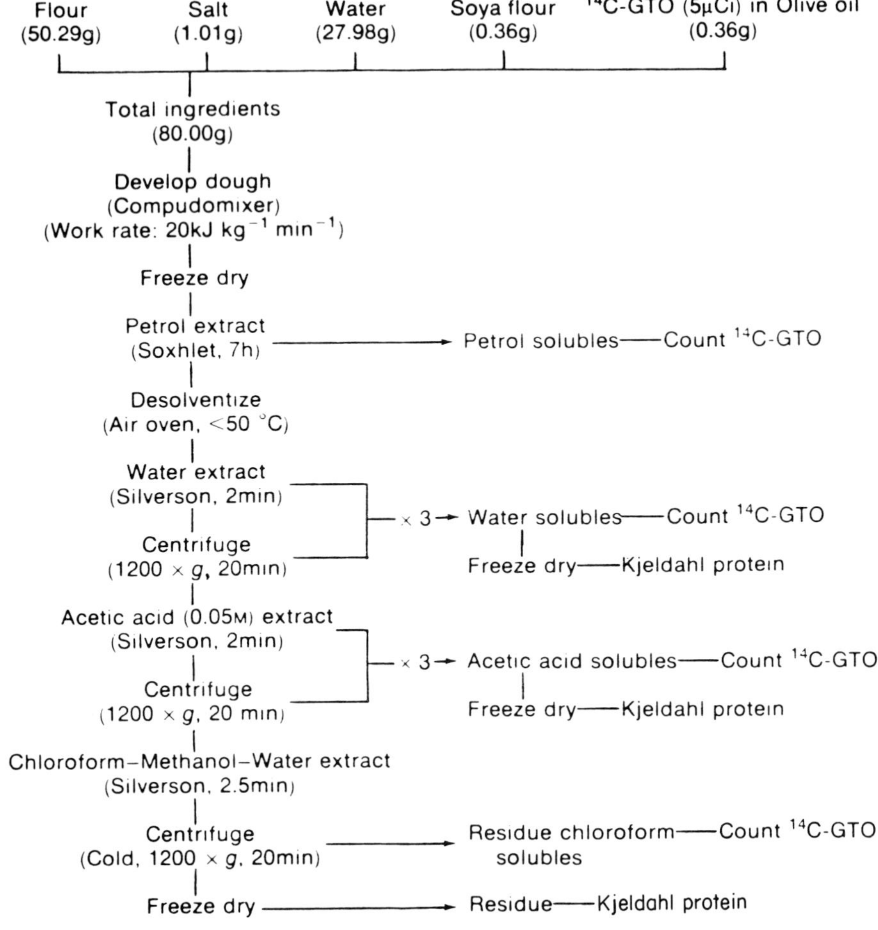

Fig. 9.5 Summarized scheme for the preparation and fractionation of doughs containing radionuclide-labelled triglyceride (from Frazier *et al.*, 1981).

0.4 μCi kg^{-1} dough at 10 kJ kg^{-1}) although this showed some increase at higher work levels.

Plotted as percentage ^{14}C-GTO recovered, Figure 9.6 illustrates the change in lipid composition of dough fractions with work input and clearly shows the acetic acid-soluble material to be almost totally responsible for the binding of lipid during dough development under nitrogen.

When doughs were mixed in air, protein recoveries were very similar to

Table 9.1 Protein and lipid (^{14}C-GTO) distribution in fractions of whole dough mixed under nitrogen (g kg^{-1} dough and μCi kg^{-1} dough respectively)

	Work level (kJ kg^{-1})					
Fraction	10	40	100	150	200	300
Protein (g kg^{-1} dough)						
Light petroleum	nd	nd	nd	nd	nd	nd
Water	13.9	14.2	15.0	15.0	15.0	18.9
Acetic acid	58.8	55.8	60.0	59.5	54.8	52.7
Residue	3.8	3.1	3.0	3.6	3.2	3.4
TOTAL	76.5	73.1	78.0	78.1	73.0	75.0
Recovery (in %)	91.5	87.4	93.3	93.4	87.3	89.7
Lipid (μCi kg^{-1} dough)						
Light petroleum	55.2	52.9	44.3	47.0	36.1	39.9
Water	0.48	0.34	0.30	0.36	0.30	0.55
Acetic acid	5.90	11.0	13.7	15.5	15.6	17.8
Residue	0.39	0.35	0.54	0.87	1.33	1.93
TOTAL	62.0	64.6	58.8	63.7	53.3	60.2
Recovery (in %)	99.1	103.4	94.1	101.9	85.3	96.3

NOTE: Total protein content of dough: 83.6 g kg^{-1} dough
Total lipid content of dough: 14.8 g kg^{-1} dough
Lipid (^{14}C-GTO) added to dough: 62.5 μCi kg^{-1} dough
nd: not determined

those obtained under nitrogen. The lipid distribution, however, was different. Most labelled lipid was extracted in the "free" fraction (approximately 90% of recovered lipid) and remained almost constant with work level. The low level of bound lipid (approximately 10%) occurred almost exclusively in the acetic acid-soluble fraction and, again, unlike the nitrogen-mixed dough, remained almost constant with work input. These results are compared with those for nitrogen-mixed doughs in Fig. 9.6. Figure 9.6 also shows the effect of letting air into the mixing bowl after developing dough under nitrogen (Frazier *et al*., 1981). The acetic acid-solubles, relatively high in "bound" lipid after 150 kJ kg^{-1} under nitrogen, rapidly released lipid during further mixing in air, ^{14}C-GTO falling from 28% to 11% of recovered lipid. These results confirmed earlier gravimetric findings (Daniels *et al*., 1967, 1970; Frazier *et al*., 1973) and clearly identified the acetic acid-soluble protein as that fraction of dough intimately involved in oxidative interactions with dough lipids.

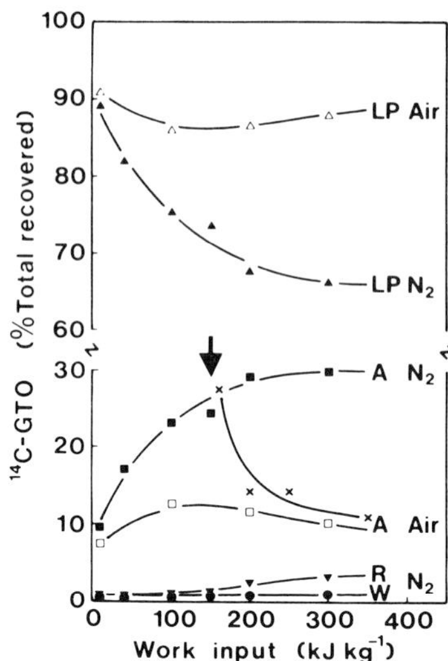

Fig. 9.6 Distribution of triglyceride lipid (^{14}C-GTO) in fractions of doughs mixed in nitrogen and in air. LP: light petroleum-soluble; W: water soluble; A: acetic acid-soluble; R: insoluble residue; ×: acetic acid-soluble after admission of air to a nitrogen mix at position shown by arrow (from Frazier *et al*., 1981).

B Fractionation of Acetic Acid-Soluble Protein

To probe further into the pattern of lipid binding, using radionuclide-labelled lipid, Frazier *et al*. (1981) sub-fractionated the acetic acid-soluble gluten proteins by different methods: the classical aqueous ethanolic extraction to separate glaidin from glutenin and the precipitation method of Wasik and Bushuk (1974).

1 *Aqueous Ethanol Extraction*

Freeze-dried, acetic acid-soluble protein from doughs mixed under nitrogen to a range of work levels in the presence of ^{14}C-GTO was exhaustively extracted with 70% ethanol until the supernatant had negligible absorbance at 280 nm wavelength. After removal of the ethanol by evaporation and freeze drying the two fractions designated F_1-glutenin residue and F_2-gliadin were examined for labelled bound lipid levels, with the results shown in Table 9.2. Recoveries of lipid averaged 96%. At low work levels about three-quarters of the bound lipid was recovered in the glutenin residue (F_1). As dough work level increased, accompanied by greater overall lipid binding in the acetic acid-soluble protein, labelled lipid increased in both gliadin and glutenin fractions. Glutenin always appeared to be the

Table 9.2: Lipid (^{14}C-GTO) distribution in aqueous ethanol fractionated acetic acid-soluble protein from doughs mixed under nitrogen (μCi kg^{-1} dry acetic acid-soluble protein)

	Work level (kJ kg^{-1})				
Fraction	10	40	100	200	300
Stock acetic acid-soluble protein	52.3	176.6	189.2	225.2	257.2
Residue (glutenin) F_1	34.9	103.6	122.1	135.1	119.8
Ethanol solution (gliadin) F_2	11.2	74.8	59.0	91.4	127.0
TOTAL	46.1	178.4	181.1	226.5	246.8
Recovery (in %)	88.1	101.1	95.7	100.6	96.0

fraction of primary importance in lipid binding, but with a significant contribution from the gliadins, particularly at high work levels.

2 *Ammonium Sulphate Precipitation*

In 1974, Wasik and Bushuk described a method for preparing pure glutenin by precipitation from the highly-dissociating AUC-solvent (acetic acid, 0.1M; urea, 3M; cetyltrimethyl-ammonium bromide, 0.01M) of Meredith and Wren (1966). Frazier *et al.* (1981) adapted this method for the fractionation of ^{14}C-GTO-labelled acetic acid-soluble protein using the procedure summarized in Fig. 9.7. The results are shown in Table 9.3. Recoveries of ^{14}C-GTO averaged 99% and the highest proportion of labelled lipid (between 55% and 72% of the recovered lipid) was found in the P_1-precipitate. This was confirmed to be reasonably pure glutenin by sodium dodecylsulphate polyacrylamide gel electrophoresis (SDS-PAGE) agreeing with the designation given by Wasik and Bushuk (1974). These workers reported the gliadin proteins to occur mainly in the P_2-precipitate with traces in P_3-precipitate. A similar distribution was observed by Frazier *et al.* (1981) and confirmation was provided by aluminium lactate PAGE. However, in marked contrast to the ethanol-soluble gliadin described above, Table 9.3 shows that very little labelled lipid was found to be associated with the P_2-gliadin and P_3 fractions. Under 4% of the total ^{14}C-GTO precipitated in P_2 and less that 1% precipitated in P_3.

A significant proportion of the bound lipid had not been accounted for in the normally accepted glutenin or gliadin fractions because of this very low

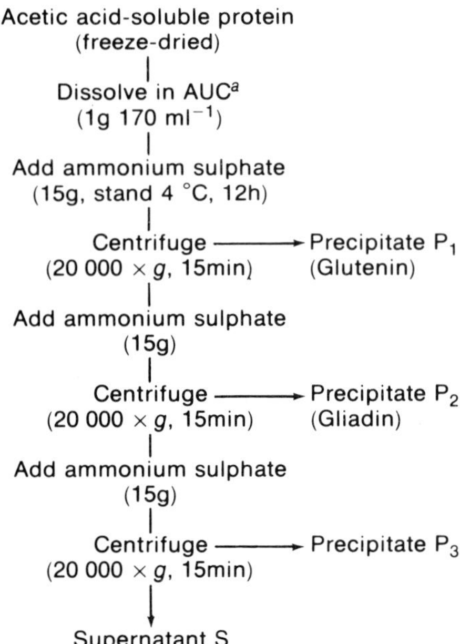

Fig. 9.7 Summarized scheme for the fractionation of acetic acid-soluble protein by ammonium sulphate precipitation from AUC solvent (from Frazier *et al.*, 1981).

Table 9.3 Lipid (^{14}C-GTO) distribution in ammonium sulphate fractionated acetic-acid-soluble protein from doughs mixed under nitrogen (μCi kg^{-1} dry acetic acid-soluble protein)

	Work level (kJ kg^{-1})					
Fraction	10	40	100	150	200	300
Stock AUC solution	95.5	186.7	192.1	222.7	197.6	251.3
Precipitate P_1	56.4	99.3	126.8	156.1	127.4	164.5
Precipitate P_2	3.07	6.22	6.24	7.72	7.38	6.32
Precipitate P_3	0.75	1.24	1.44	0.95	1.05	1.29
Supernatant S	43.4	72.0	61.8	50.8	50.8	75.7
TOTAL	103.6	178.8	196.3	215.6	186.7	247.9
Recovery (in %)	108.5	95.8	102.2	96.8	94.5	98.6

lipid activity in the P_2 and P_3-precipitates. Table 9.3 shows that between 24 and 42% of the total ^{14}C-GTO remained in solution after the ammonium sulphate precipitation procedure was complete. Only low molecular weight water-soluble proteins could normally persist in this supernatant fraction and these had all been removed from the dough by triple water extraction before the acetic acid-solubles were prepared. Furthermore, it had already been established that the water-solubles contained very little labelled lipid (see Section IIIA).

Seeking to establish precisely the status of the lipid in this unexpectedly active fraction, Frazier *et al.* (1981) found that the activity could be precipitated by 10% trichloroacetic acid (TCA), showing that the supernatant lipid was attached to protein. Adding SDS dissociated the lipid from the protein, which then precipitated without ^{14}C-activity on addition of TCA.

SDS-PAGE of the supernatant solution indicated the presence of a protein band corresponding to about 10,000 mol. wt which was not present in the P_1-glutenin or P_2-gliadin fractions. Figure 9.8 relates the labelled lipid recovery from doughs across all fractions and shows clearly that the supernatant fraction (S) contained the highest level of bound lipid after the

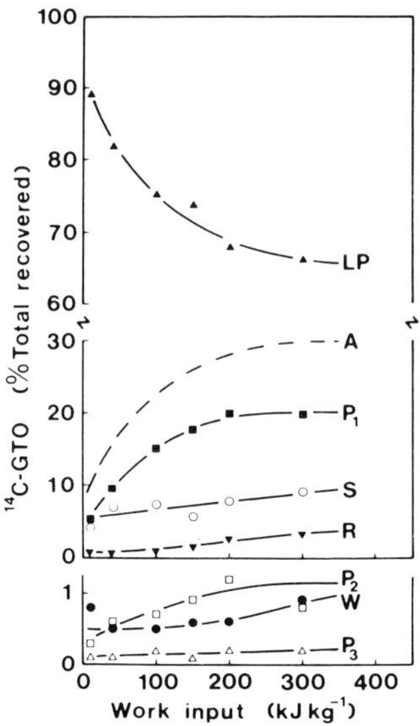

Fig. 9.8 Distribution of triglyceride lipid (^{14}C-GTO) in fractions of doughs mixed under nitrogen, showing sub-fractionation of acetic acid-soluble protein by ammonium sulphate-precipitation from AUC solvent. P_1, P_2, P_3 and S are the precipitates and supernatant shown in Fig. 9.7; other abbreviations as described in the legend to Fig. 9.6 (from Frazier *et al.*, 1981).

Table 9.4 Lipid (^{14}C-GTO) distribution in ammonium sulphate sub-fractions of aqueous ethanol fractionated acetic acid-soluble protein from dough mixed under nitrogen to 150 kJ kg^{-1} (μCi kg^{-1} dry acetic acid-soluble protein)

		Fraction		
Sub-fraction	Stock acetic acid-soluble protein	Ethanol-insoluble residue F_1	Ethanol-soluble protein F_2	Total $F_1 + F_2$
Stock AUC solution	270.6	111.3	104.1	215.4
Precipitate P_1	128.6	72.3	12.2	84.5
Precipitate P_2	8.00	1.46	5.24	6.70
Precipitate P_3	2.36	0.32	1.46	1.78
Supernatant S	145.0	22.6	81.3	103.9
TOTAL	284.0	96.7	100.2	196.9
Recovery (in %)	105.0	86.9	96.3	91.4
TCA Precipitate of supernatant S	124.4	24.6	66.0	90.6

P_1-glutenin, and both fractions showed an increase in bound lipid with increasing dough work input. By comparison, the water-soluble proteins and the gliadin fractions had only very low levels of bound lipid and showed negligible change with work input. This result suggested that the AUC-ammonium sulphate fractionation procedure was splitting off a fragment from glutenin, of very low molecular weight compared with glutenin, but which appeared particularly important for lipid binding.

3 *Aqueous Ethanol and Ammonium Sulphate Fractionation*

In an attempt to resolve the apparent anomaly concerning the extent of lipid binding between gliadin prepared by ethanol extraction and by ammonium sulphate precipitation, Frazier *et al.* (1981) used both methods on a sample of acetic acid-soluble protein from a single work level dough (150 kJ kg^{-1}). The results are shown in Table 9.4. As would be expected, the ethanol-insoluble residue glutenin (F_1), when refractionated by the AUC-ammonium sulphate procedure, precipitated mainly as P_1-glutenin and most of the F_1-bound label was also associated with this fraction. Very little material or label precipitated in P_2 or P_3 and the activity of the supernatant was also relatively low.

An examination of the F_2-gliadin fraction showed that its high content of labelled lipid arose largely because of the presence of some glutenin (P_1-

precipitate, contributing about 12% of the ^{14}C-GTO activity) together with a high level of supernatant protein (contributing over 80% of the total ^{14}C-GTO activity of the F_2-fraction). In fact, both the P_2 and P_3 gliadin protein precipitates prepared from ethanol soluble gliadin had low levels of bound lipid, similar to those found by direct precipitation from the acetic acid-soluble gluten. Adding together the ethanol-soluble and insoluble ^{14}C-GTO recoveries for each ammonium sulphate fraction produced a lipid distribution similar to that obtained directly from the acetic acid-soluble protein. Thus, it was concluded that the two fractionation methods were fundamentally in agreement regarding lipid distribution between glutenin and gliadin. Confusion can occur because of the supernatant protein which is recorded separately by the ammonium sulphate precipitation method but which appears to be part of the classical gliadin fraction owing to its solubility in ethanol. Lipid binding activity is therefore largely confined to two fractions of dough: the glutenin proteins and the supernatant proteins which have been named "ligolins" (Frazier *et al.*, 1981).

IV CHARACTERIZATION OF LIPID-BINDING PROTEINS (LIGOLINS)

A Quantitation of Protein and Lipid

Although the results discussed in Tables 9.3 and 9.4 had indicated the presence of a new protein fraction which was important in lipid binding, its full signifiance did not become apparent until the supernatant was examined for both protein quantity and molecular weight distribution. Frazier *et al.* (1981) repeated the AUC-ammonium sulphate fractionation procedure at one work level only and subjected the four fractions to extensive dialysis to allow accurate calculation of protein recoveries by Kjeldahl nitrogen determination, after removing all residual urea and ammonium sulphate.

The results are summarized in Table 9.5. Over 92% of the wet precipitates were recovered, representing 97% of the protein. The P_1-glutenin accounted for just over 50% of the recovered protein and bound almost 30 g triglyceride per kg protein (over 70% of the recovered bound lipid). The P_2-gliadin, representing 37% of the recovered protein, bound only 2 g triglyceride per kg protein (less than 4% of the recovered bound lipid). The P_3-precipitate contained too little material for nitrogen determination, so confirming the absence of albumins and globulins from the system. However, after dialysis the supernatant still contained an appreciable mass of material, accounting for almost 13% of the recovered protein, and bound

Table 9.5 Kjeldahl protein and lipid (^{14}C-GTO) distribution in ammonium sulphate fractionated aceitic acid-soluble protein from dough mixed under nitrogen to 150 kJ kg^{-1}

	Kjeldahl protein		Lipid (^{14}C-GTO)		
Fraction	(g kg^{-1} dry acetic acid-soluble protein)	(% protein recovered)	(μCi kg^{-1} dry acetic acid-soluble protein)	(% ^{14}C-GTO recovered)	(g lipid kg^{-1} Kjeldahl protein)†
Stock AUC solution	788	–	222.7	–	–
Precipitate P_1	384	50.2	156.1	72.4	29.3
Precipitate P_2	284	37.0	7.72	3.6	1.96
Precipitate P_3	–	–	0.95	0.4	–
Supernatant S	98.4	12.8	50.8	23.6	37.2
TOTAL	766.4	100	215.6	100	20.3
Recovery (%)	97.3	–	96.8	–	–

†1 μCi = 72 mg added triglyceride.

over 37 g triglyceride per kg protein (almost a quarter of the recovered bound lipid). Thus, weight for weight, the glutenins were shown to bind 15 times as much lipid as the gliadin proteins but the highest concentration of lipid was found to be associated with the supernatant proteins – over 25% greater than that bound to glutenin.

B Molecular Weight Fractionation

Using acetic acid-solubles from a dough mixed to 150 kJ kg^{-1}, containing 10 times the previous level of ^{14}C-GTO activity and omitting soya flour (Fig. 9.5), Frazier *et al.* (1981) prepared the AUC-ammonium sulphate supernatant protein as before. After dialysis and freeze-drying, some difficulty was experienced in resolubilizing the protein and a large amount was found to be present at the void volume of a Sephadex G50 column (50 000 mol. wt). Aggregation was suspected in view of the earlier failure of the protein to precipitate with ammonium sulphate and the observed low molecular weight bands during SDS-PAGE. Further runs were therefore made, avoiding dialysis and freeze drying. These revealed a large, clearly separated protein peak at a molecular weight of approximately 9000 (Fig. 9.9) with a substantial level of labelled lipid coinciding. Despite the considerable number of fractionation steps from dough mixing, through Soxhlet extraction, triple homogenization in both water and acetic

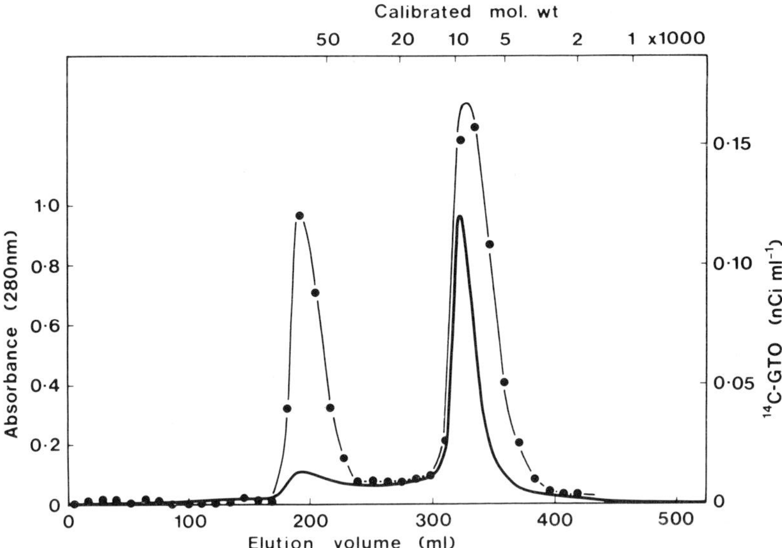

Fig. 9.9 Fractionation on Sephadex G50SF of supernatant protein from AUC-ammonium sulphate precipitation of acetic acid-soluble protein. ——: distribution of protein; -●—●- : distribution of lipid (^{14}C-GTO) (from Frazier *et al.*, 1981).

acid, freeze-drying, AUC-solubilization, ammonium sulphate precipitation and finally gel filtration, the continuing strong association of the labelled triglyceride with this low molecular weight protein was confirmed by the co-precipitation of radioactivity and protein by TCA. Adding SDS was found to disrupt the complex, the protein then precipitating without radioactivity. The small protein peak present at the void volume of the column probably represented an aggregate of the low molecular weight material and also contained bound lipid, although part of the void volume ^{14}C-GTO activity arose from dissociated lipid.

C Amino Acid Analysis

Since the supernatant 9000 mol. wt protein peak was well separated from other material, Frazier *et al.* (1981) regarded it as sufficiently pure to justify amino acid analysis. Two separate column preparations of the fraction were used and a comparison made with P_1-precipitated glutenin using similar sample weights. Table 9.6 shows the replicate amino acid determinations and compares the nearest integer values with the residue frequency of glutenin. It is evident that the supernatant protein differed considerably from glutenin. Aspartic acid frequency was much higher and

Table 9.6 Amino acid composition of AUC-supernatant protein (Ligolin) compared with P_1-glutenin (mols amino acid per 10^4g protein)

Amino acid		Ligolin			P_1-glutenin
		Experimental replicates		Nearest integers	Residue frequency
Asp	**‡	6.66	6.55	7	2
Thr		4.00	4.43	4	2
Ser		5.80	5.81	6	5
Glu	**	10.78	9.72	10	27
Pro	**	5.02	4.99	5	11
Gly		7.79	8.46	8	5
Ala	**	6.80	7.52	7	3
CySH		2.62	2.39	3	1
Val		5.09	5.24	5	3
Met		0.00	0.00	0	1
Ileu		3.28	3.23	3	3
Leu		6.84	6.60	7	5
Tyr		2.20	2.42	2	2
Phe		3.12	3.04	3	3
His		1.44	1.38	1	1
Lys	**	3.28	3.73	4	1
Trp		nd†	nd	nd	nd
Arg	**	4.68	4.52	5	2

†nd: not determined.
‡Major differences between Ligolin and P_1-glutenin are marked **

glutamic acid much lower than in glutenin so that the asp : ratio for the supernatant protein was almost 1 : 1 compared with 1 : 14 in glutenin. Supernatant proline was about half and alanine about twice the frequency found in glutenin. Cysteine frequency was higher in the supernatant protein than in glutenin, but not nearly as high as that reported for purothionin (15 residues per 10^4 g protein for α-purothionin [Redman and Fisher, 1968]). Methionine was absent, however, in common with purothionin. Lysine and arginine were both considerably higher in frequency in the supernatant protein than in glutenin, but only about half the values reported for purothionin.

The residue frequency indicated a molecular weight of 8600 for the supernatant protein excluding the lipid component. Further calculations, based on the radioactivity recovered in the supernatant protein peak, showed an average triglyceride content of 60 g kg^{-1} protein. In terms of added lipid alone, this represented a mole ratio of lipid : protein approaching 0.6. However, if the distribution of label was taken in proportion to the total triglyceride content of the original dough (i.e. including the natural flour triglycerides and approximating their molecular weights to that of glycerol triolein) then the mean mole ratio of lipid : protein was found to be 0.95. Thus, the supernatant protein appears to be a stable lipoprotein molcule formed as a result of dough development under conditions known to promote lipid binding, and probably representing the minimum molecular weight (9500) for a lipid-protein complex in dough. The name "ligolin" (from the Latin *ligare*: to bind, to tie) has been proposed for this class of highly interactive wheat proteins (Frazier *et al*., 1981). The mechanism of attachment of the triglyceride was assumed to depend on hydrophobic interaction since the complex is easily disrupted by SDS.

V CONCLUSIONS

The discovery of a low molecular weight protein fraction, responsible for lipid complexing in wheat flour doughs has made possible the clarification of a number of other observations regarding gluten behaviour. However, it has also raised questions in relation to the many earlier reports of wheat protein fractionation and it now seems probable that the highly interactive and aggregative nature of ligolin contributed greatly to its previous elusiveness.

The distribution of labelled lipid almost entirely between P_1-precipitated glutenin and the AUC-supernatant protein suggests that lipid is bound to a low molecular weight peptide chain, strongly associated with glutenin, to resist water-extraction while yielding at least in part to a more hydrophobic solvent such as 70% ethanol or to a highly dissociative solvent such as AUC. Thus, when classical ethanol-soluble gliadin, which appears to contain nearly half the bound lipid label, is sub-fractionated by ammonium sulphate precipitation from AUC, most of the lipid label is found to reside in the supernatant fraction, while the precipitated gliadin (P_2) carries negligible activity.

The possibility that ligolin could represent a fragment of glutenin, split off by AUC, is discounted by the amino acid analysis which shows it to be very different from glutenin in a number of respects, particularly the aspartic : glutamic ratio. Neither can it be considered a true sub-unit of glutenin since reduction is not required to achieve separation. However, it would normally be expected to appear in the very low molecular weight bands on SDS-PAGE of reduced glutenin. Unfortunately, in many publications using the preparation method recommended by Orth and Bushuk (1973), glutenin is precipitated from ethanolic AUC, so that all or most of the low molecular weight components are removed.Since ligolin is soluble in 70% ethanol it could also be classified with the gliadins and may be expected to be present in gliadin aluminium lactate electrophoretograms. Again, it is unfortunate that most fast moving bands would probably have been dismissed as albumin/globulin contaminants and aggregated ligolin would have remained at the origin and been confused with glutenin contamination. In the absence of a radioactive label to draw attention to it both the presence and significance of the ligolin protein fraction has evidently been previously overlooked.

The nearest reported protein, purothionin, differs considerably from ligolin in amino acid analysis and would have been removed in the light petroleum extract before the water- and acetic acid-soluble proteins had been extracted and fractionated. Furthermore, purothionin has been reported in flour only in amounts under 0.1%, very much lower than the level of supernatant protein (approximately 13% of the acetic acid-soluble protein). If the lipid content of P_1-glutenin is assumed to be due to further, undissociated ligolin attached to lipid, then the true proportion of this small protein in gluten could be even higher.

However, the total lipid binding capability of gluten could not be accounted for in this way on a 1 : 1 molar basis, since the amount of supernatant protein required would be incompatible with the amino acid analysis of glutenin. It is possible, therefore, that ligolin may be a key functional element in the formation of glutenin structure, involving both -SH and hydrophobic interactions, and that the bound lipid consequently incorporated into the gluten structure under appropriate dough mixing conditions could form a nucleus for further hydrophobic binding of lipid.

This mechanism would help to explain the reported dynamic interchange of lipid during dough-mixing (Wood *et al*., 1974). Since sodium salts of long-chain fatty acids could disrupt hydrophobic interactions between ligolin and glutenin, as well as between ligolin and lipid, the dispersibility of glutenin in soap solutions (Kobrehel and Bushuk, 1977) would also be explained. Although much further work is clearly required to explore the properties and function of ligolin, it may prove, by virtue of its interaction with other gluten proteins and with lipid, to be of fundamental importance in the formation of an optimum dough protein structure during modern breadmaking processes.

REFERENCES

Axford, D. W. E., Chamberlain, N., Collins, T. H. and Elton, G. A. H. (1963). *Cereal Sci. Today* **8**, 265–270.

Baldwin, R. R., Johansen, R. G., Keough, W. J., Titcomb, S. T. and Cotton, R. H. (1963). *Cereal Sci. Today* **8**, 273–276, 284, 296.
Balls, A. K., Hale, W. S. and Harris, T. H. (1942). *Cereal Chem.* **19**, 279–288.
Blain, J. A. and Shearer, G. (1965). *J. Sci. Food Agric.* **16**, 373–378.
Chan, H. W.-S., Levett, G. and Matthew, J. A. (1979). *Chem. Phys. Lipids* **24**, 245–256.
Daniels, N. W. R. (1975). In *Water Relations of Foods* (R. B. Duckworth, ed.), pp. 573–586. Academic Press, London and New York.
Daniels, N. W. R., Frazier, P. J. and Wood, P. S. (1971). *Bakers' Digest* **45** (4), 20–28.
Daniels, N. W. R., Richmond, J. W., Russell Eggitt, P. W. and Coppock, J. B. M. (1966). *J. Sci. Food Agric.* **17**, 20–29.
Daniels, N. W. R., Richmond, J. W., Russell Eggitt, P. W. and Coppock, J. B. M. (1967). *Chemy Ind.* 955–956.
Daniels, N. W. R., Richmond, J. W., Russell Eggitt, P. W. and Coppock, J. B. M. (1969). *J. Sci. Food Agric.* **20**, 129–136.
Daniels, N. W. R., Wood, P. S., Russell Eggitt, P. W. and Coppock, J. B. M. (1970). *J. Sci. Food Agric.* **21**, 377–384.
Davies, R. J. Daniels, N. W. R. and Greenshields, R. N. (1969). *J. Food Technol.* **4**, 117–123.
Davies, R. J. and Webb, T. (1969). *Chemy Ind.* 1138.
Fisher, N., Redman, D. G. and Elton, G. A. H. (1968). *Cereal Chem.* **45**, 48–57.
Frazier, P. J. (1974). In The Institute of Food Science and Technolgy (U.K.) 10th Anniversary Symposium on Milling and Baking Technology, pp. 10–12.
Frazier, P. J. (1979). *Baking Ind. J.* **12**, 20–27.
Frazier, P. J., Brimblecombe, F. A. and Daniels, N. W. R. (1974). *Proc. IV Int. Congress Food Sci. Technol.* **1**, 127–129.
Frazier, P. J., Brimblecombe, F. A., Daniels, N. W. R. and Russell Eggitt, P. W. (1977). *J. Sci. Food Agric.* **28**, 247–254.
Frazier, P. J., Brimblecombe, F. A., Daniels, N.W. R. and Russell Eggitt, P. W. (1979). *Getreide Mehl u Brot* **33**, 268–271.
Frazier, P. J., Daniels, N. W. R. and Russell Eggitt, P. W. (1975). *Cereal Chem.* **52**, 106r–130r.
Frazier, P. J., Daniels, N. W. R. and Russell Eggitt, P. W. (1981). *J. Sci. Food Agric.* **32**, 877–897.
Frazier, P. J., Leigh-Dugmore, F. A., Daniels, N. W. R., Russell Eggitt, P. W. and Coppock, J. B. M. (1973). *J. Sci. Food Agric.* **24**, 421–436.
Graveland, A. (1970). *Biochem. Biophys. Res. Commun.* **41**, 427–434.
Grosskreutz, J. C. (1961). *Cereal Chem.* **38**, 336–349.
Hoseney, R. C., Rao, H., Faubion, J. and Sidhu, J. S. (1980). *Cereal Chem.* **57**, 163–166.
Kobrehel, K. and Bushuk, W. (1977). *Cereal Chem.* **54**, 833–839.
Mecham, D. K. and Pence, J. W. (1957). *Bakers' Digest* **31** (1), 40–46, 79.
Meredith, O. B. and Wren, J. J. (1966). *Cereal Chem.* **43**, 169–186.
Olcott, H. S. and Mecham, D. K. (1947). *Cereal Chem.* **24**, 407–414.
Oliveto, E. P. (1972). *Chemy Ind.* 677–679.
Orth, R. A. and Bushuk, W. (1973). *Cereal Chem.* **50**, 106–114.
Patey, A. L., Shearer, G. and McWeeny, D. J, (1977). *J. Sci. Food Agric.* **28**, 63–68.
Pomeranz, Y. (1967). *Bakers' Digest* **41**(5), 48–50, 170.
Pomeranz, Y. (1971). *Bakers' Digest* **45**(1), 26–31, 58.

Redman D. G. and Fisher, N. (1968). *J. Sci. Food Agric.* **19**, 651–655.
Redman, D. G. and Fisher, N. (1969). *J. Sci. Food Agric.* **20**, 427–432.
Schaich, K. M. and Karel, M. (1975). *J. Food Sci.* **40**, 456–459.
Siddiqi, A. M. and Tappel, A. L. (1957). *J. Am. Oil Chem. Soc.* **34**, 529–533.
Sumner, R. J. (1942). *J. Biol. Chem.* **146**, 215–218.
Sumner, J. B. and Sumner, R. J. (1940). *J. Biol. Chem.* **134**, 531–533.
Tsen, C. C. (1967). *Cereal Chem.* **44**, 308–317.
Tsen, C. C., Levi, I. and Hlynka, I. (1962). *Cereal Chem.* **39**, 195–203.
Wasik, R. J. and Bushuk, W. (1974). *Cereal Chem.* **51**, 112–118.
Wood, P. S., Daniels, N. W. R. and Greenshields, R. N. (1972). *J. Food Technol,* **7**, 183–189.
Wood, P. S., Daniels, N. W. R. and Greenshields, R. N. (1974). *J. Sci. Food Agric.* **25**, 1045.
Yasumoto, K., Yamamoto, A. and Mitsuda, H. (1970). *Agr. Biol. Chem.* **34**, 1162–1168.

10 Lipoxygenase and Some Related Enzymes in Breadmaking

J. NICOLAS* **and R. DRAPRON†**

*Station de Technologie des Produits Végétaux, INRA, Montfavet, France.
†Laboratoire de Technologie Alimentaire, INRA, Massy, France

I INTRODUCTION

Although numerous papers have been published on enzyme action in food technology, only short paragraphs were devoted to the effect of

"Lipids in Cereal Technology"
ISBN 0-12-079020-3

oxidoreduction enzymes (more specifically lipoxygenase) in breadmaking (Reed, 1966; Drapron and Uzzan, 1968; Axelrod, 1974; Morrison, 1976, 1978; Frazier, 1979).

The effects of lipoxygenase in breadmaking (mainly in flour maturation and dough mixing) and its interconnections with the other oxidoreducing systems are explored in this chapter; a recent review was published in French on this subject (Nicolas, 1979). After reviewing the oxidoreduction phenomena observed during flour maturation and dough mixing, the possible effects of lipoxygenase (E.C.1.13.11.12), peroxidase (E.C.1.11.1.7), catalase (E.C.1.11.1.6), polyphenoloxidase (E.C.1.14.18.1), ascorbic acid oxidase (E.C.1.10.3.3), dehydroascorbate reductase (E.C.1.8.5.1) and protein disulphideisomerase (rearrangease; E.C.5.3.4.1) are discussed.

Most of these enzymes have been identified in the wheat grain, and in some cases, purified. The activity distribution in the different wheat grain anatomical parts and milling fractions is given in Table 10.1. The direct comparison of the figures obtained by the different authors for the same enzyme is impossible because of the different units used. However, it can be seen that the wheat germ is rich in lipoxygenase, peroxidase, catalase, dehydroascorbate (DHA) reductase and protein disulphide isomerase whereas polyphenoloxidase is concentrated in the bran fraction.

II OXIDOREDUCTION PHENOMENA DURING FLOUR MATURATION AND DOUGH MIXING

A Flour Maturation

Freshly milled flour is normally stored for short periods only and during this time the baking quality improves. A progressive loss in baking performance is reported with prolonged storage. (Bell *et al.*, 1979).

Together with an increasing amount of free fatty acids attributed to the action of fungal and flour lipases (Cuendet *et al.*, 1954; Gracza, 1965), a loss of polyunsaturated fatty acids (PUFA) and carotenoid pigments is observed (Bellenger and Godon, 1972; Warwick *et al.*, 1979). In addition to the modification of the lipid fraction during storage, the flour thiol content decreases (Cuendet *et al.*, 1954; Tsen and Dempster, 1963; Yoneyama *et al.*, 1970a) and the glutenin/gliadin ratio increases (Bellenger and Godon, 1972). A similar effect on protein was found by Rao *et al.* (1978) during storage of wheat grains.

In addition to the biochemical changes, the rheological properties, and hence the baking performance of the flour, vary during storage in air. With

Table 10.1 Distribution of enzyme activities of the wheat grain after milling or dissection†

Milling fraction or dissected part of wheat grain	Lipoxygenase		Catalase			Peroxidase	Polyphenoloxidase		Dehydro-ascorbate reductase	Protein disulfide isomerase
Whole grain		7–11		29–109		5.6–10.3	0–0.69	6–31		0–0.45
Commercial flour	2		2–22	7–67	2.3	0.7–21	0–0.3			
Bran	4.5		21–43	134–573	4.5	8.1– 9.8	0.59–1.93	27–116		
Embryo	32	62–84			83–103				70.7	0.8
Break shorts	9.5		59–110	137–879		6.2–24.7	0.38–1.7			
Reduction shorts			36.5–61	123–950		5.4–15.4	0.8 –2.2			
Red dog			30–60	76–912		3.5–17.3	0.32–2.1			
Pericarp									0.6–1	
Aleurone									9.4–11.2	
Endosperm		1–3							14.3–15.1	0–0.27
Scutellum		57–87							68.5–71	
Reference	Miller and Kummerow (1948)	Blain and Todd (1955)	Honold and Stahmann (1968)		Hawthorn and Todd (1955)	Honold and Stahmann (1968)		Milner and Gould (1951)	Carter and Pace (1964)	Grynberg (1977)

†Values may be compared within each vertical column but not between columns; this is because the authors used different assay methods and the original quoted values are tabulated here.

the Brabender farinograph, Yoneyama *et al.* (1970a,b) noticed an increase of dough tenacity. Bellenger and Godon (1972) obtained similar results using the Chopin alveograph. These phenomena were not observed when storage was in a confined atmosphere or under nitrogen or below 0 °C.

All these results highlight the major role of oxygen in these phenomena, since with the exception of lipolysis, flour maturation is due to oxidative changes in the flour components.

B Dough Mixing

Most of the oxygen absorbed by lipid oxidation during dough mixing is used in the oxidation of free and esterified (at least in the monoglyceride form) linoleic and linolenic acids (Cosgrove, 1956; Smith and Andrews, 1957; Graveland, 1970a, 1973a,b), whereas other fatty acids are not affected (Drapron and Beaux, 1969; Graveland, 1973a,b; Mann and Morrison, 1974).

Lipid binding is influenced by the dough-mixing conditions (Daniels *et al.*, 1966; Pomeranz *et al.*, 1968; Chung and Tsen, 1975). Under work-free conditions, lipid binding increases with flour moisture content. Above 30% moisture and after mixing, there is an additional 15% of bound lipids (Daniels, 1975). When oxygen tension is increased in the mixing bowl, the free lipid level increases with low work input but a higher proportion of bound lipids is released with higher work input levels whereas lipid binding increases in anaerobic dough (Daniels *et al.*, 1970).

The bleaching of carotenoid pigments of wheat flour and semolina is a well-known effect of enzymatic oxidation during dough mixing (Irvine and Winkler, 1950; Hawthorn and Todd, 1955b; Tsen and Hlynka, 1962; Daniels *et al.*, 1970; Matsuo *et al.*, 1970; Drapron *et al.*, 1971; Laignelet *et al.*, 1972; Nicolas, 1978; Laignelet, 1979).

Dough mixing affects the sulfhydryl group (SH) and disulphide groups of gluten proteins. SH oxidation is associated with aerobic mixing (Smith *et al.*, 1957; Matsumoto and Hlynka, 1959; Sokol *et al.*, 1960). In the absence of oxygen, Tsen and Bushuk (1963) and Bloksma (1963) were unable to detect any significant change in the SH content of dough. Moreover, for a longer period of mixing under nitrogen, Mecham and Knapp (1966) noticed an increasing amount of SH groups.

Several flour additives, used to improve baking quality, have oxido-reducing properties and among them, ascorbic acid is used worldwide. Its improver effect is due to DHA, its oxidation product formed from the oxygen incorporated during dough mixing (Melville and Shattock, 1938; Tsen, 1965).

We shall now examine how these phenomena are brought about by activity of these enzymes.

III THE LIPOXYGENASE (E.C.1.13.11.12) SYSTEM

For detailed review of the biochemical properties of plant lipoxygenases, the reader is referred to several recent papers (Eskin *et al*., 1977; Veldink *et al*., 1977; Galliard and Chan, 1980; Nicolas and Drapron, 1981) and Chapter 6.

A Polyunsaturated Fatty Acid Oxidation (PUFA)

It is clear that with regard to its specificity in the oxidation of PUFA, lipoxygenase is involved in the loss of linoleic and linolenic acids in dough during mixing. This loss, undetectable under anaerobic conditions, increases with mixing under pure oxygen or with added purified lipoxygenase (Graveland, 1968). In model systems, the first step of enzyme-catalysed oxidation of linoleic acid is, through free radicals, the formation of 9- or 13-hydroperoxide depending on the source of the lipoxygenase used and the reaction conditions (Hamberg, 1971; Christopher *et al*., 1972; Veldink *et al*., 1972). According to Graveland (1970b), hydroperoxy radicals formed in an aqueous extract of flour lead to 9- and 13-hydroperoxides (L_1), part of these being reduced to hydroxy acids (L_2) (Fig. 10.1). In addition, in a flour-water suspension, there is formation of trihydroxy acids (L_4). In a different case, hydroxy-epoxy acids (L_3) are present in a dough, whereas L_1 are undetectable. Kinetic studies during mixing show that L_4 is the hydrolysis product of L_3. In a flour-water suspension, L_3 do not accumulate because they are rapidly hydolysed. To explain these results, Graveland (1971) proposed the scheme shown in Fig. 10.1. In all cases the peroxy radical is the precursor. L_3 and L_4 are only obtained in the presence of the water-insoluble fraction of flour which redirects lipoxygenase-catalysed oxidation. For a model system with linoleic acid, purified lipoxygenase and glutenin, Graveland (1970b) showed that the enzyme was adsorbed on glutenin and that the $L_4/(L_1 + L_2)$ ratio was equal to the adsorbed/non-adsorbed enzyme ratio. Mann and Morrison (1975) and Markwalder *et al*. (1975) later confirmed some of these results, more particularly the formation of L_4 in dough. Graveland (1973a) also noticed the presence of small amounts of keto, keto-hydroxy and dihydroxy acids in flour-water suspensions.

Heimann and Dreisen (1973) and Heimann *et al*. (1973) purified, from oat, enzymes acting on L_1 with isomerase activity leading to L_2. Although their presence was supposed, these enzymes have never been fully characterized in wheat flour. In the wheat germ however, Zimmerman and Vick (1970) and Christianson and Gardner (1975) showed the presence of enzymes catalysing conversion of hydroperoxides into keto-hydroxy acids. Gardner *et al*. (1974, 1977, 1978) showed that the same compounds could

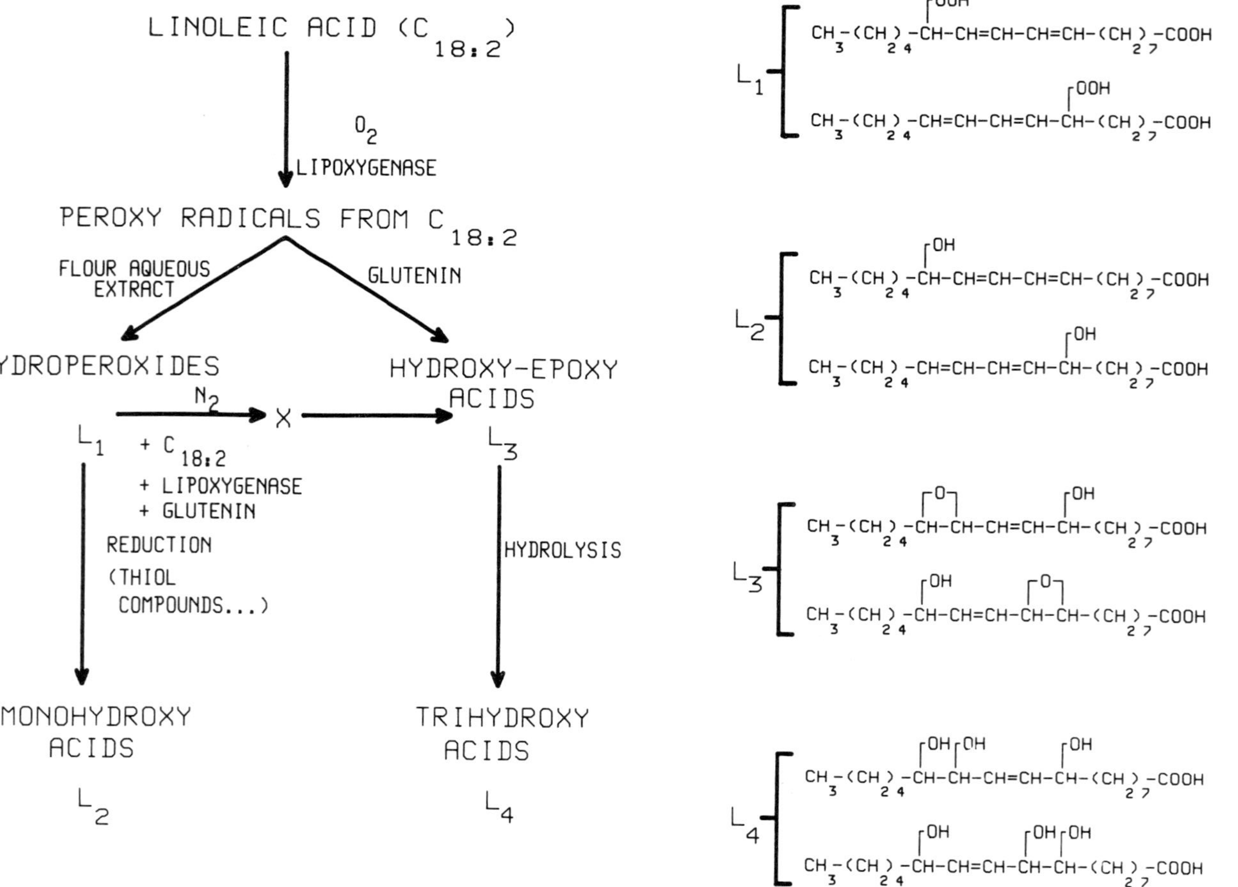

Fig. 10.1 Reaction scheme for linoleic acid oxidation by lipoxygenase in an aqueous flour extract, in a flour-water suspension or in a dough (from Graveland, 1971).

be formed in an analogous manner in the presence of cysteine and ferric ion, together with addition compounds between cysteine and hydroperoxides

In addition to free PUFA, wheat lipoxygenase is able to catalyse the oxidation of PUFA in monoglycerides (Graveland, 1970a; Morrison and Panpaprai, 1975), whereas in the presence of fully active soya flour oxidation of esterified PUFA is clearly more important (Morrison and Panpaprai, 1975). Numerous workers showed that soya flour contained enzymes able to act on esterified linoleic and linolenic acids (Koch *et al.*, 1958; Vioque and Calderon, 1967; Guss *et al.*, 1968; Christopher *et al.*, 1970; Verhue and Francke, 1972).

B Sulfhydryl Group Oxidation

To explain the loss of SH groups during mixing, different hypotheses were proposed. Pea lipoxygenase catalyses the oxidation of reduced glutathione thus forming disulphide groups (Mapson and Moustafa, 1955), and some authors have suggested that in the same way wheat lipoxygenase can catalyse oxidation of gluten proteins through co-oxidation of accessible SH groups of the proteins. Thus, Tsen and Hlynka (1963) showed that the loss of SH groups was more rapid in a normal flour than in a defatted one. They proposed that if oxygen from air can directly oxidize SH groups, the oxidation is mainly due to the enzymatically peroxidized lipids. Bloksma (1963) postulated a similar mechanism, which agrees with Graveland's scheme where SH oxidation into disulphide bridges leads to hydroxy acids formation. However, Mann and Morrison (1975) cast doubt on this scheme since they did not observe any change in the proportions of hydroxy acids in wheat flour doughs with added cysteine (increasing SH content) or added N-ethyl maleimide (decreasing SH content).

Another mechanism was suggested involving the formation of free radicals by lipoxygenase catalysis (Morrison, 1976) or by mechanical dough development (Dronzek and Bushuk, 1968; Nishiyama *et al.*, 1978). These lipid radicals would induce thiyl radicals in protein resulting in a loss of SH content by radical combination. For the reduction of hydroperoxides into hydroxy acids, Mann and Morrison (1975) postulated the presence of a lipoperoxidase (Blain and Styles, 1959; Grosch *et al.*, 1972). Later Graveland *et al.* (1978) found a decrease in the SH group content equivalent to the increase in hydroxy acids during mixing. Thus one mole of hydroperoxide oxidizes two moles of SH groups.

C Release of Bound Lipids and Rheological Effects

As early as 1970, Daniels *et al.* noticed a relationship between carotenoid bleaching and lipid release during mixing. They postulated a mechanism of release of lipids bound to protein which involved lipoxygenase (Fig. 10.2). According to Daniels *et al.*, the oxidized lipid intermediates, formed by lipoxygenase-catalysed oxidation of PUFA, are able to act on the hydrophobic sites of lipid-binding proteins resulting in oxidation of SH groups. Such an oxidation would induce a marked change in the charge distribution on the protein surface leading to inversion of the lipoprotein micelles and reversal of the previously hydrophobic nature of the lipid-binding sites. Thus, water molecules could enter the protein structure with concomitant release of bound lipids. The same authors showed that the release of lipids occurred at atmospheric oxygen concentrations below that at which a significant amount of lipid peroxides was observed. Later, Frazier *et al.* (1973) provided further evidence for the mechanism when they noticed that addition of purified soya lipoxygenase resulted in a marked increase in the release of bound lipids during dough development in air. This phenomenon did not occur after heat-denaturation of the soya flour or with a nitrogen atmosphere during dough-mixing.

The coupled oxidation of gluten proteins, suggested as early as 1956 by Koch and 1957 by Smith *et al.*, would lead to a change in the rheological properties of dough. Frazier *et al.* (1973) observed such modifications during mechanical development of wheat flour doughs; the increase in the free lipid content of aerobic dough was accompanied by an increase in dough relaxation time. The magnitude of this improvement (due to lipoxygenase) increased with increasing work input and was dependent on the rate of work input. Moreover, lipoxygenase increased dough-mixing tolerance (Koch, 1956; Frazier *et al.*, 1973; Hoseney *et al.*, 1980; Kieffer and Grosch, 1980) and it could be considered that these improvements represent oxidative effects on gluten proteins. As a general rule, oxidative chemical agents such as potassium iodate and bromate and ascorbic acid increase the dough resistance. They improve the dough handling and the bread volume. All these effects are attributed to the oxidation of flour protein SH groups (Hird and Yates, 1961; Tsen and Bushuk, 1963; Tsen, 1965). However, fast-acting agents such as potassium iodate and potassium bromate at pH 4.7 decrease dough-mixing tolerance (Sullivan *et al.*, 1961; Weak *et al.*, 1977). This different behaviour could be the result of a specific action of lipoxygenase on hydrophobic sites in the protein network (Frazier *et al.*, 1973).

Recently, Hoseney *et al.* (1980) proposed another mechanism. They believe that lipoxygenase affects mixing tolerance by creating free radicals that compete for the activated double-bond compounds (indigenous in flour or created by fast-acting oxidants) which are responsible for the rapid decrease in dough stability. By extraction of free lipids, reconstitution and

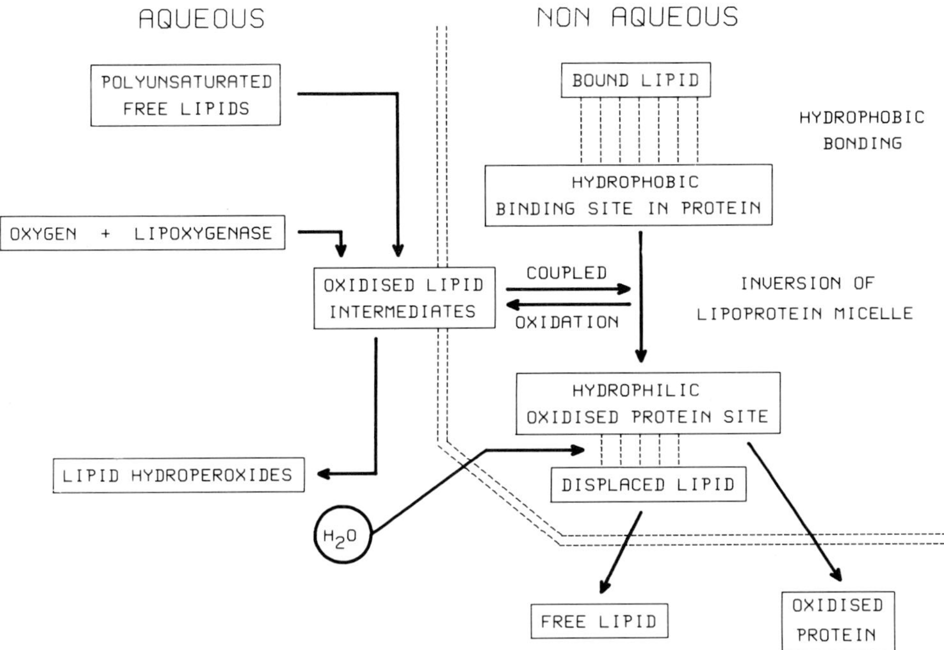

Fig. 10.2 Mechanism of lipoxygenase-induced release of bound lipids during aerobic dough-mixing (from Daniels *et al.*, 1970).

replacement experiments. Frazier *et al.* (1977) showed that the rheological effect of lipoxygenase only occurred in the presence of oxidizable, polyunsaturated, free-lipids. Adding the lipids in an already oxidized state (produced either by autoxidation or lipoxygenase-catalysed oxidation) to doughs mixed from defatted flour under nitrogen, resulted only in a small rheological improvement. Furthermore, adding nordihydroguaiaretic acid (NDGA), a strong inhibitor of lipoxygenase peroxidation (Yasumoto *et al.*, 1970; Nicolas and Drapron, 1977) in dough, although greatly reducing the peroxide formation, only weakly impaired the lipoxygenase-induced rheological improvement. According to Frazier *et al.* (1977), this improvement was achieved only if the polyunsaturated lipid oxidation occurred in dough during mixing where oxygen, PUFA and lipoxygenase were all present. Moreover, the rheological effects were only noticeable for high work input and rate of work input.

Thus the oxidation of gluten proteins, probably promoted by free-radical transfer from lipid to protein, is effective provided the contacts between lipid and protein are sufficiently frequent. This may explain why for a low rate of work input Nicolas *et al.* (1978) found, in the presence of 1%

horse-bean flour in wheat flour, a small decrease in the relaxation time of dough after a 20 min rest period. This effect was reversed after a 2 h rest period. Besides the rheological effects, the observed coupled oxidation promoted by lipoxygenase improves loaf volume and retards staling (Frazier *et al.*, 1974, 1979; Kieffer and Grosch, 1980).

D Bleaching of Carotenoid Pigments and Vitamin Losses

During dough-mixing, lipoxygenase fixes atmospheric oxygen, oxidizes PUFA and bleaches carotenoid pigments. As early as 1934, Haas and Bohn patented the use of enzyme-active soya flour as a dough-bleaching agent. Numerous workers studied the co-oxidizing power towards β-carotene, of lipoxygenases from different sources (Arens *et al.*, 1973; Grosch *et al.*, 1977; Kies *et al.*, 1969; Weber *et al.*, 1973, 1974; Zinsou, 1971). The carotene-bleaching activity of wheat lipoxygenase is much less than that of soyabean lipoxygenase type-2 (Weber *et al.*, 1973) and horse-bean lipoxygenase (Nicolas *et al.*, 1982).

Nicolas (1978) studied the influence of different mixing parameters on the bleaching of carotenoid pigment. The amount of oxidized pigments increased with increasing work input and rate of work input and dough hydration. Adding lipoxygenase in wheat flour either as horse-bean flour or as purified enzymes from horse bean or soyabean resulted in an increase in the rate of carotenoid bleaching during mixing. The reverse was found when ascorbic acid, hydrogen peroxide or an antioxidant such as butylated hydroxytoluene (BHT) was added to the dough (all of these compounds are inhibitors of the lipoxygenase activity).

In addition to carotenoid pigments, tocols were partly destroyed by coupled oxidation during mixing (Drapron *et al.*, 1971), whereas water-soluble vitamins (B_1 and B_6), although sensitive to oxidation, were not affected (Drapron *et al.*, 1974). This difference in behaviour could be the result of the relatively hydrophobic character of carotenoid pigments and tocols.

E Effect on Bread Aroma

In French breadmaking, horse-bean flour added to wheat flour used together with high-speed mixing resulted in an impairment of the bread aroma (Beaux, 1970; Etienne and Dubois, 1974; Chargelègue, 1974). Drapron *et al.* (1974) showed that this phenomenon was linked to the production of hexanal due to the horse-bean lipoxygenase action. Hexanal, the major product of the 13-hydroperoxide decomposition (Loury and Forney, 1968), is probably formed during the baking of the dough. This aldehyde also developed in unblanched peas during frozen storage (Bengtsson *et al.*, 1967). It is present in numerous fruit and vegetable flavours. A detailed review on this subject was published by Eskin *et al.* (1977).

Thus, lipoxygenase is somehow linked with the oxidoreduction phenomena during dough mixing. Although more discrete, most of these reactions (decrease in the relative content of PUFA, loss of carotenoid pigments, SH group oxidation into disulphide bridges with rheological improvement) occurred during flour maturation but occurred much more slowly. Though no experimental proof has been given, the catalysis of these reactions during maturation may be attributed to lipoxygenase. It is probable that the low-water content slows down dramatically the enzymatic reactions taking place during flour storage (Drapron, 1972).

IV RELATED ENZYMES

A The Peroxidase (E.C.1.11.1.7) and Catalase (E.C.1.11.1.6) Systems

Few papers have been devoted to the action of these enzymes in breadmaking. However, through their hematin group they are able to oxidize unsaturated lipids with intermediary free radicals, thus promoting reactions similar to those catalysed by lipoxygenase. Moreover, hydrogen peroxide is a potent inhibitor of lipoxygenase (Mitsuda *et al.*, 1967; Egmond *et al.*, 1975) and is destroyed by their catalytic action.

Their presence in wheat has long been recognized (Blish and Bode, 1935; Wallerstein *et al.*, 1948). Several peroxidase isoenzymes were isolated from durum wheat (Jeanjean *et al.*, 1975) and bread wheat germ (Shin and Nakamura, 1961; Sequi *et al.*, 1968; Zmrhal and Machackova, 1978). The distribution of peroxidase and catalase activities in different milling fractions of wheat grain was studied by Honold and Stahmann (1968) and Evans and Mecham (1971) and the catalase activity of different North American wheats was screened by Irvine *et al.* (1954a,b). Kruger and LaBerge (1974) and Kruger (1977) examined the changes in peroxidases and catalases of wheat during grain development and maturation. Recently, Fretzdorff (1980) noticed that among four cereals (wheat, rye, maize and oat), the wheat peroxidase activity showed the highest heat stability.

From a technological point of view, Hawthorn and Todd (1955a,b) showed that in dough supplemented with catalase, the rate of carotenoid bleaching increased. Nicolas (1978) obtained the same result with added horseradish peroxidase. Some authors proposed an oxidative effect on proteins, leading to rheological improvement (Irvine *et al.*, 1954a; Honold and Stahmann, 1968). In model systems involving proteins, hydrogen peroxide, hydrogen donor and peroxidase, Stahmann (1977) found such reactions. Using electrophoretic and chromatographic techniques he showed that peroxidase catalyzed protein polymerisation together with deamination of the ε-NH_2 group of lysyl residues. Noticing a similar effect with polyphenoloxidase in the presence of its substrate, he suggested that the free radicals or quinone compounds formed by these enzymes were able to react with cysteinyl, methionyl and lysyl residues. Recently, Kieffer *et al.* (1981) observed an improvement of wheat flour, which had poor baking properties by the addition of horseradish peroxidase, H_2O_2 and catechol.

B The Polyphenoloxidase (E.C.1.14.18.1) System

Polyphenoloxidase activity was detected in wheat as early as 1907 by Bertrand and Muttermilch. A weak monophenolase activity is found in immature wheat (Taneja *et al.*, 1974; Kruger, 1976) and is at a maximum 6 weeks after anthesis (Taneja *et al.*, 1974). The polyphenoloxidase activity is greater than the monophenolase activity and decreases during grain maturation. It is also claimed that mono and *o*-diphenolase activities in wheat are separable and represent different enzymes (Taneja and Sachar, 1974). The polyphenoloxidase activity is concentrated in the external parts of the grain (Kruger, 1976), thus explaining why bran, among the milling fractions, exhibits the highest activity (Milner, 1951; Honold and Stahmann, 1968; Tikoo *et al.*, 1973). The activities of different wheat classes were examined by Lamkin *et al.* (1981) who found significantly lower values for the durum wheats than for the other classes. Recently, Interesse *et al.* (1980, 1981) purified wheat *o*-diphenolase and obtained four enzyme components by column isoelectric focusing.

Polyphenoloxidase has long been associated with problems of darkening in food products. Thus, Kobrehel *et al.* (1972) found a relationship between browness index in semolina and the semolina peroxidase and polyphenoloxidase activities. However, Kuninori *et al.* (1976) recently studied the rheological effects of aqueous extracts from mushroom added to wheat dough. Using the Brabender extensigraph, a strengthening of the dough was observed similar to that due to an oxidative effect. After heat treatment, dialysis and inhibition experiments, it was demonstrated that polyphenoloxidase, present in the mushroom extract, was the factor involved. Two mechanisms were postulated. The first one involved oxidation of ferulic acid esterified to pentosan, leading to the gelation of this fraction (Fausch *et al.*, 1963; Geismann and Neukom, 1973; Neukom and Marwalder, 1978); the second assumed a change in the macroscopic structure of gluten, produced either by oxidation of tyrosyl residues or by conjugation between oxidized products of tyrosyl residues and SH groups. Later, Nishiyama *et al*. (1979) using model systems observed an addition reaction between SH groups and oxidation-products of phenol in the presence of mushroom polyphenoloxidase. Recently, Hoseney and Faubion (1981) postulated a mechanism of oxidative gelation of wheat flour pentosans. By viscometric studies, they showed that H_2O_2 in the presence of peroxidase and mushroom polyphenoloxidase was effective in increasing viscosity of wheat flour water-solubles, while common oxidants such as potassium bromate, potassium iodate and ascorbic acid, did not increase the viscosity. The postulated mechanism involved the addition of a protein thiyl radical to the activated double bond of ferulic acid esterified to the arabinoxylan fraction of pentosan.

C The Ascorbic Acid Oxidase (E.C.1.10.3.3) and Dehydroascorbate Reductase (E.C.1.8.5.1) Systems

Ascorbic acid is used world-wide as an improving agent in breadmaking. During mixing it is oxidized to dehydroascorbic acid (DHA) which is responsible for the observed effect (Melville and Shattock, 1938; Guillemet and Sonntag, 1943; Guillemet *et al*., 1942; Sandstedt and Hites, 1945; Kuninori and Matsumoto, 1963, 1964; Carter and Pace, 1964, 1965; Tsen, 1964, 1965; Nicolas *et al*., 1980). The two-step mechanism (Fig. 10.3) proposed by Tsen (1965) is widely accepted for conventional straight-dough systems. The first step (reaction A) is the oxidation of ascorbic acid to DHA by atmospheric oxygen. The second step (reaction B) is the reduction of DHA back to ascorbic acid with the formation of a disulphide at the expense of two SH groups. However, there is some controversy for chemically developed doughs. Zentner (1968) suggested that ascorbic acid could interact with the hydrogen bonds in the dough system while Dahle and Murthy (1970) proposed an effect linked to the antioxidant properties protecting certain flour lipids from oxidation (Grant and Sood, 1980a).

Tsen (1965) postulated that both reactions A and B were enzyme-catalysed. Earlier experiments indicated the probable presence of an ascorbic acid oxidase (catalysing reaction 1) in wheat flour (Sandstedt and Hites, 1945; Proskuryakov and Auerman, 1959; Meredith, 1965; Grant, 1974). More recently, the enzyme was purified simultaneously by Grant and Sood (1980b) and Pfeilsticker and Roeung (1980), with some discrepancies in the reported properties of the purified fraction. But, heat-treatment (Sandstedt and Hites, 1945; Proskuryakov and Auerman, 1959) and specific inhibition (Meredith, 1965) experiments clearly showed that non-enzymatic oxidation of ascorbic acid could occur. Moreover, besides ascorbic acid oxidase, numerous enzymes use ascorbic acid as a cofactor

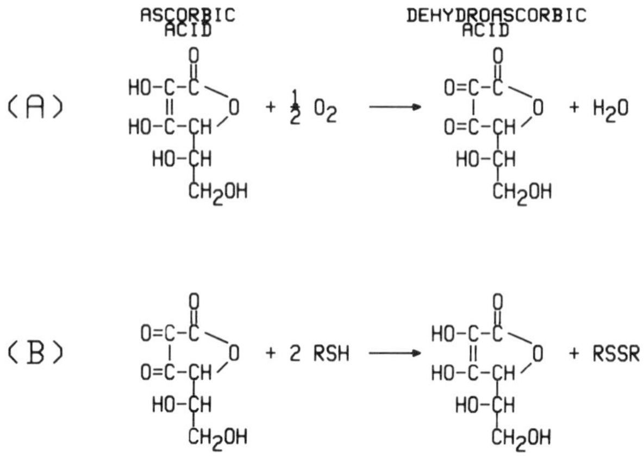

(A) ASCORBIC ACID + $\frac{1}{2} O_2$ → DEHYDROASCORBIC ACID + H_2O

(B) DEHYDROASCORBIC ACID + 2 RSH → ASCORBIC ACID + RSSR

Fig. 10.3 Oxidoreduction reactions affecting ascorbic acid in a dough (from Tsen, 1965).

(Abbott and Udenfriend, 1974). It must be emphasized that the ascorbic acid oxidation rate is very weak in aqueous flour extract (Milner, 1951; Honold and Stahmann, 1968), higher in flour-water suspensions (Proskuryakov and Auerman, 1959) and much higher in dough (Kuninori and Matsumoto, 1963).

The presence in wheat of an enzyme activity catalysing reaction B was unambiguously established (Sandstedt and Hites, 1945, Kahnt *et al.*, 1975). It was active towards reduced glutathione (GSH) (Carter and Pace, 1965) but without effect on cysteine and thioglycolic acid (Kuninori and Matsumoto, 1964). Recently, a purified preparation of GSH-DHA reductase was obtained from wheat flour by Boeck and Grosch (1976). If this enzyme represents the main mechanism of DHA reduction coupled with disulphide formation, its high specificity towards GSH could explain the tolerance of dough to overtreatment with ascorbic acid (Milatovic, 1967; Feillet, 1967; Ponte, 1971) since some SH groups which react with chemical oxidants such as iodate and bromate are not available for the coupled oxidation by the enzyme (Tsen, 1964).

D The Protein Disulphide Isomerase (E.C.5.3.4.1) System

After Goldstein (1957), numerous authors postulated that SH: disulphide exchange reactions were important to the dough rheological properties. The existence of such reactions was first demonstrated by McDermott and Pace (1961) with thiolated gelatin and later by other workers using radioactive tracers (Mauritzen, 1967; Redman and Ewart, 1967a and b; Kuninori and Sullivan, 1968; Lee and Lai, 1968, 1969; McDermott *et al.*,

1969). Though an enzyme-catalysis of SH: disulphide exchanges had never been proposed, Proskuryakov and Zueva (1964) and Gorpinchenko *et al.* (1975) found a protein disulphide reductase (E.C.1.6.4.4) requiring NADP as coenzyme. Later, Grynberg *et al.* (1977) showed the presence of a protein disulphide isomerase (PDI) in wheat embryo. Its activity was characterized by its ability to reactivate ribonuclease which had randomly cross-linked disulphide bridges. A dialysable cofactor, which could be replaced by GSH, was required by the enzyme. The optimal activity was obtained at pH 7.8 and 35 °C (Grynberg *et al.*, 1978). For different wheat varieties, Grynberg (1977) found similar activities in the germ fraction whereas the endosperm activity decreased with increasing breadmaking quality. He postulated that the PDI activity decreases the gluten strength by promoting SH: disulphide exchanges. Further work is needed to confirm this hypothesis. It is highly probable that the protein disulphide isomerase is not the single system involved since a great variety of enzymes are able to catalyse SH: disulphide interchange (Freedman, 1979).

V CONCLUSIONS

The great number of oxidoreduction reactions occurring during dough mixing (and flour maturation) have been discussed. A tentative overall scheme for the enzyme-catalysed reactions with their interrelationships is shown in Fig. 10.4. Lipoxygenase, situated at the very centre, catalyses PUFA oxidation to yield hydroperoxides with intermediate free radicals. These highly reactive products are able to co-oxidize several substances, including lipophilic pigments and vitamins and SH groups.

Hematin-containing substances, including peroxidase and catalase, catalyze oxidation of unsaturated lipids but with less specificity and at a lower rate than lipoxygenase (Tappel, 1961). Since these two enzymes destroy hydrogen peroxide, a strong inhibitor of lipoxygenase, they enhance the lipoxygenase effect.

This is also true for polyphenoloxidase since it oxidises polyphenols, compounds which often have antioxidant properties, i.e. are inhibitors of lipoxygenase activity (Tappel, 1961; Nicolas and Drapron, 1977). However, it competes with lipoxygenase for oxygen consumption. The enzymatic and non-enzymatic oxidations of ascorbic acid also participate in this competition for the available oxygen (Nicolas *et al.*, 1980). Furthermore, ascorbic acid is an inhibitor of lipoxygenase (Walsh *et al.*, 1970; Nicolas, 1978). The DHA formed is reduced back to ascorbic acid by a DHA reductase in the presence of GSH which is oxidized, resulting in

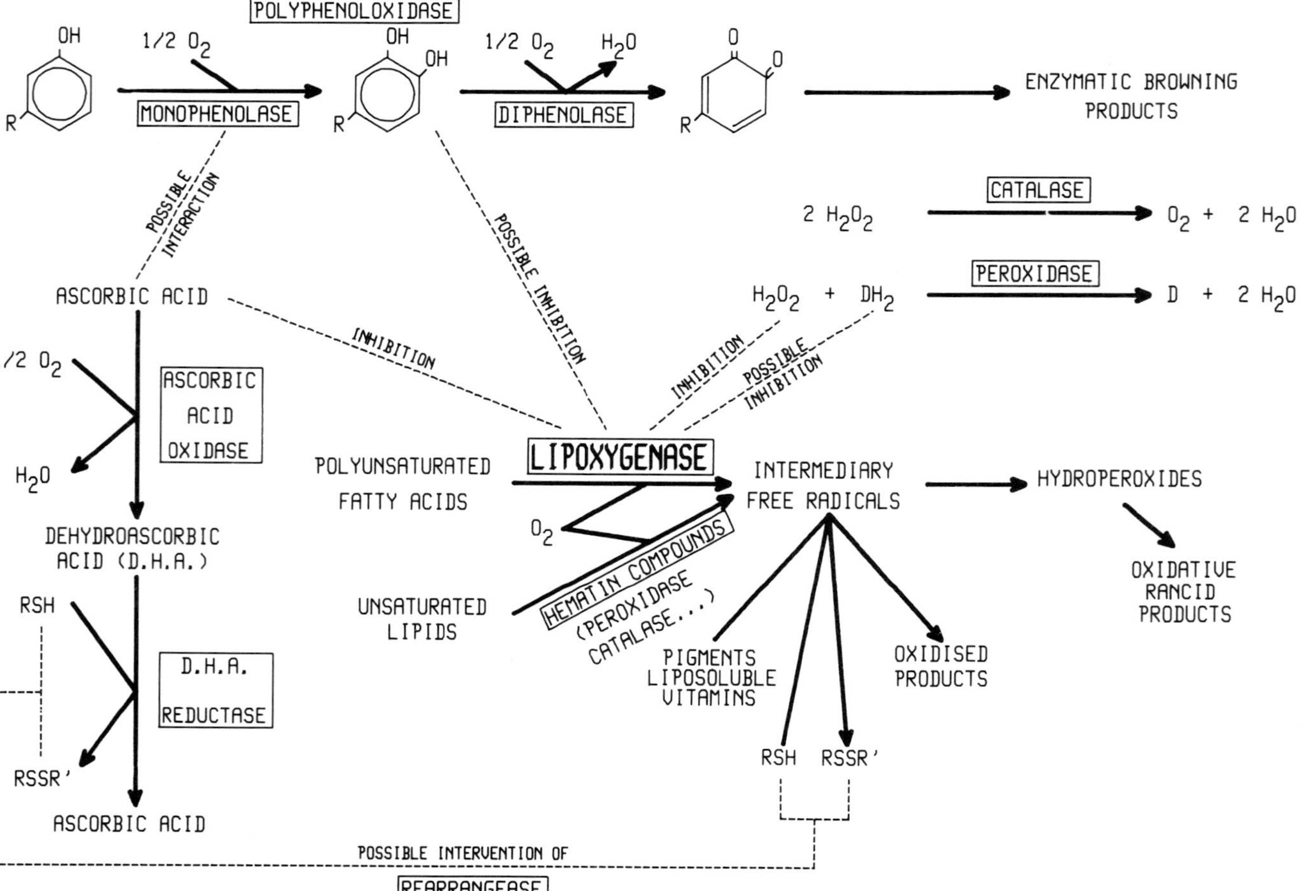

Fig. 10.4 General scheme of the different oxidoreduction systems involved in breadmaking and the possible reactions between these systems (from Nicolas, 1979).

disulphide bridge formation (GSSG) (Kuninori and Matsumoto, 1964; Carter and Pace, 1965; Boeck and Grosh, 1976).

Finally, protein disulphide isomerase also influences SH: disulphide group interchange (De Lorenzo *et al.*, 1966), which plays a prominent part in the rheological properties of dough (Bloksma, 1975).

Some results seem firmly established for lipoxygenase effects in breadmaking, i.e. carotenoid pigment bleaching, release of bound lipids during mixing and rheological effects mainly for high rate of work input. Since hematin compounds catalyse the same kind of reactions, they could play a part in these effects. If the free radicals, whose formation was demonstrated during dough mixing (Dronzek and Bushuk, 1968; Nishiyama *et al.*, 1978) have a great importance in dough development (Axford and Elton, 1960), it is clear that not only lipoxygenase but also peroxidase, catalase and polyphenoloxidase play a prominent part in their formation. Although no direct proof was given and some contradictory results were obtained by Mann and Morrison (1975), several experimental results, particularly those of Graveland, could account for the oxidation of SH groups into disulphide bridges by lipoxygenase-formed hydroperoxides, thus explaining the observed rheological effects. Several models were proposed to represent the glutenin fraction (Ewart, 1977; Kasarda *et al.*, 1976; Khan and Bushuk 1979) which is mainly responsible for the dough viscoelastic properties. Besides SH groups and disulphide bridges, these authors assigned a great importance to hydrogen and hydrophobic bonding. One could assume that oxidized lipids play a major part in hydrophobic bonding (Bernardin, 1978).

Special care must be taken when the *in vitro* results, obtained for the enzyme reactions in model systems, are applied to the *in vivo* phenomena observed during dough mixing. The dough conditions are entirely different since the water, and sometimes even the substrates, are present in limited amounts. The particular hydrophobic environment, due to the presence of gluten, might also modify the reaction mechanisms. Such a phenomenon was observed by Graveland (1970b) for lipoxygenase. Also, Rothfus and Kennel (1970) noticed that wheat β-amylase gave an insoluble complex with glutenin; this non-reversible adsorption due to hydrophobic bonding did not alter the kinetic properties of the enzyme but shifted the optimum of activity, from pH 5.5 for the free enzyme to pH 4.5 for the bound one. Wang and Grant (1969) found that the proteolytic activity of an aqueous wheat flour suspension was higher than that of an aqueous wheat flour extract, with a pH optimum shift from 4.4 to 3.8. It was postulated that interactions with the insoluble proteins of flour were responsible for these results. Such an effect was also observed with ascorbic acid oxidation, the rate of which was higher in a dough than in a suspension, and higher in a suspension than in an extract (Proskuryakov and Auerman, 1959; Kuninori and Matsumoto, 1963).

Recently, Nicolas *et al.* (1981a,b) showed the occurrence in wheat flour of an inhibitor of lipoxygenase activity. The inhibitor is a protein, soluble in 0.1M acetic acid, and the results of kinetic studies suggest that it acts on the substrate rather than on the enzyme itself. Sodium chloride enhances the inhibitory effect. According to the nature of the sustrate, it is postulated that hydrophobic interactions were involved in the phenomenon. This could be related with the recent finding, by Frazier *et al.* (1981), of a gluten protein exhibiting highly interactive properties with dough lipids. They called this protein "ligolin" (see Chapter 9). With this in mind, it is considered that a more systematic study of the influence of gluten on enzyme reactions could give valuable information on the phenomena observed during dough mixing (Nicolas, 1981).

Lastly, very few studies have been devoted to the stability of oxidoreducing enzymes during storage of wheat flour (Freimuth *et al.*, 1972; Colas and Chargelègue, 1974), and on their heat-stability (Wallace and Wheeler, 1972) in wheat protein concentrate. More research in this field is needed to achieve a closer control of oxidoreduction phenomena during flour maturation and dough-mixing.

REFERENCES

Abbott, M. T. and Udenfriend, S. (1974). In *Molecular Mechanisms of Oxygen Activation* (O. Hayaishi, ed.). pp. 167–214. Academic Press London and New York.

Arens, D., Seilmeier, W., Weber, F., Kloos, G. and Grosch, W. (1973). *Biochim. Biophys. Acta* **327**, 295–305.

Axelrod, B. (1974). In *Food Related Enzymes* (J. R. Whitaker, ed.), pp. 324–348. American Chemical Society, Washington, D.C.

Axford, D. W. E. and Elton, G. A. H. (1960). *Chem. Ind. (London)*, 1257.

Beaux, Y. (1970). Ph.D. Thesis, University of Paris.

Bell, B. M., Chamberlain, N., Collins, T. H., Daniels, D. G. H. and Fisher, N. (1979). *J. Sci. Food Agric.* **30**, 1111–1122.

Bellenger, P. and Godon, B. (1972). *Ann. Technol. Agric.* **21**, 145–161.

Bengtsson, B. L., Bosund, I. and Rasmussen, I. (1967). *Food Technol.* **21**, 478–482.

Bernardin, J. E. (1978). *Bakers' Dig.* **52**, 20–23.

Bertrand, G. and Muttermilch, W. (1907). *Compt. Rend. Acad. Sci.* **144**, 1285–1288.

Blain, J. A. and Styles, E. C. C. (1959). *Nature* **184**, 1141.

Blain, J. A. and Todd, J. P. (1955). *J. Sci Food Agric.* **6**, 471–479.

Blish, M. J. and Bode, C. E. (1935). *Cereal Chem.* **12**, 133–142.

Bloksma, A. H. (1963). *J. Sci. Food Agric.* **14**, 529–535.

Bloksma, A. H. (1975). *Cereal Chem.* **52**, 170r–183r.

Boeck, D. and Grosch, W. (1976). *Z. Lebensm. Unters. Forsch.* **162**, 243–251.

Carter, J. B. and Pace, J. (1964). *Nature* **201**, 503.

Carter, J. E. and Pace, J. (1965). *Cereal Chem.* **42**, 201–208.
Chargelègue, A. (1974). *Ann. Technol. Agric.* **23**, 375–384.
Christianson, D. D. and Gardner, H. W. (1975). *Lipids* **10**, 448–453.
Christopher, J. P., Pistorius, E. K. and Axelrod, B. (1970). *Biochim. Biophys. Acta* **198**, 12–19.
Christopher, J. P., Pistorius, E. K., Regnier, F. F. and Axelrod, B. (1972). *Biochim. Biophys. Acta* **289**, 82–87.
Chung, O. K. and Tsen, C. C. (1975). *Cereal Chem.* **52**, 533–548.
Colas, A. and Chargelègue, A. (1974). *Ann. Technol. Agric.* **23**, 323–334.
Cosgrove, D. J. (1956). *J. Sci. Food Agric.* **7**, 668–672.
Cuendet, L. S., Larson, E., Norris, G. G. and Geddes, W. F. (1954). *Cereal Chem.* **31**, 362–389.
Dahle, L. K. and Murthy, P. R. (1970). *Cereal Chem.* **47**, 296–303.
Daniels, N. W. R. (1975). In *Water Relations of Foods* (R. B. Duckworth, ed.), pp. 537–588. Academic Press, London and New York.
Daniels, N. W. R., Richmond, J. W., Russell Eggitt, P. W. and Coppock, J. B. M. (1966). *J. Sci. Food Agric.* **17**, 20–29.
Daniels, N. W. R., Wood, P. S., Russell Eggitt, P. W. and Coppock, J. B. M. (1970). *J. Sci. Food Agric.* **21**, 377–384.
De Lorenzo, F., Goldberger, R. F., Steers, E., Givol, D. and Anfinsen, C. S. (1966). *J. Biol. Chem.* **241**, 1562–1567.
Drapron, R. (1972). *Ann. Technol. Agric.* **21**, 487–499.
Drapron, R. and Beaux, Y. (1969). *Compt. Rend. Acad. Sci.* **268**, 2598–2601.
Drapron, R., Beaux, Y., Cormier, M. and Geffroy, J. (1971). *Compt. Rend. Acad. Agric. Fr.* 245–251.
Drapron, R., Beaux, Y., Cormier, M., Geffroy, J. and Adrian, J. (1974). *Ann. Technol. Agric.* **23**, 353–365.
Drapron, R. and Uzzan, A. (1968). *Ann. Nutr. Alim.* **22**, 393–436.
Dronzek, B. and Bushuk, W. (1968). *Cereal Chem.* **45**, 286.
Egmond, M. R., Finazzi-Agro, A., Fasella, P. M., Veldink, G. A. and Vliegenthart, J. F. G. (1975). *Biochim. Biophys. Acta* **397**, 43–49.
Eskin, N. A. M., Grossman, S. and Pinsky, A. (1977). *Crit. Rev. Food Sci. Nutr.* **9**, 1–40.
Etienne, J. J. and Dubois, M. (1974). *Ann. Technol. Agric.* **23**, 335–351.
Evans, J. J. and Mecham, D. K. (1971). *Cereal Sci. Today* **16**, Abstract No. 22.
Ewart, J. A. D. (1977). *J. Sci. Food Agric.* **28**, 191–199.
Fausch, H., Kündig, W. and Neukom, H. (1963). *Nature* **198**, 287.
Feillet, P. (1967). *Bull. Ec. Fr. Meunerie* **219**, 117–121.
Frazier, P. J., (1979). *Bakers' Dig.* **53**, (6), 8–10, 12–13, 16, 18, 20, 29.
Frazier, P. J., Brimblecombe, F. A., Daniels, N. W. R. (1974). *Proc. IV. Int. Congress Food Sci. Technol.* (Madrid) **1**, 127–129.
Frazier, P. J., Brimblecombe, F. A., Daniels, N. W. R. and Russell Eggitt, P. W. (1977). *J. Sci. Food Agric.* **28**, 247–254.
Frazier, P. J., Brimblecombe, P. A., Daniels, N. W. R. and Russell Eggitt, P. W., (1979). *Getreide Mehl Brot.* **33**, 268–271.
Frazier, P. J., Daniels, N. W. R. and Russell Eggitt, P. W. (1981). *J. Sci. Food Agric.* **32**, 877–897.
Frazier, P. J., Leigh Dugmore, F. A., Daniels, N. W. R., Russell Eggitt, P. W. and Coppock, J. B. M. (1973). *J. Sci. Food Agric.* **24**, 421–436.
Freedman, R. B. (1979). *FEBS' Lett.* **97**, 201–210.

Freimuth, U., Ludwig, E., Heinig, R. and Gebhardt, E. (1972). *Die Nahrung* **16**, 149–156.
Fretzdorff, B. (1980). *Z. Lebensm. Unters. Forsch.* **170**, 187–193.
Galliard, T. and Chan, H. W. S. (1980). In *The Biochemistry of Plants. A comprehensive Treatise* (P. K. Stumpf, ed.), Vol. 4 pp. 131–161. Academic Press, London and New York.
Gardner, H. W., Kleiman, R., Weisleder, D. (1974). *Lipids* **9**, 696–706.
Gardner, H. W., Kleiman, R., Weisleder, D. and Inglett, C. E. (1977). *Lipids* **12**, 655–660.
Gardner, H. W., Weisleder, D. and Kleiman, R. (1978). *Lipids* **13**, 246–252.
Geissmann, T. and Neukom, H. (1973). *Lebensm, Wissen. Technol.* **6**, 59–62.
Goldstein, S. (1957). *Mitt. Lebensm. Hyg. Bern* **48**, 87–93.
Gorpinchenko, T. V., Vakar, A. B. and Kretovich, V. L. (1975). *Biochemistry USSR* **40**, 323–330.
Gracza, R. (1965). *Cereal Chem.* **42**, 333–358.
Grant, D. R. (1974). *Cereal Chem.* **51**, 684–692.
Grant, D. R. and Sood, V. K. (1980a). *Cereal Chem.* **57**, 46–49.
Grant, D. R. and Sood, V. K. (1980b). *Cereal Chem.* **57**, 231–232.
Graveland, A. (1968). *J. Amer. Oil Chem. Soc.* **45**, 834–840.
Graveland, A. (1970a). *J. Amer. Oil Chem. Soc.* **47**, 352–361.
Graveland, A. (1970b). *Biochem. Biophys. Res. Commun.* **41**, 427–434.
Graveland, A. (1971). Ph.D. Thesis, University of Utrecht.
Graveland, A. (1973a). *Lipids* **8**, 599–605.
Graveland, A. (1973b). *Lipids* **8**, 606–611.
Graveland, A., Bosveld, P. and Marseille, J. P. (1978). *J. Sci. Food Agric.* **29**, 53–61.
Grosch, W., Hoxer, B., Stan, H. J. and Schormuller, J. (1972). *Fette Seifen Anstrichm.* **74**, 16–20.
Grosch, W., Laskawy, G., Kaiser, P. (1977). *Z. Lebensm. Unters. Forsch*, **165**, 77–81.
Grynberg, A. (1977). Ph. D. Thesis, University of Paris 7.
Grynberg, A., Nicolas, J. and Drapron, R. (1977). *Compt. Rend. Acad. Sci. Paris Série D* **284**, 235–238.
Grynberg, A., Nicolas, J. and Drapron, R. (1978). *Biochimie* **60**, 547–551.
Guillemet, P. and Sonntag, G. (1943). *Compt. Rend. Acad. Sci.* **216**, 420–422.
Guillemet, P., Sonntag, G. and Hamel, P. (1942). *Compt. Rend. Acad. Sci.* **214**, 720–722.
Guss, P. L., Richardson, T. and Stahmann, M. A. (1968). *J. Amer. Oil Chem. Soc.* **45**, 272–276.
Haas, L. W. and Bohn, R. M. (1934). *Chem. Abstracts* **28**, 4137 U.S. Patents 1957333-4-5-6-7.
Hamberg, M. (1971). *Anal. Biochem.* **43**, 515–526.
Hawthorn, J. and Todd, J. P. (1955a). *Chem. Ind. (London)* 446–447.
Hawthorn, J. and Todd, J. P. (1955b). *J. Sci. Food Agric.* **6**, 501–511.
Heimann, W. and Dresen, P. (1973). *Helv. Chim. Acta* **56**, 463–469.
Heimann, W., Dresen, P. and Schreier, P. (1973). *Z. Lebensm. Unters. Forsch.* **152**, 147–151.
Hird, F. J. R. and Yates, J. R. (1961). *Biochem. J.* **80**, 612–616.
Honold, G. R. and Stahmann, M. A. (1968). *Cereal Chem.* **45**, 99–108.
Hoseney, R. C. and Faubion, J. M. (1981). *Cereal Chem.* **58**, 421–424.

Hoseney, R. C., Rao, H., Faubion, J. M. and Sidhu, J. S. (1980). *Cereal Chem.* **57**, 163–166.

Interesse, F. S., Ruggiero, P., D'Avella, G. and Lamparelli, F. (1980). *J. Sci. Food Agric.* **31**, 459–466.

Interesse, F. S., Ruggiero, P., Lamparelli, F. and D'Avella, G. (1981). *Z. Lebensm. Unters. Forsch.* **172**, 100–103.

Irvine, G. N., Bushuk, W. and Anderson, J. A. (1954a). *Cereal Chem.* **31**, 256–266.

Irvine, G. N., Bushuk, W. and Anderson, J. A. (1954b). *Cereal Chem.* **31**, 267–270.

Irvine, G. N. and Winkler, C. A. (1950). *Cereal Chem.* **27**, 205–218.

Jeanjean, M. F., Kobrehel, K. and Feillet, P. (1975). *Biochimie* **57**, 145–153.

Kahnt, W. D., Mundy, V. and Grosch, W. (1975). *Z. Lebensm. Unters. Forsch.* **158**, 77–82.

Kasarda, D. D., Bernardin, J. E. and Nimmo, C. C. (1976). In *Advances in Cereal Science and Technology* (Y. Pomeranz, ed.), Vol. 1, pp. 158–236. American Association of Cereal Chemists, St. Paul, Minnesota.

Khan, K. and Bushuk, W. (1979). *Cereal Chem.* **56**, 63–68.

Kieffer, R. and Grosch, W. (1980). *Z. Lebensm. Unters. Forsch.* **170**, 258–261.

Kieffer, R., Matheis, G., Hofman, H. W. and Belitz, H. D. (1981). *Z. Lebensm. Unters. Forsch.* **173**, 376–379.

Kies, M. W., Haining, J. L., Pistorius, E. K. and Axelrod, B. (1969). *Biochem. Biophys. Res. Commun.* **36**, 312–315.

Kobrehel, K., Laignelet, B. and Feillet, P. (1972). *Compt. Rend. Acad. Agric. Fr.* 1099–1106.

Koch, R. B. (1956). *Bakers' Dig.* **30**, 2, 48–71.

Koch, R. B., Stern, B. and Ferrari, C. G. (1958). *Arch. Biochem. Biophys.* **78**, 165–179.

Kruger, J. E. (1976). *Cereal Chem.* **53**, 201–213.

Kruger, J. E. (1977). *Cereal Chem.* **54**, 820–826.

Kruger, J. E. and LaBerge, D. F. (1974). *Cereal Chem.* **51**, 345–354.

Kuninori, T. and Matsumoto, H. (1963). *Cereal Chem.* **40**, 647–657.

Kuninori, T. and Matsumoto, H. (1964). *Cereal Chem.* **41**, 39–46.

Kuninori, T., Nishiyama, J. and Matsumoto, H. (1976). *Cereal Chem.* **53**, 420–428.

Kuninori, T. and Sullivan, B. (1968). *Cereal Chem.* **45**, 486–495.

Laignelet, B. (1979). Ph. D. Thesis, University of Montpellier.

Laignelet, B., Kobrehel, K. and Feillet, P. (1972). *Ind. Agric. Alim.* **27**, 695–713.

Lamkin, W. M., Miller, B. S., Nelson, S. W., Traylor, D. D. and Lee, M. S. (1981). *Cereal Chem.* **58**, 27–31.

Lee, C. C. and Lai, T. S. (1968). *Cereal Chem.* **45**, 627–630.

Lee, C. C. and Lai, T. S. (1969). *Cereal Chem.* **46**, 598–606.

Loury, M. and Forney, M. (1968). *Rev. Fr. Corps Gras* **15**, 663–673.

Mann, D. L. and Morrison, W. R. (1974). *J. Sci. Food Agric.* **25**, 1109–1119.

Mann, D. L. and Morrison, W. R. (1975). *J. Sci. Food Agric.* **26**, 493–505.

Mapson, L. W. and Moustafa, E. M. (1955). *Biochem. J.* **60**, 71–80.

Markwalder, H. U., Scheffeldt, P. and Neukom, H. (1975). *Lebensm. Wissen. Technol.* **8**, 234–235.

Matsumoto, H. and Hlynka, I. (1959). *Cereal Chem.* **36**, 513–521.

Matsuo, R. R., Bradley, J. W. and Irvine, G. N. (1970). *Cereal Chem.* **47**, 1–5.

Mauritzen, C. M. (1967). *Cereal Chem.*, **44**, 170–182.

McDermott, E. E. and Pace, J. (1961). *Nature* **192**, 657.
McDermott, E. E., Stevens, D. J. and Pace, J. (1969). *J. Sci. Food Agric.* **20**, 213–217.
Mecham, D. K. and Knapp, C. (1966). *Cereal Chem.* **43**, 226–236.
Melville, J. and Shattock, H. T. (1938). *Cereal Chem.* **15**, 201–205.
Meredith, P. (1965). *J. Sci. Food Agric.* **16**, 474–480.
Milatovic, L. (1967). In *La Vitamina C nelle Tecnologia dei Cereali*, pp. 1–169. Tecnica Molitoria, Pinerolo.
Miller, B. S. and Kummerow, F. A. (1948). *Cereal Chem.* **25**, 391–398.
Milner, M. (1951). *Cereal Chem.* **28**, 435–448.
Milner, M. and Gould, M. R. (1951). *Cereal Chem.* **28**, 473–478.
Mitsuda, H., Yasumoto, K. and Yamamoto, A. (1967). *Agric. Biol. Chem.* **31**, 853–860.
Morrison, W. R. (1976). *Bakers' Dig.* **50**, 29–34, 36, 47–49.
Morrison, W. R. (1978). In *Advances in Cereal Science and Technology* (Y. Pomeranz, ed.), Vol. **2**, pp. 221–348. American Association of Cereal Chemists, St. Paul, Minnesota.
Morrison, W. R. and Panpaprai, R. (1975). *J. Sci. Food Agric.* **26**, 1225–1236.
Neukom, H. and Markwalder, H. U. (1978). *Cereal Foods World* **23**, 374–376.
Nicolas, J. (1978). *Ann. Technol. Agric.* **27**, 695–713.
Nicolas, J. (1979). *Ann. Technol. Agric.* **28**, 445–468.
Nicolas, (1981). Ph. D. Thesis, University of Paris 6.
Nicolas, J., Autran, M. and Drapron, R. (1981a). *Sciences des Aliments* **1**, 213–232.
Nicolas, J., Autran, M. and Drapron, R. (1982). *J. Sci. Food Agric.* **33**, 365–372.
Nicolas, J. and Drapron, R. (1977). *Riv. Ital. Sost. Grasse* **54**, 284–288.
Nicolas, J. and Drapron, R. (1981). *Sciences des Aliments* **1**, 91–168.
Nicolas, J., Gustafsson, S., Autran, M. and Drapron, R. (1981b). *Sciences des Aliments* **1**, 203–212.
Nicolas, J., Gustafsson, S. and Drapron, R. (1980). *Lebensm. Wissen. Technol.* **13**, 308–313.
Nicolas, J., Launay, B., Roussel, P. and Drapron, R. (1978). *Bull. Ec. Fr. Meunerie* **287**, 21–34.
Nishiyama, J., Kuninori, T. and Matsumoto, H. (1978). *J. Sci. Food Agric.* **29**, 267–273.
Nishiyama, J., Kuninori, T. and Matsumoto, H. (1979). *J. Ferment. Technol.* **57**, 387–394.
Pfeilsticker, K. and Roeung, S. (1980). *Z. Lebensm. Unters. Forsch.* **171**, 425–429.
Pomeranz, Y., Tao, R. P. C., Hoseney, R. C., Shogren, M. D. and Finney, K. F. (1968). *J. Sci. Food Agric.* **16**, 974–978.
Ponte, J. G. (1971). In *Wheat Chemistry and Technology* (Y. Pomeranz, ed.), 2nd. Ed., pp. 675–742. American Association of Cereal Chemists, St. Paul, Minnesota.
Proskuryakov, N. I. and Auerman, T. L. (1959). *Biokhimiya* **24**, 297–302.
Proskuryakov, N. I. and Zueva, E. S. (1964). *Dokl. Akad. Nauk SSSR* **158**, 232–234.
Rao, V. S., Vakil, U. K. and Sreenivasan, A. (1978). *J. Sci. Food Agric.* **29**, 155–164.
Redman, D. G. and Ewart, J. A. (1967a). *J. Sci. Food Agric.* **18**, 15–18.
Redman, D. G. and Ewart, J. A. (1967b). *J. Sci. Food Agric.* **18**, 520–523.
Reed, G. (1966). In *Enzymes in Food Processing* (G. Reed, ed.), pp. 221–254. Academic Press, London and New York.

Rothfus, J. A. and Kennel, S. J. (1970). *Cereal Chem.* **47**, 140–146.
Sandstedt, R. M. and Hites, B. D. (1945). *Cereal Chem.* **22**, 161–187.
Sequi, P., Marchesini, A. and Lanzani, G. A. (1968). *Atti VII Simp. Int. Agrochim.* **13**, 75–80.
Shin, M. and Nakamura, W. (1961). *J. Biochem (Tokyo).* **50**, 500–507.
Smith, D. E. and Andrews, J. S. (1957). *Cereal Chem.* **34**, 323–336.
Smith, D. E., Van Buren, J. P. and Andrews, J. S. (1957). *Cereal Chem.* **34**, 337–349.
Sokol, H. A., Mecham, D. K. and Pence, J. K. (1960). *Cereal Chem.* **37**, 739–748.
Stahmann, M. A. (1977). In *Protein Crosslinking. Nutritional and Medical Consequences* (M. Friedman, ed.), Vol. 86B, pp. 285–298. Plenum Press, New York.
Sullivan, B., Dahle, L. and Nelson, O. R. (1961). *Cereal Chem.* **38**, 281–291.
Taneja, S. R., Abrol, Y. P. and Sachar, R. C. (1974). *Cereal Chem.* **51**, 457–465.
Taneja, S. R. and Sachar, R. C. (1974). *Phytochem.* **13**, 1367–1371.
Tappel, A. L. (1961). In *Autoxidation and Antioxidants* (W. O. Lundberg, ed.), pp. 325–366. Interscience Publishers, New York.
Tikoo, S., Singh, J. P., Abrol, Y. P. and Sachar, R. C. (1973). *Cereal Chem.* **50**, 520–528.
Tsen, C. C. (1964). *Bakers' Dig.* **38**, 44–47.
Tsen, C. C. (1965). *Cereal Chem.* **42**, 86–96.
Tsen, C. C. and Bushuk, W. (1963). *Cereal Chem.* **40**, 399–408.
Tsen, C. C. and Dempster, C. J. (1963). *Cereal Chem.* **40**, 586–589.
Tsen, C. C. and Hlynka, I. (1962). *Cereal Chem.* **39**, 209–219.
Tsen, C. C. and Hlynka, I. (1963). *Cereal Chem.* **40**, 145–153.
Veldink, G. A., Garssen, G. J., Vliegenthart, J. F. G. and Boldingh, J. (1972). *Biochem. Biophys. Res. Commun.* **47**, 22–26.
Veldink, G. A., Vliegenthart, J. F. G. and Boldingh, J. (1977). *Prog. Chem. Fats Other Lipids* **15**, 131–166.
Verhue, W. M. and Francke, A. (1972). *Biochim. Biophys. Acta* **285**, 43–53.
Vioque, E. and Calderon, M. (1967). *Grasas Y Aceites* **18**, 296–302.
Wallace, J. M. and Wheeler, E. L. (1972). *Cereal Chem.* **49**, 149–156.
Wallerstein, J. S., Hale, M. G. and Alba, R. T. (1948). *Wallerstein Lab. Commun.* **11**, 221–227.
Walsh, D. E., Youngs, V. L. and Gilles, K. A. (1970). *Cereal Chem.* **47**, 119–125.
Wang, C. C. and Grant, D. R. (1969). *Cereal Chem.* **46**, 537–544.
Warwick, M. J., Farrington, W. H. H. and Shearer, G. (1979). *J. Sci. Food Agric.* **30**, 1131–1138.
Weak, E. D., Hoseney, R. C., Seib, P. A. and Biag, M. (1977). *Cereal Chem.* **54**, 794–802.
Weber, F., Laskawy, G. and Grosch, W. (1973). *Z. Lebensm. Unters. Forsch.* **152**, 324–331.
Weber, F., Laskawy, G. and Grosch, W. (1974). *Z. Lebensm. Unters. Forsch.* **155**, 142–150.
Yasumoto, T., Yamamoto, A. and Mitsuda, H. (1970). *Agric. Biol. Chem.* **34**, 1162–1168.
Yoneyama, T., Suzuki, I. and Muroashi, M. (1970a). *Cereal Chem.* **47**, 19–26.
Yoneyama, T., Suzuki, I. and Muroashi, M. (1970b). *Cereal Chem.* **47**, 27–33.
Zentner, H. (1968). *J. Sci. Food Agric.* **19**, 464–467.
Zimmerman, D. C. and Vick, B. A. (1970). *Plant Physiol.* **46**, 445–453.
Zinsou, C. (1971). *Physiol. Végét.* **9**, 149–167.
Zmrhal, and Machackova, I. (1978). *Phytochem.* **17**, 1517–1520.

11 Physical State of Lipids and their Technical Effects in Baking

K. LARSSON

University of Lund, Sweden

I INTRODUCTION

Lipids are the most surface-active molecules in nature, and are, therefore, found in most biological interfaces. It would thus be expected that lipids from cereals also will accumulate at interfaces when given such opportunities. The production of a dough from cereal flour and water is a homogenization process involving the creation of new interfaces; the critical interfacial phenomenon during fermentation and oven spring is the expansion of gas–liquid interfaces.

Cereal lipids are of two types with regard to their aqueous interaction. About half of the non-starch lipids are non-polar (mainly triglycerides); they form no aqueous phases and give monomolecular films at the

"Lipids in Cereal Technology"
ISBN 0-12-079020-3

air–water interface with quite low-spreading pressure. The rest of the lipids are of the polar type and form aqueous phases. They can also form monomolecular films at the air–water interface which have the ability to displace all other types of biomolecules. It seems rather obvious that these properties must have significance to breadmaking, and this chapter concerns such surface-chemical aspects of wheat lipids.

II PHASE EQUILIBRIA AND STRUCTURES IN BINARY LIPID–WATER SYSTEMS

There are two main types of binary lipid–water system. The first is represented by lipids which form a micellar solution in water. Lysophosphatidylcholine (lysolecithin) the dominant lipid component in starch, is an example of this type. The main features of such a binary system are shown in Fig. 11.1 together with the different lipid–water liquid-crystalline structures. Starting from the water side of the system a micellar solution is formed above the critical concentration for micelle formation (c.m.c.). At increasing concentration, the spherical micelles are transformed into rod-shaped micelles. When the length of these rods is increased and the amount of water between them is reduced to a critical value, it is natural that an ordered structure is formed: the common hexagonal close-packing of

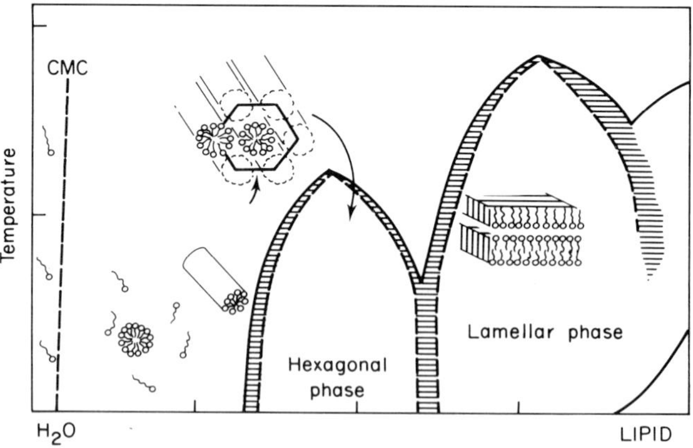

Fig. 11.1 Binary lipid–water system characteristic of polar lipids which form micellar solutions. The horizontal axis defines the composition and the vertical axis the temperature. The phase diagram thus illustrates the temperature range and composition range of existence of the phases. In the indication of structures, each molecule is illustrated by the polar head (a circle) and one attached chain tail.

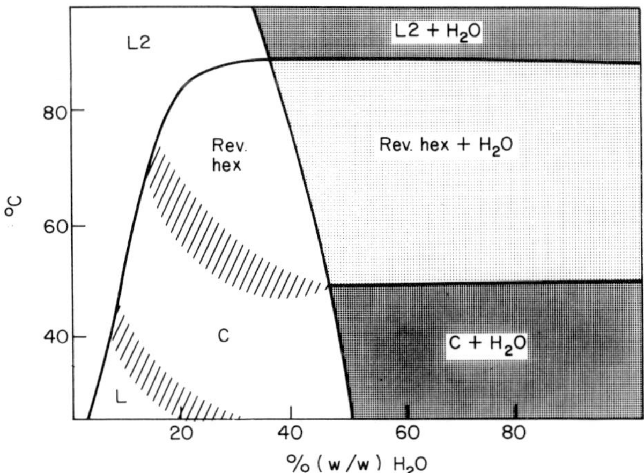

Fig. 11.2 Phase diagram of the binary system of sunflower oil monoglycerides-water (Larson *et al*., 1980). The structure of the lamellar phase is shown in Fig. 11.1. The *reverse* hexagonal phase in this system has the same geometry as the hexagonal phase in Fig. 11.1, with the difference that water forms the rods and the hydrocarbon chains form the matrix between the rods. The structure of the cubic phase has recently been determined (Larsson *et al*., 1980), but will not be discussed here as it is not of interest from a cereal point of view.

infinite rods. This hexagonal liquid-crystalline phase is transformed into the lamellar liquid-crystalline phase when a lower water concentration is reached. The characteristic feature of liquid-crystalline lipid–water phases is the occurrence of long-range order (perfect repetition in one or two dimensions), although the molecules are disordered as in liquids when considered on the atomic level. Thus the hydrocarbon chains possess a chaotic mobility, similar to that of paraffins in the liquid state. These structural features were first revealed by Luzzati and co-workers (1960).

The other type of lipid–water system occurs in the case of non-soluble lipids. Their phase behaviour will be demonstrated here in the example of fatty acyl monoglycerides, which represent a dominant group of functional additives in cereal products. Figure 11.2 shows such a binary system.

The transitions with increasing chain length and increasing temperature, from lamellar to cubic, and to the reverse hexagonal phase, are a direct consequence of changes in the molecular geometry. The thermal mobility of the hydrocarbon chains increases from the polar ester groups to the methyl end groups and this mobility will result in a tendency toward a wedge-shaped space for the molecules. The cubic phase and the reverse hexagonal phase allow such a wedge-shape.

When the lamellar phase exists in equilibrium with excess water,

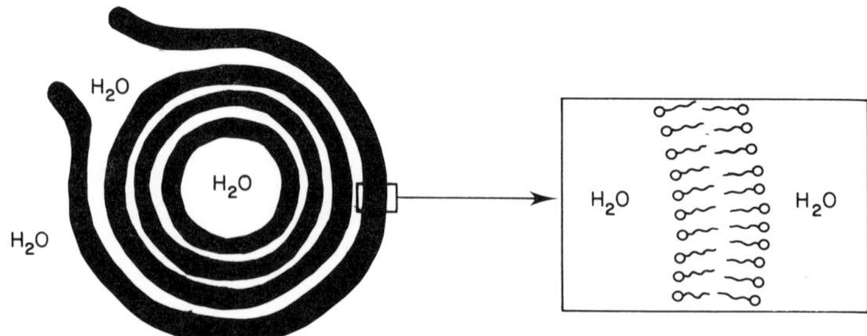

Fig. 11.3 Structure of a liposomal dispersion in water. The liposomes are multilamellar particles, whereas unilamellar ones are termed vesicles.

mechanical fragmentation of the lipid bilayers results in the formation of a liposomal dispersion. In these spherical particles the closed lipid bilayers alternate with water layers, as shown in Fig. 11.3.

If the temperature is increased to the region above that of the liquid-crystalline phases (see Fig. 11.2), a liquid phase of so-called "L2-type" is formed. This is a reverse micellar phase with water aggregates in a con-

Fig. 11.4 Structure of an L2-phase in monoglyceride-water systems (Larsson, 1979). The black regions indicate water lamellae, which are separated by a lipid bilayer.

tinuous lipid medium. The structure of this phase is close to that of the liquid-crystalline phase from which it was formed (see Fig. 11.4). A loss of long-range order is involved in the transition (Larsson, 1979). Although the structural units of the L2-phase are asymmetric, it exhibits Newtonian flow properties, which illustrates the highly dynamic nature of this liquid phase.

III SOLUBILIZATION OF NON-POLAR LIPIDS IN LIQUID-CRYSTALLINE PHASES – TERNARY SYSTEMS

The introduction of fat molecules into the binary system of polar lipids and water results in solubilization of the triglyceride molecules in the liquid-

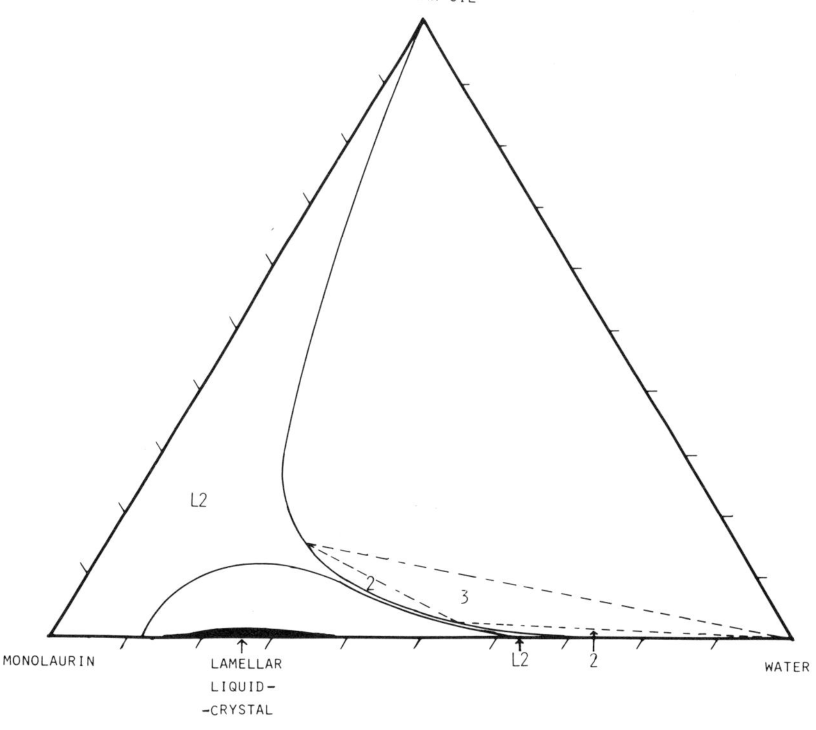

Fig. 11.5 Phase equilibria at 40 °C of the ternary system monododecanoin-water-soyabean oil. The two-phase and three-phase regions at excess of water are indicated by "2" and "3", respectively (Pilman *et al.*, 1979). Monolaurin = monododecanoin.

crystalline phases, and the interaction may even result in the formation of new phases. Such phase equilibria, involving three components, can be illustrated by a ternary phase diagram as shown in Fig. 11.5. This phase diagram was selected since it bears a resemblance to the ternary systems of cereal lipids, which will be discussed in Section IV. In this system the solubilization of soyabean oil triglycerides in the liquid-crystalline phase of monododecanoin (monolaurin) results in the formation of an L2-phase, and in the region of excess of water there are either two (L2 + H_2O) or three (L2 + H_2O + oil) phases.

IV AQUEOUS INTERACTION OF CEREAL LIPIDS

The chemical composition of lipids extracted from wheat flour is described in Chapters 2 and 7. The lipids can be separated on a chromatography column into various polar and non-polar fractions. The two dominating polar components are monogalactosyldiglycerides (MGDG) and digalactosyldiglycerides (DGDG). The ternary system of wheat lipid DGDG, MGDG, and water is shown in Fig. 11.6 (Larsson and Puang-Ngern,

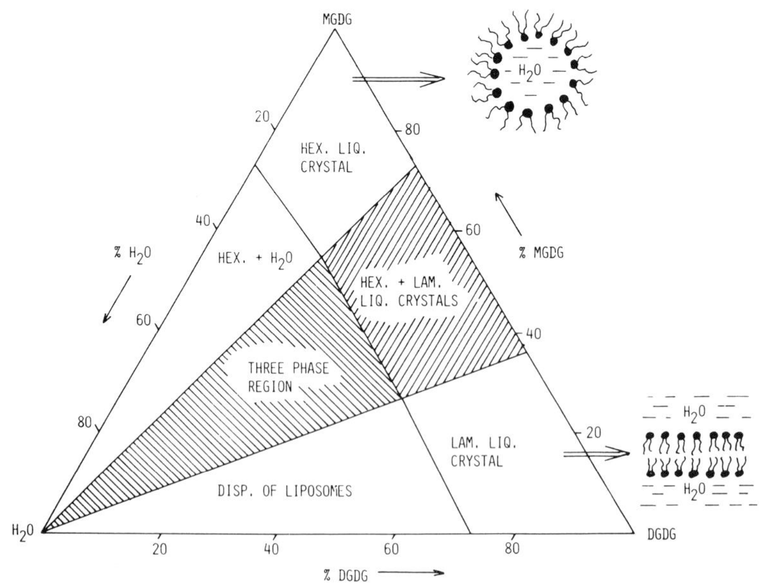

Fig. 11.6 Phase equilibria in the ternary system of MGDG-DGDG-water at 20 °C.

1979). MGDG forms the reverse hexagonal phase with water while DGDG forms the lamellar phase. From the phase diagram it can be seen that when a critical weight ratio of MGDG : DGDG of about 1.5 is exceeded there is a transition from the lamellar to reverse hexagonal phase. As will be discussed later these two phases have quite different functional properties: the lamellar one is crucial for good baking properties. The phase properties are thus sensitive to changes in lipid composition.

The aqueous interaction of all lipids obtained from a wheat flour is shown in Fig. 11.7 (Carlson *et al.*, 1978). The phase behaviour is best illustrated by using a ternary phase diagram where the components are polar lipids, non-polar lipids and water (the polar lipids are the ones which interact with water to form liquid-crystalline phases, contrary to the non-polar ones). In the phase diagram, the lipid composition corresponding to the *Amy* wheat flour used in the extraction is indicated by a broken line from the water corner to a point near the centre of the line between the polar and non-polar lipid corners. The phase equilibrium along this broken line represents

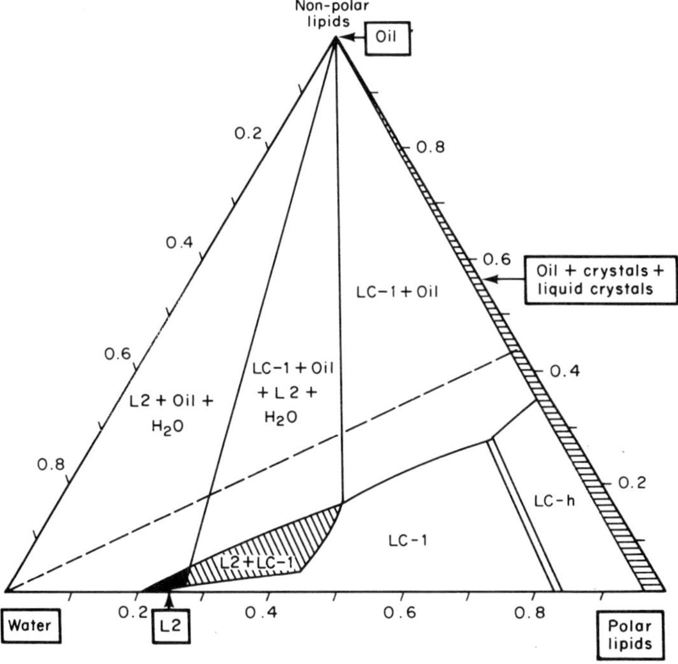

Fig. 11.7 Aqueous interaction of wheat flour lipids (extracted from *Amy* wheat flour by water-saturated butan-1-ol) and water illustrated as phase equilibria at 20 °C of a ternary system.

the equilibrium state of the lipid–water interaction in a wheat flour dough. Equilibrium will not be reached during the dough-mixing process, but this phase equilibria information is also relevant to stages before equilibrium. This will be discussed in the following paragraphs.

If a small amount of water is added to the total lipids (the broken line in Fig. 11.7), the reverse hexagonal phase is formed. At a higher water content the lamellar liquid-crystalline phase is formed, and this phase exists in equilibrium with a non-polar oil and free water. When proceeding to even higher water content the inverse micellar phase (the L2-phase) is formed, and in the region of the phase diagram close to the water corner there are three liquid phases in equilibrium (oil, water and L2). Ultra-centrifugation is needed to separate the phases. After centrifugation the oil phase floats on top, the water phase is found in the middle and the L2-phase at the bottom. The phases are identified by X-ray diffraction, which also gives the dimension of the structural units (Carlson *et al.*, 1978).

The polar lipids of the wheat flour originate from the endosperm cell membranes, and the non-polar ones from the spherosomes of the endosperm, germ and aleurone. During dough-mixing the non-polar lipids will be solubilized and emulsified by polar lipid bilayers (cell membrane residues). Simultaneously, the polar lipids will interact with water to give liquid-crystalline, and finally liquid (L2), lipid-water phases. These successive changes represent the kinetic lipid transitions which can occur during dough-making, and can be compared with the ultimate equilibrium state shown in Fig. 11.7. Figure 11.8 shows the phase diagram of lipids extracted from the corresponding wheat gluten. The general features of the phase diagram are the same, and the small differences can be attributed to the minor amount of starch lipids which occur in the extract of wheat flour (Carlson *et al.*, 1979).

In this connection it is natural to comment upon the frequently used classification into "free" and "bound" lipids, and the changes with water content and dough-mixing.

In dough-mixing studies the terms "free" and "bound" are usually applied to the lipids extracted from *freeze-dried* dough samples. Based on such experiments, a specific wheat protein fraction has recently been shown to be responsible for a significant, if not the major, part of lipid binding activity in dough (Chapter 9). However, this classification of lipids is not applicable to the solvent extraction, or ultra-centrifugation of *wet* dough. In solvent extraction, the proportion of lipids extractable from dough into a non-polar solvent is affected by the partition of the different lipids between the water phase and the organic solvent phase (the partition is influenced by the formation of liquid-crystalline phases and inverse micellar solutions). When wet dough is subjected to ultra-centrifugation,

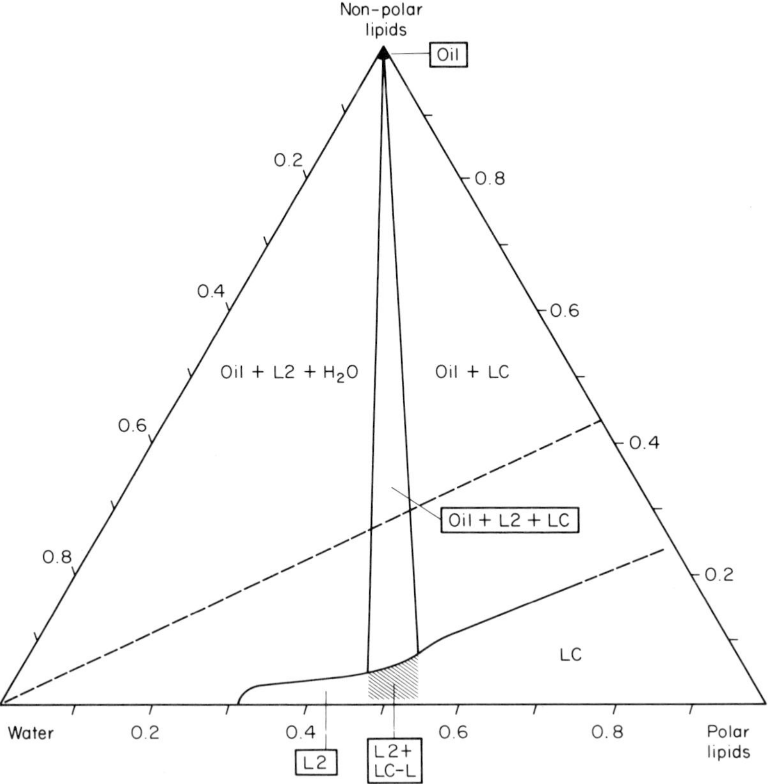

Fig. 11.8 Aqueous interaction of lipids extracted from *Amy* gluten illustrated as a ternary phase diagram at 20 °C.

an oil phase can be recovered; the lipids associated with the remaining fraction will include lipids retained by formation of lipid–water phases. Thus, for aqueous systems the failure to recover lipids by these procedures does not necessarily indicate a physical bonding, or association, of the lipids with protein.

Less is known about the aqueous interaction of lipids from other cereals. The ternary system of rye lipids has been studied (Carlson *et al*., 1980), and the characteristic feature is that there is *no* region where the well-defined lamellar liquid-crystalline phases exist. Most of the diagram exhibits free oil, water and an inverse micellar phase (L2). Only very small aggregates with liquid-crystalline character exist, and they are dispersed in the L2-phase. The phase properties of lipids from triticale are intermediate in relation to those of wheat lipids and of rye lipids (Carlson *et al*., 1980).

V LIPID MONOLAYERS AT THE AIR–WATER INTERFACE

Only insoluble monolayers of polar lipids relevant to cereal products will be considered here.

Lipid molecules orient at the air–water interface, and the spreading pressure of the film provides a force which has an opposite direction to the surface tension of the liquid. The surface tension is thus reduced. The reason for the surface activity of a lipid is the dualistic nature of the molecule, with a hydrophilic head group and a hydrophobic hydrocarbon chain region.

It is possible to record the surface film pressure when a lipid monolayer is compressed at the air–water interface and the monolayer behaviour can be illustrated by a pressure-area isotherm. An example of such an isotherm is shown in Fig. 11.9. This isotherm is characteristic for lipids forming liquid-crystalline phases, such as phosphatidylcholine (lecithin), MGDG or DGDG. At a high molecular area, the film is in the gas state, with quite low surface film pressure, and when a certain surface concentration is reached a liquid state is formed. The molecules in this state have the same type of structure as in the lamellar liquid-crystalline state. Thus at the collapse point they have an area per hydrocarbon chain of about 30 Å^2, which also is the characteristic cross-section area of the lamellar liquid-crystalline phase. After the collapse point, lipid aggregates are formed by the excess of lipids and displaced into the bulk phase. If an excess of lipids is present (forming a liquid-crystalline phase) the pressure will be the same as that

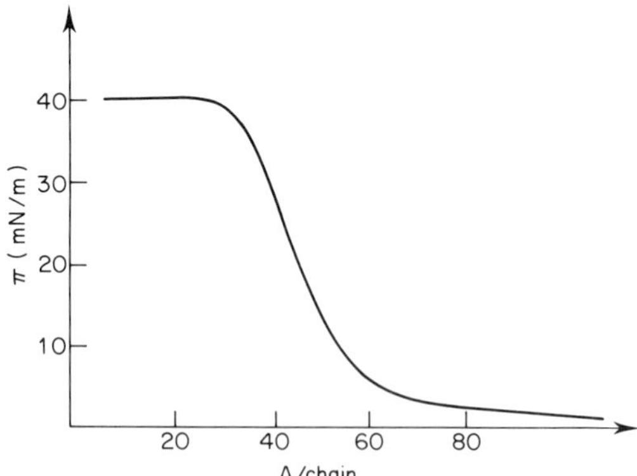

Fig. 11.9 Pressure-area isotherm characteristic of a polar lipid. From the cross-sectional area per hydrocarbon chain of the molecule it can be seen that a maximal surface film pressure of about 40 mN/m is reached when the area per chain is about 30 Å^2.

after collapse (see Fig. 11.9) and this pressure is also termed equilibrium-spreading pressure.

A lipid monolayer can contain a mixture of lipids, provided that there is molecular solubility. Phosphatidylcholine and DGDG, for example, can form such mixed films. Triglycerides or proteins, however, do not form mixed films with polar lipids. The equilibrium-spreading pressures will determine the composition of the film, and the component with the highest equilibrium-spreading pressure will displace the other components from the monolayer (so that maximal reduction in surface energy is achieved).

The equilibrium-spreading pressure for triglycerides is about 15 mN/m, and for proteins it is about 20 mN/m. Carlson (1981) has demonstrated that the equilibrium-spreading pressure obtained from wheat flour or gluten is about 41 mN/m. He has also examined mixed gliadin : DGDG films and showed that there is no molecular solubility of gliadin in the DGDG monolayer. It can therefore be stated that the monolayer at gas cell interfaces in a dough consists of polar lipids in the liquid-condensed state, provided that equilibrium is reached. The question of whether this equilibrium is obtained or not is related to the bulk structure of the polar lipids, and will be discussed in Section VI.

VI WHEAT LIPIDS AND BREAD VOLUME

One critical requirement for the baking performance of a cereal flour is the gas-holding capacity of the dough during fermentation and oven spring. On the basis of the lipid functions discussed in Section V it is possible to explain the gas-holding capacity as an effect due to the polar lipids in the dough.

First we have to consider the bulk structure of the lipids in a dough, how the lipid–water phases are dispersed during dough mixing and then how the lipid monolayers at the surface of the gas cells can be formed.

X-ray data for wheat gluten in different degrees of hydration were first reported by Hess (1954) and later by Traub *et al.* (1957) and by Grosskreutz (1960). In our laboratory we have followed the X-ray low-angle scattering as a function of water content of gluten from different wheat varieties. The scattering curve shows a pronounced maximum, and the peak shape is comparable to that obtained from a liposomal dispersion. No maximum is obtained if the lipids have been removed from the gluten and we have, therefore, reasons to assume that this is due to the lipids, and the spacing of the maximum as the dominating first-order spacing of a liquid-crystalline phase. The X-ray spacing observed for dry gluten is 45 ± 1 Å, when the water content is increased this spacing is progressively shifted towards a maximum value of 55 ± 1 Å. These data are consistent with the lamellar liquid-crystalline phase discussed above. Of the possible phases from the phase diagram (Fig. 11.7), the L2-phase gives a diffuse "liquid" type of X-ray scattering, such that it should not be expected to give any peak in the scattering curve. The hexagonal phase is not spontaneously dispersible; therefore it should not be expected to give small domains and thus line-broadening. The lamellar liquid-crystalline phase, however, spontaneously forms a liposomal dispersion in an aqueous environment, and the size of the particles gives rise to a line broadening consistent with that observed.

The rapid formation of condensed lipid monolayers at the new gas–liquid interfaces, formed during fermentation and during oven spring, means that lipid depots must be available all through the dough structure. The lamellar liquid-crystalline phase in the form of liposomes has unique properties in this respect.

Cereal flours – for example, rye – that do not form a liposomal type of dispersion of the lipids when mixed with water, should not be expected to be able to form a stable lipid monolayer at the surface of gas-cells, thus resulting in low bread volume. Conversely, it should be possible to supply

Fig. 11.10 Bread baked from a 1 : 1 rice-wheat mixed flour (to the right) compared to the corresponding wheat bread. The bread volume of the rice-wheat mixed flour was achieved by using 2% lecithin added as a liposomal dispersion.

such a liposomal dispersion and in that way increase the bread volume. This can be demonstrated from baking tests using rice-wheat flour mixtures. When part of the wheat flour is substituted for rice flour in baking ordinary bread, there is a striking reduction in bread volume, and even at a ratio of rice : wheat of 2 : 8, the bread is hardly acceptable. If a liposomal dispersion of lecithin is added, however, it is possible to obtain a bread with the same volume as a wheat bread up to a ratio of rice : wheat of 1 : 1 (Fig. 11.10). This method was also found to work in sorghum-wheat mixed flour baking (Rajapaksa *et al.*, 1983). The lecithin can be the crude aqueous sludge obtained in the degumming process in the refining of vegetable oils. The requirement for function is the formation of a liposomal dispersion, which can be easily identified in the polarizing microscope (*cf.* Carlson *et al.*, 1978).

Knowledge of the role of lipids in influencing bread volume may also be utilized in plant-breeding. Thus the ratio of (phospholipids + DGDG) to MGDG is directly related to the tendency to form the lamellar liquid-crystalline phase, which in turn favours the formation of a condensed lipid monolayer at the gas-cell interface.

VII LIPID–STARCH INTERACTION

The cereal lipid functionality discussed in Section VI is based on macroscopic association structures. However, the interaction between lipids and starch is based on lipid monomers. The driving force behind this interaction is the hydrophobic effect (*cf.*, Tanford, 1973), which causes the inclusion complex of mono-acyl lipids in amylose in the helical *V*-conformation. Optimal conditions for formation of this amylose complex are directly related to the monomer concentration. The most effective physical state of the lipid is the micellar solution (see Fig. 11.1). The reason is that the monomer concentration in equilibrium with a micellar solution is higher than that in equilibrium with a liquid-crystalline or crystalline phase Furthermore, the kinetics of supplying monomers from micelles to the solution are far superior to those of any other association state of the lipids. The lamellar liquid-crystalline phase is superior to the other liquid-crystalline phases because of its ability to form fine dispersions.

There are mainly two technical effects which can be obtained from formation of the lipid-amylose complex. The first is a reduction in stickiness of a starch gel by "inactivation" of amylose molecules in the continuous aqueous phase, since they form an insoluble precipitate. The other main application of complex formation is to reduce bread-staling. On the basis of

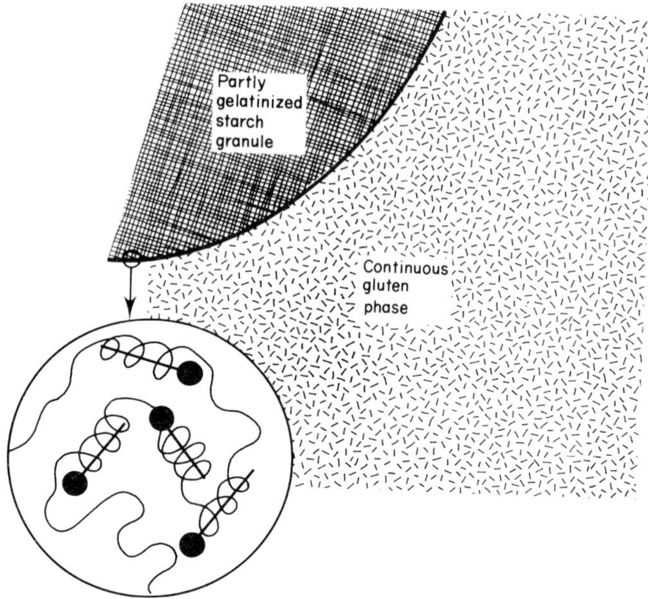

Fig. 11 11 Proposed anti-staling mechanism of mono-acyl lipids in bread. An interfacial film formed by the insoluble amylose-lipid complex acts as barrier against water transport from the gluten matrix to the swollen starch granules.

some recent starch-lipid interaction studies (*cf.* Eliasson *et al.*, 1981), the following mechanism is proposed. An insoluble amylose-lipid film is formed on the surface of the starch granules, in connection with starch gelatinization, when mono-acyl lipid monomers are present to give complex formation. Such a film is water-insoluble provided that it contains sufficient lipid monomers; it must therefore be expected to act as a barrier against water transport involved in the staling process. The proposed effect of the amylose-lipid complex is illustrated in Fig. 11.11 (see also Chapter 8).

REFERENCES

Carlson, T. L-G. (1981). Law and Order in Wheat Flour Dough. Colloidal Aspects of the Wheat Flour Dough and its Lipid and Protein Constituents in Aqueous Media. PhD. Thesis. University of Lund, Sweden.

Carlson, T., Larsson, K. and Miezis, Y. (1978). *Cereal Chem.* **55**, 168–179.

Carlson, T. L-G., Larsson, K. and Poovarodom, S. (1979). *Cereal Chem.* **56**, 417–419.

Carlson, T. L-G., Larsson, K. and Miezis, Y. (1980). *J. Disp. Sci. Techn.* 1, 197–208.
Eliasson, A.-C., Larsson, K. and Miezis, Y. (1981). *Stärke (Starch)* **33**, 231–235.
Grosskreutz, J. C. (1960). *Biochim. Biophys. Acta* **38**, 336–349.
Hess, K. (1954). *Kolloid-Z.* **136**, 84–99.
Larsson, K. (1979). *J. Colloid Interface Sci.* **72**, 152–153.
Larsson, K. and Puang-Ngern, S. (1979). In *Advances in the Biochemistry and Physiology of Plant Lipids* (L.-Å. Appelqvist and C. Liljenberg, eds), 27–33. Elsevier/North-Holland Biomedical Press, Amsterdam.
Larsson, K., Fontell, K. and Krog, N. (1980). *Chem. Phys. Lipids* **27**, 321–328.
Luzzati, V., Mustacchi, H., Skoulios, A. and Husson, F. (1960). *Acta Cryst.* **13**, 660–670.
Pilman, E., Tornberg, E. and Larsson, K. (1980). *J. Disp. Sci. Techn.* **1**, 267–281.
Rajapaksa, D., Eliasson, A.-C. and Larsson, K. (1983). *J. Cereal Science* **1**, 53–61.
Tanford, C. (1973). *The Hydrophobic Effect*, pp. 126–142. J. Wiley & Sons Inc., New York.
Traub, W., Hutchinson, J. B. and Daniels, D. G. H. (1957). *Nature* **179**, 769–770.

12 The Effect of Storage on the Lipids and Breadmaking Properties of Wheat Flour

G. SHEARER and M. J. WARWICK
Ministry of Agriculture, Fisheries and Food, Norwich, U.K.

I INTRODUCTION

The changes which occur in the breadmaking value of wheat flour during storage depend upon several factors. Flour leaving a mill will vary in composition depending on the composition of the grist, on extraction rate and flour grade (Inglett, 1973). It has been established that flours of high extraction rate have lower storage stability (Cuendet *et al*., 1954), as do flours which have been fortified with wheat protein concentrate, wheat germ (Inglett, 1973) or soya flour (Bean, 1977). The moisture content of freshly milled flour is fairly constant and usually in the range 12–15%. Flour kept under unsatisfactory storage conditions will vary from this figure depending on temperature and humidity, and may change in breadmaking quality. Fungal growth occurs at moisture contents above 15% resulting in extensive deterioration in breadmaking quality, hydrolysis of lipids and loss of fatty acids (Daftary and Pomeranz, 1965; Daftary *et al*.,

"Lipids in Cereal Technology"
ISBN 0-12-079020-3

1970). Oxidative deterioration is significant below 8% moisture, resulting in the production of objectionable odours and off-flavours (Halton and Fisher, 1937; Cuendet *et al*., 1954) and loss of breadmaking quality (Sullivan *et al*., 1936; Sullivan, 1940).

Flours of normal moisture content often show a temperature-dependent improvement in baking quality during the initial stages of storage, followed by deterioration when storage is prolonged.

II THE EFFECT OF STORAGE-INDUCED CHANGES ON BREADMAKING QUALITY

One of the features of work on flour storage is the very variable results that have been obtained, often reflecting the different objectives and approaches of the investigators. Greer *et al*. (1954) found that flours stored at 10–20 °C had a storage life of 10 years or so, whereas McCalla *et al*. (1939) found a 3-months storage life at 16–28 °C. Flour at 14% moisture content and at 37.8 °C had a storage life of only three weeks (Cuendet *et al*., 1954). Fisher *et al*. (1937) showed that flour from an English wheat deteriorated over 2 years but after this time regained its quality. Many of the differences can probably be attributed to different flour grades, wheat varieties and methods of assessment, but until recently there has been no systematic examination of these problems. However, it is a consistent observation of baking with stored flour that the doughs tend to be tougher and shorter than usual, but always prove as normal doughs. The difference becomes apparent in the initial phase of baking, when doughs made from stored flour fail to "spring", resulting in a reduced final loaf volume.

Results of baking experiments using stored flours have recently been published by Bell *et al*. (1979a) who collaborated with us in an attempt to define more closely the changes in baking quality and flour components during prolonged storage. Three white flours, of approximately 70% extraction rate, were chosen for these experiments: a weak flour milled from a grist containing English cultivars only; a strong flour milled from Manitoba wheat; and a medium flour which was milled from a 1 : 1 mixture of the grists used for the weak and strong flours. To maximize storage stability, no bleaching agents or improvers were added, and the flours were of 12% moisture content. Test flours were stored at ambient temperatures (mean of 12 °C) while controls were canned under an inert gas atmoshere and stored at −20 °C. Three test-baking procedures were used to follow the changes in quality of the flours during storage, the Chorleywood bread process (CBP), a traditional long fermentation procedure (LFP) and activated dough development (ADD). These were chosen because approximately 75% of bread in the UK is made using the CBP and the other two procedures account for the rest in approximately equal proportions (DHSS, 1981). Since stored flour baked by the CBP gives loaves with characteristics of fat failure (added fat is an essential ingredient using CBP) all flours were baked both with and without bakery fat in all procedures.

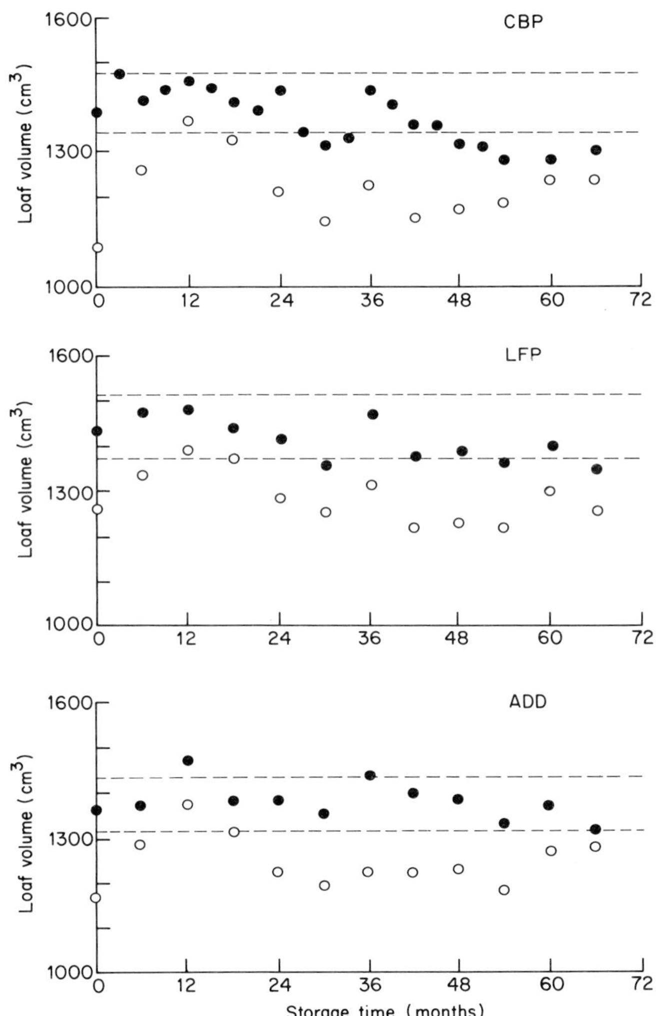

Fig. 12.1 Loaf volume plotted against storage time for the weak flour. ●, with fat; ○, without fat. Values between the dotted lines do not differ significantly from respective control means (from Bell *et al.*, 1979).

As shown in Fig. 12.1, the changes in loaf volume with flour storage time for the weak flour were not markedly different in the three test-baking procedures. However, greater loaf volumes were obtained when bakery fat was included in the recipe (the "fat effect"). When fat was omitted, an initial peak at 12 months was very noticeable and there appeared to be a

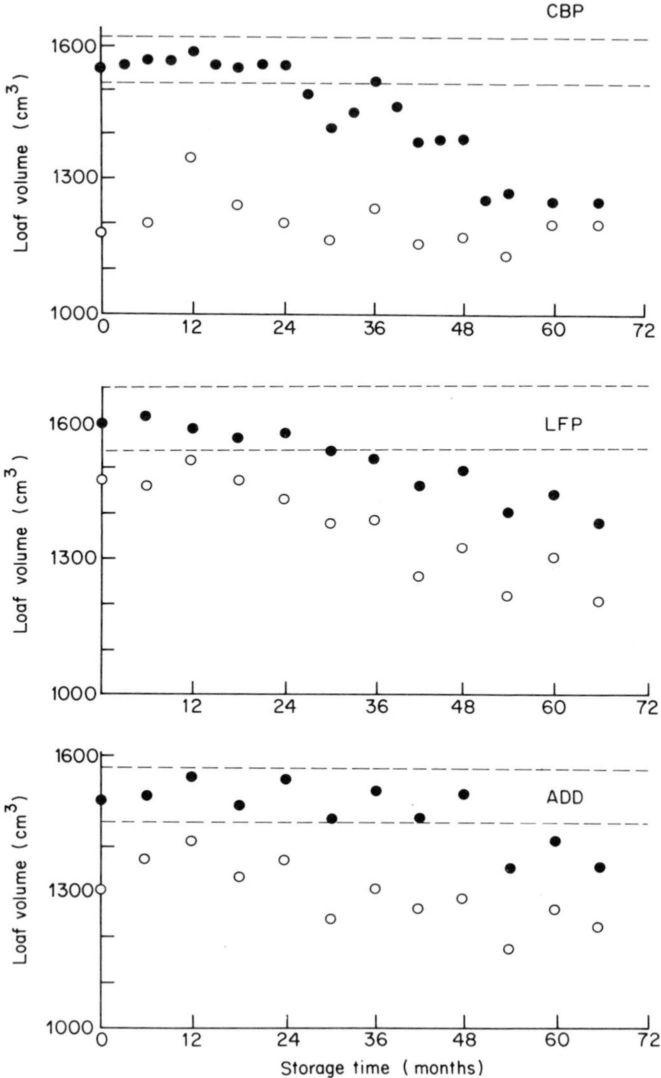

Fig. 12.2 Loaf volume plotted against storage time for the strong flour. Details as for Fig. 12.1.

second increase in loaf volume towards the end of the storage period. The magnitude of the fat effect varied considerably during storage. In strong flour, the decrease in loaf volume with flour storage time was much more pronounced than for weak flour in all three baking procedures (Fig. 12.2). Marked differences between the different baking procedures were observed when fat was omitted. Using the CBP without added fat, loaf volumes were consistently low throughout the storage period and, consequently, the observed fat effect decreased considerably during storage. In contrast, the changes in loaf volume obtained with the LFP and ADD without added fat more closely followed the changes obtained using bakery fat in the recipe. The results obtained with the medium flour using added fat more nearly resembled those for the strong than those for the weak flour. However, the results obtained when fat was omitted more closely resembled those observed with the weak flour.

The importance of fat in the recipe is very dependent on the baking process employed. The CBP has an absolute requirement for fat that has been described by Chamberlain *et al.* (1965). In all combinations of the three flours and the three test-baking procedures, greater loaf volumes were obtained with added fat than without; however, the fat effect was most pronounced for loaves baked by the CBP from unstored flour. The decrease in fat effect during storage was also greatest in the CBP for all three flours. It was concluded from the baking results that three major changes were taking place during storage. First, the baking quality increased during the early stages of storage (maturation). Second, the baking quality subsequently decreased from the level reached at the end of maturation, and third, the fat effect observed in the CBP decreased during flour storage.

III PHYSICAL AND CHEMICAL CHANGES IN WHEAT FLOUR LIPIDS DURING STORAGE AND THEIR POSSIBLE RELATIONSHIP TO CHANGES IN BREADMAKING QUALITY

A Changes in the concentration of individual lipid classes

During the storage of wheat flour, the lipid fraction undergoes extensive changes which, like changes in baking quality, are dependent on flour moisture content, storage temperature and atmosphere and flour type. Increases in the free lipid acidity and free fatty acid (FFA) content of stored flours have been repeatedly demonstrated (Cuendet *et al.*, 1954; Greer *et al.*, 1954; Morrison, 1963; Clayton and Morrison, 1972; Lin *et al.*, 1974; Warwick *et al.*, 1979; Bell *et al.*, 1979a) and are caused by the slow hydrolysis of other lipids, even at 12–14% moisture content. Warwick *et al.* (1979) examined changes in the lipid composition during storage of the weak, medium and strong flours whose baking characteristics were described in Section IIA. Quantitative and qualitative results were comparable to those previously reported for shorter term studies (Clayton and Morrison, 1972; Lin *et al.*, 1974). All results are for the non-starch lipid fraction; no lipolysis of starch lipids during storage has been demonstrated (Morrison, 1976).

The rate of accumulation of FFA in all three flours is rapid at the beginning of storage but gradually decreases thereafter (Fig. 12.3). The concentration of FFA differed between flours at all times and was always in the order weak flour > medium flour > strong flour; this is the same order

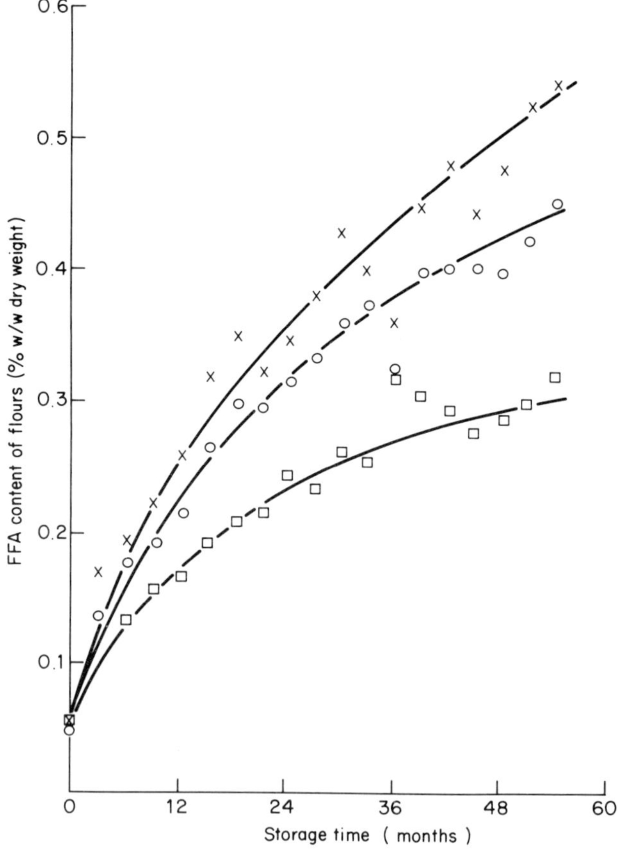

Fig. 12.3 Increase in FFA with flour storage time. ×, weak flour; ○, medium flour; □, strong flour (from Warwick *et al.*, 1979).

as the lipolytic enzyme activities of the flours (Bell *et al.*, 1979a). In Table 12.1 the composition is shown for the other lipids in the medium flour during storage; the lipid composition of the weak and strong flours showed similar trends. The greatest change occurred in the concentration of the triglycerides (TG) fraction, the magnitude of the decrease being sufficient to account for the quantity of FFA accumulated by hydrolysis. The decrease in concentration of galactosyldiglycerides (MGDG, DGDG) was accompanied by increased concentrations of the galactosylmonoglycerides (MGMG, DGMG). Relatively large decreases in the concentration of 1,2-diacylphospholipids were observed with no increases in the corresponding lysophospholipids. From the results of short-term storage experiments, Clayton and Morrison (1972) suggested that phosphatidylcholine, and probably phosphatidylethanolamine, were specifically hydrolyzed at the 2-position while galactosyldiglycerides were apparently randomly hydrolyzed. TG appeared to undergo both random and specific hydrolysis.

Changes occurring in the lipid fraction during long-term storage were not confined to hydrolysis. Farrington *et al.* (1981) showed that in the weak, medium and strong flours described previously, significant acylation of free sterol occurred, giving rise to increased sterol ester concentration after five years' storage. Acylation of free carotenoid was also indicated, but significant differences were not obtained with these compounds. It is probable that this esterification is the result of acyl transferase activity as demonstrated for potato tuber tissue (Galliard and Dennis, 1974).

It is generally accepted that most storage studies on wheat flour have been unable to correlate changes in baking quality with increasing FFA content (Morrison, 1963; Bell *et al.*, 1979a) though there are exceptions (Cuendet, 1954). However, it has been shown that FFA are detrimental to baking quality (de Stephanis and Ponte, 1976). Bell *et al.* (1979b) showed that, using the CBP and added fat, a linear decrease in loaf volume occurred with increasing concentration of added *cis* unsaturated FFA. Both linoleic and oleic acid were equally effective in decreasing volume, so that the effect was not related to either the number of double bonds or to the action of lipoxygenase for which oleic acid is not a substrate. Saturated FFA did not have a deleterious effect on loaf volume.

Daniels and colleagues (D. G. H. Daniels, unpublished results) added increasing concentrations of wheat germ FFA to the control weak, medium and strong flours previously described and determined the effect on loaf volume using the CBP, LFP and ADD baking procedures. The results for the CBP are shown in Fig. 12.4 with loaf volume/FFA relationships of the same flours where the FFA have been generated by lipolysis during storage. For all the control flours in the CBP with both added fatty acid and added fat a linear decrease in loaf volume occurred with increasing FFA.

Table 12.1 Changes in the content of individual lipid classes in medium flour during storage at ambient temperature (mg kg^{-1})†

	Storage time (months)					Storage time (months)			
Lipid class	0	18	36	54	Lipid class	0	18	36	54
TG	6300	3720	2240	1190	DGMG	100	540	520	420
DG	650	900	700	590	APE	800	720	590	420
AMGDG	480	540	570	370	ALPE	470	420	430	370
ASG	240	210	210	160	PC	840	580	450	280
MGDG	640	570	530	420	PE	190	160	120	n.d.§
MGMG	170	260	320	380	LPC + LPE	900	1 060	920	900
DGDG	2460	1970	1830	1330	Total Lipid‡	14 740	14 450	13 230	11 330

†From Warwick *et al.* (1979); based on flour at 13% moisture content.
‡Including FFA.
§Not determined.

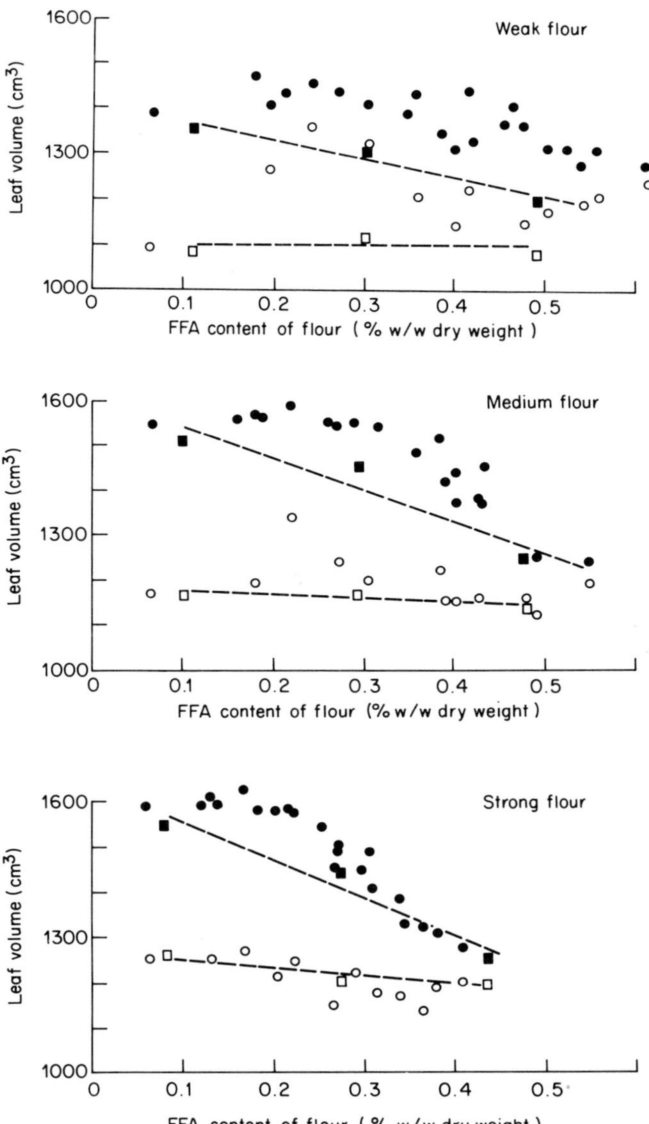

Fig. 12 4 Loaf volume plotted against FFA content using the CBP. ●, stored flour with added fat; ○, stored flour without added fat; ■, control flour with added FFA and added fat; □, control flour with added FFA without added fat (from D. G. H. Daniels, unpublished results).

The decrease was greater in the strong flour than in the weak one. In the absence of added fat, added FFA had no significant effect. These results suggest that in the CBP the decrease in the fat effect may be largely due to the increase in FFA during flour storage. The relationship between CBP loaf volume and FFA content of the stored flours was similar to that between loaf volume and flour storage time. For the LFP and ADD baking procedures, the addition of wheat germ FFA to control flours caused a relatively small decrease in loaf volume with increasing FFA concentration, independent of the presence of added fat. The loss of volume was less than that found for the CBP with added fat and the effect of added FFA was largely independent of flour type.

It is notable that the relationship between loaf volume and FFA concentration for the stored strong flour and medium flour was non-linear in all three baking procedures. This contrasts with the linear relationship between loaf volume and FFA added to the control flours. It is possible that the non-linear relationship found for stored flours is a result of the increased volume due to maturation compensating for decreased volume due to FFA accumulation. After the initial maturation stage of flour storage, the effect on loaf volume of the increasing FFA content is greater than would be predicted from the experiments using control flour with added FFA. Thus, another factor as well as the FFA concentration must be involved in the deterioration of baking quality during prolonged storage.

A major difference between the stored flours and the control flours with added fatty acids is the decreased concentration of other lipid classes in the former. Bell *et al.* (1979a) reported that the effect on loaf volume of added fat does not occur if the flour TG fraction has been removed. It is thus possible that the decrease in flour triglyceride content during storage contributes to the deterioration in baking quality. Certain polar lipids have an important role in baking quality (Chapter 8) and work has shown that addition of digalactosyldiglyceride (DGDG) in particular gives definite increases in loaf volume. The addition of DGDG has been shown to improve the baking quality of stored flour (Bell *et al.*, 1976), and during storage this lipid class also shows the greatest absolute decrease in concentration compared with the other polar lipids. However, the losses of these polar lipids on storage are relatively small and, in the three flours described, the extent of the losses was in reverse order to the decrease in baking quality. The present evidence suggests that increases in FFA and decreases in TG during flour storage contribute to decreased baking quality but that changes in lipid composition alone cannot fully account for the observed changes in quality.

B Oxidation of Flour Lipids during Storage

In a study of flour stored at 15 °C for 6 months, Clayton and Morrison (1972) found no significant decrease in the total lipid content and no change in the fatty acid composition. However, Warwick *et al*. (1979), in the long-term storage trial of the weak and medium flours described found a large decrease in total extractable non-starch lipid after 36 months and a significant decrease in concentration of total fatty acids from 24 months onwards. The decrease in total esterified and unesterified fatty acids was largely accounted for by a fall in the content of linoleic acid, equivalent to 12–15% of the total linoleic acid content of the non-starch lipid fraction, and is a strong indication that significant oxidation had occurred. Further evidence for oxidation is provided by the decrease in carotenoid content of the three flours (Farrington *et al*., 1981). Carotenoids are known to be involved in co-oxidation reactions during the lipoxygenase-catalysed oxidation of linoleic acid (Chapters 6 and 10).

A search for the characteristic, non-volatile oxidation products of lipoxygenase was made in the FFA fractions of the three flours. Two such products were identified as monohydroxyoctadecadienoic acid and dihydroxyoctadecenoic acid (Warwick and Shearer, 1980), both of which have already been identified among the products derived from lipoxygenase action in slurries of wheat flour in water (Graveland, 1973). The rate of accumulation of these acids with flour storage time is shown in Fig. 12.5. The hydroxy acids seem to accumulate approximately linearly with time and the concentrations, which are nearly identical in the weak and medium flours, are twice the level of that in the strong flour. It is notable that the strong flour exhibited less accumulation of hydroxy acids, a lower decrease in total linoleic acid and less FFA formation than did the weak and medium flours. The total hydroxy acid concentration, accumulated over the storage period, does not account for more than one-third of the amount of total linoleic acid lost during the same period. The difference between linoleic acid lost and hydroxy acids formed was not accounted for by the hydroperoxides present, nor were any oxidation products of acyl esters detected. The formation of hydroperoxides that decomposed to undetected products may be responsible for the discrepancy. Although this evidence is fully in accord with lipoxygenase action, it does not exclude the possibility of autoxidation; in fact, it is probable that a combination of both was responsible for the observed loss of polyenoic acid. The strong flour, which showed the least improvement in loaf volume during the early stages of storage (maturation), also exhibited the lowest accumulation of hydroxy acids, suggesting a possible relationship between lipid oxidation and the

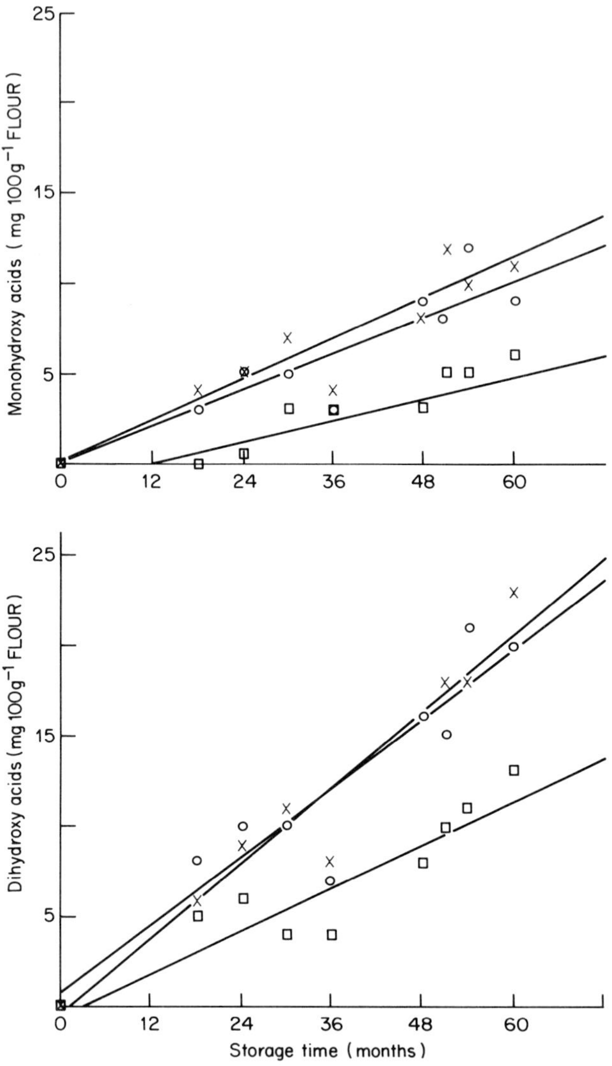

Fig. 12.5 Increase in concentration of hydroxy acids during flour storage. ×, weak flour; ○, medium flour; □, strong flour (Warwick and Shearer, 1980).

maturation effect. However, the major proportion of the hydroxy acids accumulated after the maturation effect had reached a maximum.

A further possibility is that lipid oxidation may be involved in the decrease in baking quality which takes place on prolonged storage, subsequent to the maturation stage. Oxidized lipids have been shown to cause deterioration in baking quality (Sullivan *et al.*, 1936; Sullivan, 1940) and could, therefore, augment the deleterious effect of FFA formation and loss of acyl lipids. However, in the experiments previously described the greatest decrease in baking quality during storage was found for the strong flour, but this flour also exhibited the lowest accumulation of hydroxy acids. It would thus be necessary to postulate that the baking quality of the strong flour was much more sensitive to the presence of oxidized lipids than was that of the weak flour. It is apparent that although lipid oxidation during flour storage has been demonstrated, there is as yet no definite evidence that lipid oxidation plays a role in the changing baking quality.

C Effect of Flour Storage on Lipid-Protein Interaction during Dough-Mixing

Storage-deteriorated flour shows increased "binding" of flour lipids and added fat when a dough is made (see Chapter 9 for a discussion of lipid binding). It has been established that the addition of oleic and linoleic acids to fresh flour also increases lipid binding, and the acids themselves become bound (Bell *et al.*, 1979b). The increased binding of model fat consisting of glyceryl trioleate (GTO) and glyceryl tristearate (GTS) was found to be due entirely to increased binding of GTO, whereas the saturated TG component (GTS) was not significantly affected. It was concluded that the adverse effect of unsaturated fatty acids was likely to be due mainly to their influence on the association of other lipids in the dough with non-lipid flour components.

In the previous studies, the extent of lipid binding was assessed on the basis of extraction into polar and non-polar solvents. More recently, an aqueous fractionation procedure has been applied to the study of lipid-protein associations during dough mixing (Warwick and Shearer, 1982). The results suggest that a proportion of the saturated TG component of the added fat becomes associated with protein, and that this association is affected by the addition of unsaturated FFA. It thus appears possible that the relatively high concentrations of FFA in flours after prolonged storage may exert an adverse influence on baking quality through the effect of FFA on lipid-protein associations in dough mixing.

ACKNOWLEDGEMENTS

The authors would like to thank staff at FMBRA particularly Miss B. Bell, Dr D. G. H. Daniels and Dr N. Fisher for their collaboration during this work, for valuable discussion of the data and providing unpublished work for inclusion in this chapter.

REFERENCES

Bean, M. M., Hanamolo, M. M., Nishita, K. D., Mecham, D. K. and Fellers, D. A. (1977). *Cereal Chem.* **54**, 1159–1170.

Bell, B. M., Chamberlain, N., Daniels, D. G. H. and Fisher, N. (1976). Flour Milling and Baking Research Association, Research Report No. 70, Chorleywood, UK.

Bell, B. M., Chamberlain, N., Hylton Collins, T., Daniels, D. G. H. and Fisher, N. (1979a). *J. Sci. Food Agric.* **30**, 1111–1122.

Bell, B. M., Daniels, D. G. H. and Fisher, N. (1979b). *J. Sci. Food Agric.* **30**, 1123–1130.

Bennett, R. and Coppock, J. B. M. (1957). *J. Sci. Food Agric.* **8**, 261–270.

Chamberlain, N., Collins, T. H. and Elton, G. A. H. (1965). *Cereal Sci. Today,* **10**, 415–419, 490.

Clayton, T. A. and Morrison, W. R. (1972). *J. Sci. Food Agric.* **23**, 721–736.

Cuendet, L. S., Larson, E., Norris, C. G. and Geddes, W. F. (1954). *Cereal Chem.* **31**, 363–389.

Daftary, R. D., and Pomeranz, Y. (1965). *J. Agric. Food Chem.* **13**, 442–446.

Daftary, R. D., Pomeranz, Y. and Sauer, D. B. (1970). *J. Agric. Food Chem.* **18**, 613–616.

Department of Health and Social Security (UK) (1981). Report of the Panel on Bread, Flour and Cereal Products Committee on Medical Aspects of Food Policy.

Farrington, W. H. H., Warwick, M. J. and Shearer, G. (1981). *J. Sci. Food Agric.* **32**, 948–950.

Fisher, E. A., Halton, P. and Carter, R. H. (1937), *Cereal Chem.* **14**, 135–141.

Galliard, T. and Dennis, W. (1974). *Phytochemistry.* **13**, 1731–1735.

Graveland, A. (1973). *Lipids.* **8**, 599–605.

Greer, E. N., Jones, C. R. and Moran, T. (1954). *Cereal Chem.* **31**, 439–450.

Halton, P. and Fisher, E. A. (1937). *Cereal Chem.* **14**, 267–275.

Inglett, G. E. and Anderson, R. A. (1973). In *Wheat: Production and Utilization* (G. E. Inglett, ed.), AVI, Connecticut.

Lin, M. J. Y. L., Youngs, V. L. and D'Appolonia, B. L. (1974). *Cereal Chem.* **51**, 17–33.

McCalla, A. G., McCaig, J. D. and Paul, A. D. (1939). *Can. J. Res.* **17C**, 452–459.

Mecham, D. K. (1971). In *Wheat: Chemistry and Technology* (Y. Pomeranz, ed.). American Association of Cereal Chemists, St. Paul, Minnesota.

Morrison, W. R. (1963). *J. Sci. Food Agric.* **14**, 870–873.

Morrison, W. R. (1976), *Bakers Digest.* **50**, 29–36, 47–48.

Pomeranz, Y. (1974). In *Storage of Cereal Grains and Their Products* (C. M. Christensen, ed.), pp. 56–114, 2nd ed., American Association of Cereal Chemists, St. Paul, Minnesota.

de Stephanis, V. A. and Ponte, J. G. (1976). *Cereal Chem.* **53**, 636–642.

Sullivan, B., Near, C. and Foley, G. H. (1936). *Cereal Chem.* **13**, 318–331.

Sullivan, B. (1940). *Cereal Chem.* **17**, 661–668.
Warwick, M. J., Farrington, W. H. H. and Shearer, G. (1979). *J. Sci. Food Agric.* **30**, 1131–1138.
Warwick, M. J. and Shearer, G. (1980). *J. Sci. Food Agric.* **31**, 316–318.
Warwick, M. J. and Shearer, G. (1982). *J. Sci. Food Agric.* **33**, 918–924.

13 Lipids in Pasta and Pasta Processing

B. LAIGNELET

Laboratoire de Technologie des Céréales, INRA, Montpellier, France

I INTRODUCTION

The lipids of pasta are a quantitatively minor component compared with proteins and carbohydrates and were not extensively studied in the earlier investigations. More recently, there has been an increased interest in the lipids and pigments involved in pasta quality. In this chapter the term "pasta" will be restricted to products made from durum wheat semolina or bread wheat flour or a mixture of these two raw materials (with the inclusion of egg when specified). Pasta products made from gluten-free cereals such as rice, maize and sorghum will not be considered. Canneloni, ravioli and canned pasta are also excluded.

"Lipids in Cereal Technology"
ISBN 0-12-079020-3

Pasta is made by mixing semolina and water, with or without the addition of egg, to give a very stiff dough of 50% water content. The dough is forced into the extrusion head by a kneading worm at pressures as high as 100 kg cm^{-2} and passes through a die which is chosen to give the type of pasta product desired. At this stage, vacuum may be applied to the product to remove occluded air bubbles and to minimize oxidation. The pasta is finally dried to a moisture content of 14%; the drying stage is a critical and time-consuming operation which must be carefully controlled to prevent deterioration of the final pasta quality. High-temperature drying may be used, with air temperatures of 70 °C or more, to prevent or minimize enzymatic reactions.

The aspects of pasta quality that will be considered are composition, appearance and cooking quality. According to Feillet (1977) the minimum requirements are that the surface of the raw pasta is smooth and brilliant, free from cracks and free from black or white specks; a yellow to amber colour is generally preferred by consumers rather than brown or cream. The cooking time must be short, with low cooking losses and a high degree of swelling. The cooked pasta must be free from stickiness and retain firmness and elasticity even after cooking. In terms of composition and nutritional quality, the pasta should retain the basic nutritional factors present in the wheat endosperm.

II LIPIDS OF TECHNOLOGICAL AND NUTRITIONAL IMPORTANCE IN PASTA PRODUCTS

A higher lipid content is found in durum wheat and durum semolina compared with bread wheat and bread flour respectively, although the fatty acid (FA) compositions are similar. Typical values currently accepted for total lipid content would be 33 g kg^{-1} for whole durum grain and 18 g kg^{-1} for the semolina (Weihrauch and Matthews, 1977). The content of polyunsaturated FA and tocols of the wheat grain is of nutritional significance but the tocol content is influenced by the growing conditions (Davis *et al.*, 1980).

The addition of egg to semolina results in a pasta with a much higher lipid content and a different FA composition (Table 13.1). The content of polyunsaturated FA (18 : 2 + 18 : 3) increases but the greatest increase is shown in the concentration of 16 : 0 and 18 : 1 (Weihrauch and Matthews, 1977). The egg also provides an important increase in vitamin A content (Matthews and Workman, 1977).

Table 13.1 Fatty acid composition of macaroni and noodles†

Fatty acid	Fatty acid content (g kg^{-1})‡	
	Macaroni§	Egg noodles§
16 : 0	2	8
18 : 0	1	2
18 : 1	2	11
18 : 2	6	10
18 : 3	< 1	1

†Modified from Weihrauch and Matthews (1977).
‡Expressed as fatty acid methyl esters after transesterification.
§Total lipid content of macaroni and noodles was 15 and 46 g kg^{-1} respectively.

A Acyl Lipids

There are relatively few published studies of the effects of acyl lipids on pasta quality. Brokaw *et al.* (1960) noted an improvement in the quality of cooked spaghetti caused by the addition of lipids such as monoglycerides (MG); the effect may be related to the ability of MG to form a complex with the amylose fraction of starch. According to Thoren (1972), the addition of 2% of MG hydrate prevented stickiness and increased the resistance of the pasta to overcooking. The formation of a complex between MG and starch delayed gelatinization with the result that cooking time was not a critical point. The commercial use of MG and diglycerides is the subject of many patents especially in the manufacture of Japanese noodles.

To study the significance of the endogenous semolina lipids and flour lipids, Dahle and Muenchow (1968) extracted the lipids and proteins from raw spaghetti and observed the effects after cooking. Removing the lipids alone resulted in increased stickiness after cooking, while removing both lipids and proteins resulted in greater leaching of amylose into the cooking water and decreased firmness of the cooked pasta in addition to the increased stickiness. It was concluded that the proteins were essential for retention of high cooking quality but that the lipids improved the functional properties of the proteins and minimized the secondary cooking effects such as stickiness. The lipids were not analysed but probably comprised the major proportion of the non-starch lipids.

Laignelet (1979a) showed that the firmness of cooked pasta was slightly decreased when the semolina had been defatted with light petroleum, but was considerably increased when the water-soluble fraction had been extracted from the semolina (Table 13.2). Defatting the semolina which

Table 13.2 Effect of light petroleum (LP)-soluble fraction and water-soluble fraction of semolina on the quality of macaroni†

Characteristic of macaroni	Fraction removed from semolina			
	None	LP-soluble	Water-soluble	LP/water-soluble
Yellow index	16.25	14.10	16.65	14.35
Swelling (%)				
at 15 min	180	185	180	175
at 25 min	225	240	235	225
Firmness (kg)				
at 15 min	3.6	3.3	5.0	5.0
at 25 min	2.3	2.2	3.1	3.3

†Laignelet (1979a).

already had the water-soluble fraction removed had little effect on firmness and decreased the extent of swelling in the cooked pasta. It appears that the functional properties of the petroleum-soluble lipids could be related to the presence of water-soluble components.

Semolina lipids have been extracted, fractionated and used in reconstitution experiments (Lin *et al.,* 1974). Defatting the semolina changed the colour by the removal of carotenoid pigments but the colour of petroleum-defatted semolina was restored by adding the non-polar lipid (NL) fraction to a level of 0.6%. Water absorption of the untreated semolina and firmness of the spaghetti were slightly increased by the addition of NL and monogalactosyldiglyceride and decreased by phospholipids and digalactosyldiglyceride. Unlike that of Dahle and Muenchow (1968), the report by Lin *et al*. (1974) does not consider cooking losses and change in stickiness, and concludes that neither polar lipids nor NL were factors of great importance in cooking quality. However, it is difficult to make a direct comparison of the two studies because Lin *et al*. used light petroleum to defat semolina, whereas Dahle and Muenchow used the much more polar water-saturated butan-1-ol and considered the interaction between lipids and proteins. Extraction with light petroleum would have less effect on gluten proteins and this may explain why Lin *et al*. observed less change in pasta cooking quality.

B Carotenoids

Although present at relatively low concentration, the carotenoid pigments are important because they are the major factor determining the yellow colour of pasta, and thus influence acceptability to the consumer. The general aspects of carotenoids in cereal grains are discussed in Chapter 3.

1 *Nature of the Carotenoid Pigments*

Historically, carotene has been believed to be the major yellow carotenoid pigment and, for this reason, the carotenoid content has usually been expressed in carotene units. However, Markley and Bailey (1935) showed that the predominant carotenoid pigments in the extracts of ground durum wheat grains are xanthophylls, and Munsey (1938) confirmed that durum wheat grains contain only a small proportion of carotene but a relatively large quantity of xanthophylls (see Chapter 3). According to Zechmeister and Cholnoky (1940), lutein represented most of the xanthophyll present in wheat flour, the other carotenoid pigments being unidentified isomers.

It is now agreed that β-carotene is the main carotene isomer present in wheat grain as shown by Chen and Geddes (1945), and later confirmed by other workers (Bouguerra, 1964; Wildfeuer and Acker, 1968; Tykheeva *et al.*, 1971; Farre-Rovira and Costes, 1974). The presence of α-carotene has occasionally been reported (Manunta, 1946b; Bouguerra, 1964).

The major carotenoid pigment in the wheat grain is the xanthophyll lutein which occurs unesterified or as mono- and diacyl esters (Chapter 2) and represents 90% of the total carotenoids. The relative proportions of unesterified lutein and lutein esters vary between different types of wheat (Table 13.3). Improvements in the methods for carotenoid analysis has permitted the detection of other xanthophylls in flour and semolina, including triticoxanthin (Manunta, 1946b), taraxanthin (Irvine and Anderson, 1949), flavoxanthin (Bouguerra, 1964; Farre-Rovira and Costes, 1974) and canthaxanthin (Wildfeuer and Acker, 1968). In addition, phytofluene, lycopene and violaxanthin have been suggested to be present (Farre-Rovira and Costes, 1974). It has not proved possible to identify a

Table 13.3 Lutein composition of wheat grain carotenoids†

	Percentage composition of total lutein fraction	
	Durum wheat	Hard red spring wheat
Unesterified lutein	84.8	21.6
Lutein monoester	9.8	46.5
Lutein diester	5.3	31.9

†Lepage and Sims (1968).

carotenoid pigment specific for durum wheat or bread wheat in spite of the technical and commercial significance of distinguishing between products from the two species.

2 *Factors affecting Carotenoid Content*

A number of factors influence the carotenoid content of wheat grain, especially the species, variety and growing conditions. Durum wheats are generally higher in carotenoid content than bread wheats; for example, durum flour and bread flour contained 3.7 and 2.8 mg kg^{-1} respectively (Lepage and Sims, 1968). In Australian bread wheat flours the carotenoid content varied from 1.45 to 4.3 mg kg^{-1} (Moss, 1967), whereas for the semolina of eight American durum varieties it ranged from 4.4 to 6.9 mg kg^{-1} (Irvine and Anderson, 1953).

Irvine and Anderson (1953) have demonstrated a highly significant effect of the variety and the growing location on carotenoid content of durum grain; this has been confirmed by Grignac (1970) who found even greater variation among varieties grown in France (Table 13.4). Recently it has been shown that grains of winter-sown wheat are lower in carotenoid content than those of spring-sown varieties (Vdovkin *et al.*, 1980).

It is well established that high carotenoid content has a high degree of

Table 13.4 Variation in carotenoid content of durum semolina and pasta according to the variety and the stage of processing†

	Carotenoid content (mg kg^{-1})		
Variety	semolina	pasta	Percentage loss
Mandon	10.1	7.0	30.7
Adur	10.0	7.1	29.0
Lakota	8.4	6.4	23.8
Leeds	8.3	6.3	24.1
Sentry	8.4	6.4	23.8
Wells	8.2	6.1	25.6
Steward 63	8.2	5.8	29.3
Nugget	8.2	6.0	26.8
Montferrier	8.4	2.1	75.0
Bidi 17	6.1	1.9	68.9
Oued Zenati 368	5.8	1.8	69.0
Chili 931	5.7	2.0	64.9
Candeal	5.5	1.9	34.5
Lez	4.8	1.7	35.4
Alcala la real	3.3	–	–

†Modified from Grignac (1970).

heritability (Clark and Smith 1925; Worzella and Cutler, 1935; Braaten *et al.*, 1962; Grignac, 1970). In a study of eight Italian bread wheats, it was considered that the total carotenoid content is a genetic characteristic independent of growing location, but that the ratio of carotene content to xanthophyll content varies with the location (Manunta, 1946a). The results of experiments in Tunisia exhibited negative correlations between carotenoid content and grain yield, and between carotenoid content and thousand grain weight; it is possible that only a fixed quantity of carotenoids can be synthesized in the grain regardless of the rate of formation of dry matter (Ferchat and Bouguerra, 1962). Variations have been observed in the nature of the grain pigments according to the ripening conditions (Bouguerra, 1964). Total carotenoid content can be increased by the application of chlorophenylthiotriethylamine (CPTA) one week after anthesis (Lier and Lacroix, 1974); the lutein content was not changed and the increase was mainly due to a greater quantity of carotene (Lacroix and Lier, 1975).

Climatic and pathological effects have a strong influence over semolina colour. Heavy rains during ripening result in grain which appears more white and less vitreous (Grignac, 1965) and pigment loss can occur if the grain begins to germinate or is infected with fungus before harvesting (Braaten *et al.*, 1962).

III CHANGES IN LIPID COMPOSITION DURING PASTA PROCESSING

Since colour is an important aspect of pasta quality the changes in carotenoid composition during processing have been studied in detail and reviewed previously (Irvine, 1971; Laignelet *et al.*, 1972). In contrast, the behaviour of acyl lipids during processing has received less attention.

A Acyl Lipids

Burini and Damiani (1978) reported the composition and distribution of FA in the triglycerides of flour streams from the milling of durum wheat, but gave no indication of the total lipid content of the streams. Recently, Ramirez Rico and Laignelet (unpublished results) have studied the lipid composition of flour streams from a pilot-scale mill at the author's institute in Montpellier. A French durum wheat (variety Kidur) was milled to yield six semolina fractions, four break roll flours and four reduction roll flours. The lipid content was found to vary widely among the mill streams and was

Table 13.5 Lipid content and fatty acid composition of durum wheat milling fractions (variety Kidur)†

Milling fraction	Ash (% dry weight)	Lipid content‡ (g kg^{-1} dry weight)	Fatty acid composition (%)				
			16 : 0	18 : 0	18 : 1	18 : 2	18 : 3
Semolina							
S 1	0.80	13.0	18.6	0.6	16.1	62.8	1.9
S 2	0.95	14.8	15.9	0.6	16.8	64.7	2.0
S 3	0.72	9.9	19.6	0.8	13.7	63.4	2.5
S 4	0.86	15.2	17.3	0.7	15.9	63.7	2.4
S 5	1.21	16.0	16.9	0.6	16.8	63.2	2.6
S 6	1.36	11.0	16.2	0.7	18.0	62.4	2.8
Flour: break rolls							
B 1	1.93	14.4	14.5	0.8	19.7	62.5	2.5
B 2	1.96	19.4	15.9	0.8	19.3	61.6	2.5
B 3	2.33	31.9	15.4	0.7	20.5	60.9	2.5
B 4	2.96	37.2	14.7	0.6	20.2	61.7	2.7
Flour: reduction rolls							
D 1	1.78	19.4	16.1	0.9	20.2	59.4	3.4
D 2	2.24	26.9	15.8	0.8	20.7	60.1	2.6
D 3	2.45	31.0	15.1	0.7	20.8	60.7	2.7
D 4	2.85	35.0	15.9	0.8	20.4	60.3	2.7

†C. Ramirez Rico and B. Laignelet (unpublished results).
‡Expressed as fatty acid methyl esters after transesterification.

correlated with the ash content (Table 13.5), which provides an indication of purity of the streams. Among the semolina fractions, those with the higher purity had the greater proportion of 16 : 0 in the FA and the lower proportion of 18 : 1. Fractions S1 and S3 (Table 13.5) are the purest fractions and may have a relatively high polar lipid content; these two fractions yield pasta products with the lowest cooking index (Houliaropoulos *et al.*, 1981). Similar results were found by Dahle and Muenchow (1968) and Lin *et al.* (1974).

There is less known about the changes in lipid composition during pasta processing than during breadmaking. The content of acetone-extractable triglycerides and phospholipids, the yield of acetone-extractable lipid and the acidity were all lower on a dry weight basis in spaghetti or farfalle compared with the corresponding semolina (Fabriani *et al.*, 1968). It was suggested that the difference was due to a chemical change resulting in the loss of a proportion of the lipids, or to binding or complexing of the lipids with another component of the dough. However, no data were given for total lipid content and it was impossible to determine whether the changes represented destruction of lipid or decreased extractability.

More recently, Barnes *et al.* (1981) showed that there was no significant decrease in total lipid content during pasta production but that the most notable change occurred in the extractability of the lipids. About 50% of the lipids in the semolina were extractable in hexane (free lipids) compared with 10% in the dry pasta; the remaining lipids required extraction with water-saturated butan-1-ol (bound lipids). The change took place mainly during the pre-drying stage, which is when most of the freezable water is lost from the pasta, but there is no definite evidence to show that the change in lipid binding and the disappearance of freezable water are related. There is also no information to show whether lipid binding influences processing characteristics or product quality.

During storage of pasta there is an increase in acidity and this is more pronounced in pasta products containing egg (Fabriani and Quaglia, 1973; Yakovenko and Antyukhova, 1977). It has been suggested that wheat phospholipase D is involved in the degradation of phospholipids in stored egg pasta products (Nolte *et al.*, 1974).

B Carotenoids

1 *Production of Semolina*

The destruction of carotenoids may begin during grain storage, then increase further during and after milling when the cellular structure has been disrupted. Abnormal flavours in semolina may also arise in parallel with the loss of carotenoids. These changes are generally influenced by temperature and moisture and by the presence of micro-organisms during storage (Drapron and Uzzan, 1968). The method of drying grain after harvest has been shown to influence fat acidity, lipoxygenase activity and carotenoid content (Lasseran and Barthelemy, 1968).

Although there was no major variation in the pigment content between different mill streams, the extent of pigment loss during pasta making was related to the ash content of the semolina (Matsuo and Dexter, 1980). The loss of pigment was greater in the semolina streams having a higher ash content. Heat treatment of semolina at temperatures between 45 ° and 70 °C lowered the lipoxygenase activity, decreased pigment loss and maintained the desirable yellow colour of the pasta (Menger *et al.*, 1969). The low thermal stability of durum wheat lipoxygenase has recently been confirmed (MacDonald, 1979).

2 *Pasta production*

There have been many studies on the destruction of carotenoids during the pasta-making process and it is known that the extent of carotenoid loss is influenced by wheat variety, ranging from 30% up to as much as 60% or more in some varieties (Irvine and Anderson, 1953; Schettino and Di Lieto, 1968; Grignac, 1970; Laignelet, 1979a; MacDonald, 1979). It is even possible to differentiate between certain types of durum wheat according to the extent of carotenoid loss; for instance, American varieties exhibit a smaller decrease in pigment content during processing than do varieties from the Mediterranean region (Laignelet, 1979b).

Irvine and Winkler (1950) and Irvine and Anderson (1953) considered that the destruction of carotenoids was due to coupled oxidation reactions catalysed by the lipoxygenase present in semolina. Later studies have confirmed that the lipoxygenase activity of semolina is involved in pigment destruction (Matsuo *et al.,* 1970; Walsh *et al.,* 1970), although Burov (1974) and Burov *et al.,* (1973, 1974) have suggested that carotenoid pigments became bound to gluten and, in this condition, were not susceptible to lipoxygenase-catalysed co-oxidation. In the latter context, it is notable that the carotenoid pigment in egg pasta products, derived mainly from the egg, would not be oxidized by this mechanism, according to Fernandes *et al.* (1979).

Nevertheless, it is known that lipoxygenase is involved in carotenoid oxidation in bread dough (see Chapter 10) and the early studies of Irvine and Winkler (1950) showed that the addition of lipoxygenase led to destruction of pigments during pasta making. The loss of pigment was greater with unesterified 18 : 2 and 18 : 3 FA as substrates than with esters of these acids; the rate of pigment loss with the ester substrates increased with time, possibly due to lipase-catalysed hydrolysis of the esters to yield unesterified FA (Matsuo *et al*., 1970). Some attention is now being given to the formation of FA hydroperoxides by lipoxygenase activity in durum wheat; MacDonald (1979) was unable to demonstrate linoleic acid hydroperoxide isomerase activity; Biermann *et al.* (1980) suspected the presence of lipoperoxidase activity, lower than that in bread wheat, leading to formation of hydroxy FA. It is now thought that lipoxygenase activity is necessary but by itself it is not sufficient to catalyse oxidation of the carotenoid pigments (Buré, 1974; Laignelet, 1979a).

Irvine and Winkler (1950) suggested that three stages could be distinguished in the destruction of carotenoids during processing. First, there was a rapid degradation during the first minute of mixing that resulted from the addition of water to the semolina. It did not require atmospheric oxygen and sufficient oxygen was already present in the mixture, even under a nitrogen atmosphere. Second, a slower rate of degradation occurred from the second minute of mixing and the introduction of molecular oxygen was obligatory at this stage. Finally, the degradation continued at a very slow but still significant rate. The destruction of pigments also continued during the drying stage, and a further 20% of the pigments may thus be lost (Matsuo *et al.,* 1970). The use of relatively high temperatures for drying might limit pigment degradation but promotes the onset of browning (Laignelet, 1976).

3 *Prevention of Pigment Destruction during Pasta Manufacture*

Many investigations have been carried out to find methods of preventing bleaching of the yellow colour during pasta manufacture. Both physical and chemical methods are available.

The most common physical approach is to use vacuum, which removes the molecular oxygen necessary for extensive oxidation and also prevents air bubble formation in the dough (Irvine, 1971; Nazarov *et al.,* 1971; Nazarov and Egorova, 1972). Nazarov *et al.* (1971) have shown that the use of vacuum is more efficient at preventing carotenoid oxidation than is the addition of chemical inhibitors of oxidation. Burov *et al.* (1974) claimed that vacuum had no effect on pigment loss based on their belief that lipoxygenase is not involved in carotenoid oxidation.

One of the chemical methods for avoiding the problem of low pigment content in pasta is to add purified carotenoid pigments during dough-mixing (Counsell and Webb, 1971; Kläui, 1976; Kläui and Raunhardt, 1976; Burov *et al.,* 1979). However, the added pigments might also be oxidized and, for an equivalent level of pigment addition, it was found that it was better to use maize gluten feeds with a high carotenoid content; it is possible that the binding of pigments to protein in these fractions minimized the extent of carotenoid oxidation (B. Laignelet, unpublished results). In this context, it should be noted that the carotenoid composition of maize grain is different from that of wheat grain; maize contains a major proportion of a second xanthophyll, zeaxanthin, in addition to the lutein of wheat (see Chapter 3).

A second chemical approach to avoiding pigment loss is the use of chemical inhibitors of the oxidation. In tests carried out by Irvine and Winkler (1950), the activity of chemical inhibitors varied according to the stage during the mixing process as already explained above. In practise, a number of substances have been shown to limit pigment degradation, including sodium chloride (Menger, 1965), ascorbic acid at low concentration (Irvine, 1963; Menger, 1965; Walsh *et al.,* 1970), phospholipids and butyl-hydroxyanisole (BHA) (Nazarov *et al.*, 1971), cysteine (Menger, 1972), propylene glycol monostearate and monoglycerol stearate (Nazarov *et al.,* 1972) and modified maize starch and tannol or both (Nazarov *et al.,* 1973, 1974). Sodium chloride also inhibits pasta browning (Menger, 1965) and cysteine improves the cooked pasta consistency (Menger, 1972), but ascorbic acid at high concentrations promotes browning (Dahle, 1970). Walsh *et al.* (1970) have shown that ascorbic acid is a competitive inhibitor of lipoxygenase.

From the information available it appears that the physical methods for preventing pigment degradation are probably superior to the chemical methods. However, to obtain high pigment content in the pasta it is also desirable that the best raw materials are chosen and that the wheat should be genetically improved to enhance the carotenoid content.

IV THE LIPID ASPECTS OF QUALITY CONTROL

Lipid analysis has been and still is used in some laboratories to determine the colour, egg content or bread wheat content of pasta products. However, the lipid methods are only a few among the many others used for quality control.

A Colour Determination

Since the yellow colour of pasta depends on the content of carotenoid pigments, many investigations have been carried out on the determination of pigment content as an indicator of yellow colour.

Since the work of Fifield *et al.* (1937) it has been known that determining raw material carotenoid content alone is not enough to predict the colour of the final pasta and therefore lipoxygenase activity must also be measured. Several prediction equations have been published, the most recent is that of MacDonald (1979). According to Dahle (1965), it is also necessary to consider the content of polyunsaturated FA, which may be a limiting factor in the extent of pigment co-oxidation.

Progress made in colorimetric instruments and in the standard CIE colorimetry methods has allowed the instrumental measurement of not only yellow but of the other colours exhibited by raw materials and pasta. Thus, the measurement of pigment concentration is now of only limited interest in the pasta industry, although it may still be useful to the plant breeder.

Determining carotenoid content involves extracting the pigments with water-saturated butan-1-ol followed by measuring the absorption of light at 440 nm wavelength on a spectrophotometer; the results are expressed in units of β-carotene (Fortmann and Joiner, 1971).

B Determination of Egg Content

Determining the egg content of pasta products is used to control the quality during processing and to detect subsequent adulterations. The methods used for the determination of egg content may be based on analysis for yolk lipids or for egg proteins.

A number of components of the yolk lipids have been considered for detection of egg, but the most effective lipid analysis at present is the determination of cholesterol content. The methods for measuring the cholesterol extracted from pasta include colorimetry, spectrophotometry, fluorimetry, gravimetry, gas-liquid chromatography (glc) and enzymology (Hadorn and Jungkunz, 1951; Chioffi, 1966; Pelagatti, 1966; Muntoni *et al*., 1966; Laskowski and Bolibok, 1967, 1968; Eeckhaut, 1968; Roberts, 1968; Beutler and Michal, 1976). Methods based on glc are now preferred because of the higher specificity and greater reproducibility (Toth, 1978). However, the value of the methods is limited by the significant variability of cholesterol content in eggs (Calet, 1959; Edwards *et al.*, 1960; Eeckhaut, 1968; Wenker and Herrmann, 1975) and by the influence of the degree of milling, duration of lipid extraction (Briski, 1975) and the extraction solvent used (Cauderay, 1978).

C Determination of Bread Wheat Content

Detecting bread wheat (*Triticum aestivum*) in durum wheat (*Triticum durum*) and in durum products (semolina and pasta) is not only of scientific and technological interest, but is also important commercially because of the risk of accidental mixing or adulteration. Investigations have been concerned with the search for differences in lipid and protein composition between the wheat species. The search for differences in protein composition is essentially based on the electrophoretic behaviour of specific enzymes associated with the D-genome of bread wheats, and on immunological analysis; there are now several official methods of this type.

Methods of differentiation based on lipid analysis were the earliest to be used. They are less frequently used now but could still be important where the application of high temperatures during pasta processing may cause denaturation of proteins. A brief review of all the methods has been published (Canuti, 1974). The methods utilizing differences in lipid composition may be divided into two classes, one is based on sitosterol palmitate determination, and the other on infra-red spectrophotometry.

1 *Sitosteryl Palmitate Determination*

As described in Chapter 2, the FA in the steryl esters of endosperm have a high content of linoleic acid (18 : 2) in wheats lacking the D-genome (including durum wheat) and in a few bread wheats, but in the majority of bread wheats the endosperm steryl esters have a high content of palmitic acid (16 : 0). A number of investigations have been made to find methods of using this difference to differentiate bread wheat from durum wheat. A short description of the investigation follows.

Walde and Mangels (1930) first recorded the presence in bread wheat flour of a sterol lipid that was undetectable in durum wheat flour. Shortly afterwards it was identified as sitosteryl palmitate (Dangoumau, 1933; Spielman, 1933). From these observations, Matveef (1952) proposed a method for detecting bread wheat based on the determination of sitosteryl palmitate content. Many attempts were made to improve upon this method and to increase the precision of the calibration curves (Montefredine and Laporta, 1955; Fabriani and Fratoni, 1955; Alliot, 1957; Zoubovsky, 1958; Guilbot, 1959). Sitosteryl palmitate contents of bread wheats were found to range from 16 to 400 mg kg^{-1} (Guilbot, 1959) and 50% of the bread wheat acreage in Italy was claimed to have a negligible content when measured by Matveef's method (Fabriani and Fratoni, 1955). The distinction between the wheats appeared less well-defined because the content of the palmitate was found to vary from 6 to 60 mg kg^{-1} in durum wheat (Guilbot, 1959).

In later investigations Gilles and Youngs (1964, 1969) used thin-layer chromatography to separate and measure the sitosteryl esters and found only very low concentrations of sitosteryl palmitate in durum wheats. In another study using the same method, the palmitate could not be detected in any of the durum wheat varieties analysed, but it was also undetectable in 40% of the French bread wheats (Fruchard *et al.*, 1967). Using a modification of the method, Gruener and Bernaerts (1968) showed that sitosteryl palmitate was undetectable in durum wheat, but also confirmed that numerous varieties of bread wheat contained only very low contents of the ester. This observation has been confirmed by Dimopoulos (1976), Ahmed (1977) and accepted by Bure and Guilbot (1972). The method is reported to also allow detection of bread wheat in pasta products containing egg (Bernaerts and Gruener, 1968; Salvioni, 1969).

Since many bread wheats respond similarly to durum wheats in the test, the absence of sitosteryl palmitate is not a guarantee of the purity of a durum product (Garcia-Faure *et al.*, 1965). The methods have only a qualitative or semi-quantitative value and bread wheat can be detected only when it is present in large proportion (Salvioni, 1976). The more recent application of high-performance liquid chromatography allows a more rapid analysis and a better quantification of wheat sterol esters (Artaud *et al.*, 1979). However, it is necessary to have some knowledge of the bread wheat involved and therefore the interest in the method is decreased.

2 *Infra-red Spectrophotometry*

To resolve the difficulties found with the sitosteryl palmitate methods, Brogioni and Franconi (1963) proposed a test based on infra-red (IR) spectrophotometry of acetone extracts of flour and semolina. Extracts from bread wheat flour yield spectra with a characteristic absorption whose maximum is between 1075 and 1098 cm^{-1}; extracts of durum wheat semolina yield spectra with a much less significant absorption in this region and the method can be used to quantify the proportions of bread wheat flour and durum wheat semolina in a mixture by reference to a calibration curve.

The method is empirical and is generally considered precise and valid. The differences in the spectra of the extracts are not a result of different sterol compositions (Custot *et al*., 1966; Fruchard *et al*., 1967), although the sterols of durum wheat do yield different spectra to those of bread wheat (Fabriani and Quaglia, 1968). It is possible that the differences are related to the glycolipid compositions (Cavallaro, 1966; Fruchard *et al.*, 1967). Sturm *et al.* (1966) have suggested that it is necessary to have some knowledge of the varieties involved when using the method due to the variability of response for different wheats. The method can be used only for flours, semolina and pasta and not for the whole grain (Custot *et al.*, 1966). The pasta must not be older than 12 months and it is also possible to detect the addition of fat used to conceal adulteration with bread wheat (Brogioni, 1969).

REFERENCES

Ahmed, S. (1977). *Macaroni J.* **59**, 30–34.

Alliot, R. (1957). *Ann. Fals. Fraud.* (579), 1–9.

Artaud, J., Iatrides, M. C. and Estienne, J. (1979). *Ann. Fals. Exp. Chim.* **72**, 153–157.

Barnes, P. J., Day, K. W. and Schofield, J. D. (1981). *Z. Lebensm. Unters. Forsch.* **172**, 373–376.

Bernaerts, M. J. and Gruener, M. (1968). *Ann. Fals. Exp. Chim.* **61**, 297–308.

Beutler, H. O. and Michal, G. (1976). *Getreide Mehl Brot* **30**, 116–118.

Biermann, V., Wittmann, A. and Grosch, W. (1980). *Fette Seifen Anstrichmittel,* **82**, 236–240.

Bouguerra, M. (1964). *Ann. Inst. Nation. Rech. Agron. Tunisie* **37**, 141–192.

Braaten, M. O., Lebsock, K. L. and Sibbit, L. D. (1962). *Crop Sci.* **2**, 277–281.

Briski, B. (1975). *Hrana I Shrana* **16**, 149–157.

Brogioni, M. (1969). *Ann. Ist. Super. Sanita* **5**, 402–403.

Brogioni, M. and Franconi, U. (1963). *Boll. Lab. Chim. Provinciali* **14**, 135–160.

Brokaw, G. Y., Lee, L. J. and Neu, G. D. (1960). *Food Processing* **21**(7), 47–54.

Buré, J. (1974). *Tec. Molitoria* **25**, 35–45.

Buré, J. and Guilbot, A. (1972). *Techn. Industr. Céréalières* (139), 8–13.

Burini, G. and Damiani, P. (1978). *Tec. Molitoria* **29**, 97–108.

Burov, L. A. (1974). *Lebensmittelindustrie* **21**, 270–271.

Burov, L. A., Ilias, A., Medvedev, G. M. and Popov, M. P. (1973). *Khlebopek. Konditer. Prom.* (10) 24–25.

Burov, L. A., Medvedev, G. M. and Ilias, A. (1974). *Khlebopek. Konditer. Prom.* (11) 25–26.
Burov, L. A., Medvedev, G. M. and Semko, V. T. (1979). *Khlebopek. Konditer. Prom.* (6), 33–34.
Calet, C. (1959). *Ann. Nutr. Alim.* **13**, A163–A203.
Canuti, A. (1974). *Tec. Molitoria* **25**, 95–101.
Cauderay, P. (1978). *Mitt. geb. Lebensm. Hyg.* **69**, 550–555.
Cavallaro, A. (1966). *Tec. Molitoria* **17**, 553–559.
Chen, K. T. and Geddes, W. F. (1945). Ph.D. Thesis, University of Minnesota, St. Paul, Minnesota.
Chioffi, V. (1966). *Boll. Lab. Chim. Provinciali* **17**, 788–800.
Clark, J. A. and Smith, R. W. (1925). *J. Amer. Soc. Agron.* **20**, 1097–1304.
Counsell, J. N. and Webb, G. K. (1971). *Flavour Industr.* **2**, 519–524.
Custot, F., Mezonnet, R. and Caley, M. (1966). *Ann. Fals. Exp. Chim.* **59**, 300–316.
Dahle, L. K. (1965). *J. Agr. Food Chem.* **13**, 12–15.
Dahle, L. K. (1970). US Patent, 3, 503753.
Dahle, L. K. and Muenchow, H. L. (1968). *Cereal Chem.* **45**, 464–468.
Dangoumau, A. (1933). *Bull. Soc. Chim. Biol.* **15**, 1083–1093.
Davis, K. R., Litteneker, N., Le Tourneau, D., Cain, R. F. Peters, L. J. and MacGinnis, J. (1980). *Cereal Chem.* **57**, 178–184.
Dimopoulos, J.S. (1976). *Hernika Hronika* **41**, 31–35.
Drapron, R. and Uzzan, A. (1968). *Ann. Nutr. Alim.* **22**, B393–B436.
Edwards, H., Friggers, C., Dean, R. and Carmon, J. (1960). *Poultry Sci.* **39**, 487–489.
Eeckhaut, R. (1968). *Fermentatio* **2**, 66–83.
Fabriani, G. and Fratoni, A. (1955). *Quadr. Nutr.* **15**, 130–141.
Fabriani, G. and Quaglia, G. (1968). *Quadr. Nutr.* **28**, 487–492.
Fabriani, G. and Quaglia, G. (1973). *Sci. Tecnol. Alimenti* **3**, 165–169.
Farre-Rovira, R. and Costes, C. (1974). *Physiol. Veg.* **12**, 251–288.
Feillet, P. (1977). *Cahiers Nutr. Diététique* **12**, 299–310.
Ferchat, A. and Bouguerra, M. (1962). *Ann. Inst. nation. Rech. Agron. Tunisie* **35**, 27–50.
Fernandes, J. L. A., Shuey, W. C. and Maneval, R. D. (1979). *Cereal Chem.* **56**, 520–524.
Fifield, C. C., Smith, G. S. and Mayes, J. F. (1937). *Cereal Chem.* **14**, 661–673.
Fortmann, K. L. and Joiner, R. R. (1971). In *Wheat: Chemistry and Technology* (Y. Pomeranz, ed.), pp. 493–522. American Association of Cereal Chemists, St. Paul, Minnesota.
Fruchard, C., Poma, J. and Buré, J. (1967). *Ann. Fals. Exp. Chim.* **60**, 193–205.
Garcia-Faure, R., Garcia Olmedo, F., Sotelo Aboy, I. and Salto Andreu, Y. M. (1965). *I.N. Investigaciones Agronomicas* **25**, 395–408.
Gilles, K. A. and Youngs, V. L. (1964). *Cereal Chem.* **41**, 502–513.
Gilles, K. A. and Youngs, V. L. (1969). *Macaroni J.* **51**, 12–22.
Grignac, P. (1965). *Producteur Agric. France* (1), 5.
Grignac, P. (1970). *Ann. Amélior. Plantes* **20**, 159–188.
Gruener, M. and Bernaerts, M. (1968). *Getreide Mehl* **18**, 69–74.
Guilbot, A. (1959). In *Aktuelle Problem über Durum und Teigwaren*, pp. 101–114 Granum-Verlag, Detmold.
Hadorn, H. and Jungkunz, R. (1951). *Mitt. geb. Lebensm. Hyg.* **42**, 452–458.

Houliaropoulos, E., Abecassis, J. and Autran, J. C. (1981). *Industr. Céréales,* **12**, 3–13.
Irvine, G. N. (1963). Grain Research Lab. Report p. 11.
Irvine, G. N. (1971). In *Wheat: Chemistry and Technology* (Y. Pomeranz, ed.) pp. 777–798. American Association of Cereal Chemists, St. Paul, Minnesota.
Irvine, G. N. and Anderson, J. A. (1949). *Cereal Chem.* **26**, 507–512.
Irvine, G. N. and Anderson, J. A. (1953). *Cereal Chem.* **30**, 334–342.
Irvine, G. N. and Winkler, C. A. (1950). *Cereal Chem.* **27**, 205–218.
Kläui, H. (1976). *Intern. Flavours Food Additives* **7**, 165–172.
Kläui, H. and Raunhardt, O. (1976). *Alimenta* **15**, 37–45.
Lacroix, L. J. and Lier, J. B. (1975). *Canadian J. Plant Sci.* **55**, 579–684.
Laignelet, B. (1976). *Getreide Mehl Brot* **30**, 277–280.
Laignelet, B. (1979a). Ph.D. Thesis, Université des Sciences et Techniques du Languedoc, Montpellier, pp 75–77.
Laignelet, B. (1979b). *Industr. Alim. Agr.* **96**, 1243–1254.
Laignelet, B., Kobrehel, K. and Feillet, P. (1972). *Industr. Alim. Agr.* **89**, 413–427.
Laskowski, K. and Bolibok, E. (1967). *Roczn. Paustw. Zahl. Hig.* **13**, 749–753.
Laskowski, K. and Bolibok, E. (1968). *Przegl. Zboswo. Mlyn.* **12**, 9–10.
Lasseran, J. C. and Barthelemy, P. (1968). *Industr. Alim. Agr.* **85**, 823–835.
Lepage, M. and Sims, R. P. A. (1968). *Cereal Chem.* **45**, 600–604.
Lier, J. B. and Lacroix, L. J. (1974). *Cereal Chem.* **51**, 188–194.
Lin, M. J. Y., D'Appolonia, B. L. and Youngs, V. L. (1974). *Cereal Chem.* **51**, 34–45.
MacDonald, C. E. (1979). *Cereal Chem.* **56**, 84–89.
Manunta, C. (1946a). *Genet. Agr. (Roma)* **1**, 19–44.
Manunta, C. (1946b). *Genet. Agr. (Roma)* **1**, 167–186.
Markley, M. C. and Bailey, C. H. (1935). *Cereal Chem.* **12**, 33–39.
Matsuo, R. R. and Dexter, J. E. (1980). *Cereal Chem.* **57**, 117–122.
Matsuo, R. R., Bradley, J. W. and Irvine, G. N. (1970). *Cereal Chem.* **47**, 1–5.
Matthews, R. H. and Workman, M. Y. (1977). *Cereal Chem.* **54**, 1115–1123.
Matveef, M. (1952). *C. R. Acad. Agriculture France,* **39**, 658–663.
Menger, A. (1965). *Getreide Mehl,* **15**, 60–61.
Menger, A. (1972). *Getreide Mehl Brot* **26**, 307–311.
Menger, A., Cleve, M. and Ander, E. (1969). *Brot u. Gebäck,* 179–183.
Montefredine, A. and Laporta, L. (1955). *Italia Cereali,* **10** (2), 95 and **10** (3), 175.
Moss, H. J. (1967). *Austral. J. of Exper. Agric. and Anim. Husbandry* **7**, 462–464.
Munsey, V. E. (1938). *J. Assoc. Offic. Agr. Chemists* **21**, 331–351.
Muntoni, F., Tiscornia, E. and Tassi-Mico, C. (1966). *Quadr. Nutr.* **26**, 26–34.
Nazarov, N. I. and Egorova, N. I. (1972). *Khlebopek. Konditer. Prom.* (5), 24–25.
Nazarov, N. I., Egorova, N. I. and Kondratenko, S. S. (1972). *Khlebopek. Konditer. Prom.* (2) 20.
Nazarov, N. I., Kondratenko, S. S. and Egorova, N. I. (1971). *Izv. Vyssh. Ucheb. Zaved., Pishch. Tekhnol.* (4), 34–35.
Nazarov, N. I., Luk'yanov, A. B., Kondratenko, S. S. and Egorova, N. I. (1973). USSR Patent 385 572.
Nazarov, N. I., Rozantsev, R. G., Kondratenko, S. S. and Egorova, N. I. (1974). *Izv. Vyssh. Ucheb. Zaved., Pishch. Tekhnol.* (1) 148–150.
Nolte, D., Rebmann, H. and Acker, L. (1974). *Getreide Mehl Brot*, **28**, 189–191.
Pelagatti, G. (1966). *Tec. Molitoria* **17**, 337–340.
Roberts, L. (1968). *J. Assoc. Off. Anal. Chem.* **51**, 1220–1224.

Salvioni, C. (1969). *Tec. Molitoria* **20**, 129–130.
Salvioni, C. (1976). *Boll. Lab. Chim. Provinciali* **27**, 251–269.
Schettino, O. and Di Lieto, A. (1968). *Tec. Molitoria* **19**, 557–565.
Spielman, M. A. (1933). *Cereal Chem.* **10**, 239–242.
Sturm, P. A., Parkhurst, R. M. and Skinner, W. A. (1966). *Cereal Sci. Today* **11**, 523–528.
Thoren, I. (1972). *Getreide Mehl Brot* **26**, 340–343.
Toth, L. (1978). *Getreide Mehl Brot* **32**, 116–119.
Tykheeva, E. B., Nechaev, A. P. and Denisenko, J. I. (1971). *Isv. Vyssh. Ucheb. Zaved. Pishch. Teckhnol.* (3) 20–22.
Vdovkin, E., Pershakova, N., Kalugina, N. and Kostin, V. (1980). *Isv. Vyssh. Ucheb. Zaved. Pishch. Tekhnol.* (4), 28–29.
Walde, A. W. and Mangels, C. E. (1930). *Cereal Chem.* **7**, 480–486.
Walsh, D. E., Youngs, V. L. and Gilles, K. A. (1970). *Cereal Chem.* **47**, 119–125.
Weihrauch, J. L. and Matthews, R. H. (1977). *Cereal Chem.* **54**, 444–453.
Wenker, K. and Herrmann, H. (1975). *Mitteilungsblatt GDCH-Fachgruppe Lebensmittelchemie Gerichtliche Chem.* **29**, 253–257.
Wildfeuer, I. and Acker, L. (1968). *Mitt. Geb. Lebensm. Hyg.* **59**, 392–400.
Worzella, W. W. and Cutler, G. H. (1935). *Cereal Chem.* **12**, 708–713.
Yakovenko, V. A. and Antyukhova, L. I. (1977). *Khlebopek. Konditer. Prom.* (1) 33–24.
Zechmeister, L. and Cholnoky, L. (1940). *J. Biol. Chem.* **135**, 31–36.
Zoubovsky, P. (1958). *Pâtes Alim. Semoules Industr. Annexes* (77) 4.

14 Lipids in Cereal Products

Y. POMERANZ and O. K. CHUNG

U.S. Grain Marketing Research Laboratory, U.S. Department of Agriculture, Manhattan, Kansas 66502 U.S.A.

I INTRODUCTION

This chapter discusses lipids in cereal products. To the best of our knowledge no comprehensive compilation on lipids in cereal products has been published recently elsewhere. Generally, data on lipids in commercial products include information on gross composition for comparative purposes and for product characterization.

The mention of a trademark, proprietary product, or vendor does not constitute a guarantee or warranty of the product by the U.S. Department of Agriculture and does not imply its approval to the exclusion of other products or vendors that may also be suitable.

"Lipids in Cereal Technology"
ISBN 0-12-079020-3

II WHEAT

A Lipids in Wheat Starch Preparations

About 1.1% of wheat starch preparations consists of lipids; these are listed in Table 14.1 (from Acker *et al.*, 1967). About 75% of these lipids are represented by the lysophospholipids, with lysophosphatidylcholine (LPC) as the main component (60%). No phosphatidylcholine is present and the LPC should be considered a native, integral starch granule component. Among the starch fatty acids, palmitic acid amounts to 56% of the total, compared with 60% linoleic acid in non-starch fatty acids. Seventy-one per cent of the free fatty acids of starch are saturated. A major proportion of the lipids in the starch preparations are assumed to be present as inclusion compounds in the starch granule and largely unavailable to affect dough processing before starch gelatinization. They can, however, influence pasting characteristics (as determined by the Amylograph) and overall flour properties in the baked bread. Compositions of lipids in wheat, rye, barley, and oat starch preparations are compared in Table 14.2 (from Acker, 1974, Acker and Becker, 1971). Further information concerning wheat starch lipids will be found in Chapter 2.

B Wheat Gluten Lipids

Starch-bound lipids are essentially unavailable for interaction with gluten proteins during dough formation (Acker *et al.*, 1967). Other lipids are

Table 14.1 The lipid composition of wheat starch preparations†

Lipid type	Content (% of total lipids)
Lysophospholipids	
-choline	62.3
-ethanolamine	8.5
-serine	4.2
-inositol	1.5
Free fatty acids	11.6
Mono- and diglycerides	0.7
Triglycerides	0.8
Sterols	1.1
Steryl esters	0.7
Galactolipids	1.0
Miscellaneous	7.6
(hydroxy-fatty acids, hydrocarbons, unidentified)	

†Acker *et al.* (1976).

Table 14.2 Lipids in cereal starch preparations†

Lipids	Wheat (%)	Rye (%)	Barley (%)	Oats (%)
Total (%)	1.1	0.5	1.0	1.3
of which:				
Lysophosphatidylcholine	62.3	51.9	62.4	51.6
Lysophosphatidylethanolamine	8.3	n.d.	6.0	5.1
Lysophosphatidylinositol	1.5	n.d.	3.1	7.0
Free fatty acids	11.6	2.1	4.4	7.7

†Acker (1974) and Acker and Becker (1971).

bound to differing degrees by various gluten fractions and, once bound, are difficult to extract into solvents. Gluten lipids are compared with wheat flour lipids in Table 14.3. The lipids were extracted with water-saturated butan-1-ol at room temperature, separated on a silicic acid columnn into non-polar and polar fractions, sub-fractionated by thin-layer chromatography and assayed densitometrically after spraying. The results were semi-

Table 14.3 Comparison of the lipid compositions of wheat flour and gluten†

Lipid type	Content (% of total lipids) in	
	Flour	Gluten
Total non-polar	35.6	37.9
Total polar	64.4	62.1
Total phospholipids	19.5	14.0
Phosphatidic acid	2.3	4.8
Phosphatidylglycerol	4.0	2.7
Phosphatidylcholine	2.1	3.6
Phosphatidylethanolamine	traces	1.2
Phosphatidylserine	1.2	–
Lysophosphatidylcholine	9.1	1.7
Lysophosphatidylethanolamine	0.8	–
Galactolipids, *total*	19.4	25.1
Monogalactosyldiglyceride	6.0	8.8
Digalactosyldiglyceride	13.4	16.3
Other lipids (free fatty acids, steryl glycosides, glycosidesters, cerebrosides, phytoglycolipids)	25.5	23.0

†The total lipid content of the Type 405 flour was 1.28%, dry matter basis (Acker *et al.*, 1967).

Table 14.4 Typical analysis of vital wheat gluten†

Moisture (%)	5.0– 8.0
Protein (%, d.b.)	75.0–80.0
Diethyl ether extractable (%, d.b.)	0.5– 1.5
Ash (%, d.b.)	0.8– 1.2

†From "Wheat Gluten — A Natural Protein for the Future — Today"; International Wheat Gluten Association, Shawnee Mission, KS 66208, U.S.A.

quantitative. The following facts are important:

(1) The polar : non-polar lipid ratios in wheat gluten and flour were about 1.6 : 1 and 1.8 : 1, respectively.

(2) Gluten contains more glycolipids than phospholipids; the amounts of LPC (the main starch lipid) in wheat flour, and especially gluten, are low.

(3) Although the gluten contains more than 95% of wheat-flour glycolipids, it contains only 53% of flour phospholipids.

(4) Determination of lipids in the wedge protein fraction and in a mixture of starch and adhering proteins showed no unusual or specific distribution patterns.

Table 14.5 Lipids in flour and gluten from wheats that vary widely in breadmaking potential

Component	Manitoba	Jubilar	Wimax
A. Wheat Flour			
Protein (%)	16.8	7.8	9.2
Total lipids (%)	2.05	1.62	1.39
Non-polar lipids (%)	0.71	0.42	0.38
Polar lipids (%)	1.34	1.20	1.01
of which:†			
Monogalactosyldiglyceride	13.2	16.3	18.8
Digalactosyldiglyceride	16.3	18.5	20.5
Lysophosphatidylcholine	24.2	24.9	23.5
Phosphatidylcholine	3.9	3.4	3.0
B. Wheat Gluten			
Total lipids (%)	5.70	7.63	7.11
Non-polar lipids (%)	2.16	1.97	1.98
Polar lipids (%)	3.54	5.66	5.13
of which:†			
Monogalactosyldiglyceride	17.7	20.3	21.3
Digalactosyldiglyceride	24.2	25.6	24.3
Lysophosphatidylcholine	3.8	3.0	3.9
Phosphatidylcholine	1.2	1.3	1.4

†As a percent of polar lipids (Acker, 1974).

According to Bartolome (1981) commercial vital wheat gluten contains (on a moisture-free basis) at least 75% protein ($N \times 5.7$), 1.0% diethyl ether extractable fat and 6.5 to 7% fat by acid hydrolysis. A typical analysis of commercial vital gluten is given in Table 14.4.

Lipids were determined in flours and gluten from three wheats that varied widely in breadmaking potential: Manitoba (high); Jubilar (low); and Wimax (low) (Acker, 1974). The results are summarized in Table 14.5. Though some differences are indicated, much more work would be required to determine whether the differences are related to protein contents and protein quality or both (in terms of functional, breadmaking properties).

III MAIZE

The term corn is used for specific products; in all other cases, the term maize is used.

A Dry Milling

Dry-milled maize can be produced from either whole or degermed grain. In milling whole maize, preferably white dent, the product has a rich, oily flavour because of its high fat content. Meals with a soft-texture-feel are produced on mill stones run slowly at low temperatures. Two types of products can be produced: first, meals that are essentially ground whole maize, not bolted (not sifted), and which are produced in small mills; second, meals bolted, with about 5% of coarse hull and germ particles removed. Whole and bolted meals have short shelf lives because of the high fat content of the ground material which has a large surface area and contains active lipases. Approximate compositions of dry-milled maize products are given in Table 14.6 (from Brockington, 1970).

A typical composition of dry-milled products from degermed maize is listed in Table 14.7 (from Brekke, 1970). Grits and meal are largely produced from the horny, or vitreous, endosperm; they contain less than 1.0% and 1.5% fat, respectively. Flour produced by grinding the starchy (or mealy) endosperm contains from 2 to 3% fat derived from broken germ during processing. The large surface area and relatively high fat content of maize flour lower its shelf life.

The Food and Drug Administration (FDA-USA) has established standards of identity for dry-milled corn products used for food. According to those standards (Code of Federal Regulations, Title 21, part 15), the fat

Table 14.6 Dry-milled maize products composition†

Product	Protein (%)	Fat (%)	Starch (%)	Crude fibre (%)	Ash (%)
Whole-meal	9.2	3.9	73.5	1.6	1.2
Meal-bolted	9.0	3.4	74.5	1.0	1.1
Meal-degermed	7.9	1.2	78.4	0.6	0.5
Grits	8.7	0.8	78.1	0.4	0.4
Flour	7.8	2.6	76.8	0.7	0.8

†Brockington (1970); 12% moisture basis.

content of corn meal may not differ by more than 0.3% from that of cleaned corn; the fat content of bolted corn meal should be not less than 2.25%, nor more than 0.3% greater than that of cleaned corn; the fat content of degerminated corn meal should be less than 2.25%, of corn grits not more than 2.25%, and the fat content of corn flour may not exceed that of cleaned corn.

The composition of commercial maize products is listed in Table 14.8. The meals and flours are produced from maize ground to typical granulations. The cooked flour hydrates readily in cold water to form a stable paste. The toasted germ is a food-grade product in flake form and contains all the original oil of the germ. The germ cake is a feed product from maize germ from which most of the oil has been removed. It is used as a carrier for vitamins and antibiotics in animal feed formulations. Massa Harina (yellow regular grind, or yellow coarse grind, or white) is a food-grade product. It is milled from maize that has been steeped, ground and dried to produce a stable flour for the production of Mexican foods.

B Wet Milling

The main products of wet-maize milling are starch (unmodified and modified, including syrups and dextrose) and several co-products. The co-products, used mainly as feed ingredients, include gluten meal, gluten feed, corn germ meal, and condensed, fermented corn extractives (about 50% solids). Processing maize germ yields refined oil (together with fatty acids from crude oil refining) and corn germ meal (Harness, 1978).

While maize starch contains only 0.04% fat as determined by diethyl ether extraction, total lipids amount to about 0.54%. The lipids are almost entirely associated with amylose and are predominantly free fatty acids (Morrison, 1978). Commercial maize starch products contain less than 0.1% diethyl ether extractives, typically 0.03% (Harris, 1981).

Table 14.7 Typical yields and analyses of products from a degerming-type dry-maize mill

Products	Yield (%)	Particle size range†	Moisture (%)	Fat (% d.b.)‡	Crude fibre (% d.b.)	Ash (% d.b.)	Protein (% d.b.)
Maize	100		15.5	4.5	2.5	1.3	9.0
			Primary products				
Cereal flaking (hominy grits)	12	−3.5 + 6	14.0	0.7	0.4	0.4	8.4
Coarse grits	15	−10 + 14	13.0	0.7	0.5	0.4	8.4
Regular grits	23	−14 + 28	13.0	0.8	0.5	0.5	8.0
Coarse meal	3	−28 + 50	12.0	1.2	0.5	0.6	7.6
Dusted meal	3	−50 + 75	12.0	1.0	0.5	0.6	7.5
Break flour	4	−75 + pan	12.0	2.0	0.7	0.7	6.6
Oil	1						
Hominy feed	35		13.0	6.3	5.4	3.3	12.5
			Alternative products				
Brewers' grits	30	−12 + 30	13.0	0.7	0.5	0.5	8.3
100% meal	10	−28 + pan	12.0	1.5	0.6	0.6	7.2
Fine meal	7	−50 + pan	12.0	1.6	0.6	0.7	7.0
Germ fraction§	10	−3.5 + 20	15.0	18.0	4.6	4.7	14.9

†U.S. standard sieve.
‡Dry basis.
§Yield is distributed between maize oil and hominy feed (Brekke, 1970).

Table 14.8 Composition of typical maize products†

Maize product	Moisture (%)	Protein (%)	Fat (%)	Crude fibre (%)	Ash (%)
		(Dry matter basis)			
Yellow maize					
Meal‡	8.2–12.6	8.0–9.0	0.9–2.3	0.3–0.7	0.3–0.7
Flour‡	8.5–12.1	7.6–8.8	1.4–3.0	0.6–1.0	0.6–1.0
Fine flour	7.1–11.3	6.4–7.4	2.2–3.3	0.04–0.60	0.66–1.12
Cooked flour	6.7–11.5	9.0–9.2	0.6–0.7§	–	0.55–0.73
White maize					
Meal‡	11.0–13.0	8.0–9.0	1.7–2.2	0.5–1.0	0.3–0.7
Flour‡	8.0–12.0	7.0–8.0	2.0–2.7	0.5–1.0	0.4–0.9
Fine flour	10.0–13.0	8.0–8.8	4.0–5.2	0.5–1.2	0.8–1.0
Toasted corn germ	4.2	17.0	25.4§	4.2¶	7.2
Corn germ cake	5.0 max	14.0 min	3.5 min	8.5 max	–
Massa harina	10.0–12.0	7.0–9.0	3.5–4.5	1.8–2.6	1.2–1.7

†Courtesy of the Quaker Oats Co., Chicago, Ill., U.S.
‡From degermed maize.
§Diethyl ether extract.
¶Dietary fibre = 20.8%.

The co-products of maize starch wet milling amount to about one-third of the total output. With the exception of maize oil and steep liquor (used in industrial fermentations) the co-products are mainly sold as feed ingredients. Listed in decreasing value they are: corn gluten meal, corn gluten feed; spent germ meal; corn starch molasses or hydrol; steep liquor (condensed corn fermentation extractives); corn bran; fatty acids. The composition of the main maize wet milling feeds is summarized in Table 14.9.

Corn gluten meal is a high-protein product, used as a protein balancing ingredient in feed formulations. It is used widely in broiler and layer rations because of its high content of carotenoid pigments. Among the three carotene isomers only β-carotene has significant vitamin A activity. The dihydroxy xanthophylls are potent pigments for colouring poultry skin and egg yolks. The major isomer, lutein, is slightly superior to zeaxanthin in producing colour. The monohydroxy pigments, zeinoxanthin and cryptoxanthin, have less than half the pigmenting value of the dihydroxy pigments. Xanthophyll levels in gluten meal are highest in winter months and gradually drop to half the original value by the end of the summer.

The linoleic acid content (as-is basis) is 3.2% in corn gluten meal, 2.2%

Table 14.9 Composition of feeds from maize wet milling†

Component	Corn gluten feed	Corn gluten meal		Corn germ meal	Condensed fermented corn extractives (about 50% solids)
	Guaranteed analysis (%)				
Protein (minimum)	21.0	60.0	41.0	20.0	23.0
Fat (minimum)	1.0	1.0	1.0	1.0	0.0
Crude fiber (maximum)	10.0	3.0	6.5	12.0	0.0
	Typical analysis (%)				
Fat – average	2.5	2.5	2.5	1.9	0.0
Fat (AOAC) – range	1.4–3.5	1.0–5.2	1.2–4.4	1.0–2.9	0.0
Total fat – average‡	3.8	5.7	4.8	4.6	0.0
Total fat – range	2.7–4.7	4.4–7.9	3.6–6.4	4.1–5.3	0.0

†Adapted from "Corn Wet-Milled Feed Products"; Corn Refiners Association Inc., Washington, D.C. (1975) and S. A. Watson, AACC Short Course, April, 1980.

‡As determined by extraction with a chloroform-methanol mixture 4 : 1; widely used in Europe.

in corn gluten feed and about 0.5% in corn germ meal (Rapp, 1978). Corn gluten meal is relatively rich in xanthophylls (220 to 500 mg kg^{-1}); in corn gluten feed 22 mg kg^{-1} are present and practically none in corn germ meal and concentrated steepwater. Corn gluten meal contains 30 to 65 vitamin A equivalents as retinol (0.15 mg retinol = 5000 I.U. vitamin A) and 45–65 mg kg^{-1} of β-carotene.

C Maize Oil (Corn Oil)

Maize contains about 4.5% oil, 85% of which is present in the germ (Reiners, 1978). The germ fraction separated from maize by the wet milling process contains about 50% oil, the fraction separated by the dry milling process contains about 25% oil. Germ oil can be extracted by a continuous screw press (expeller) to yield a meal with a residual oil content of 7 to 10%; solvent extraction (directly or following expeller extraction) produces a meal with a residual oil content of 1 to 3%. About 2.25 kg of oil can be recovered from a hectolitre of maize (1.75 lb per U.S. bushel) by solvent extraction of the germ. The products are crude oil and corn germ meal. Crude and refined maize oils are compared in Table 14.10 (from Reiners, 1978).

Table 14.10 Composition of corn oil (%)†

Component	Grade of oil	
	Crude	Refined
Triglycerides	95.6	98.8
Free fatty acids	1.7	0.03
Phospholipids	1.5	—
Sterols	1.2	1.1
Tocols	0.06	0.05
Waxes	0.05	—
Carotenoids	0.0008	—

†Reiners (1978).

Refining maize oil removes free fatty acids, phospholipids, waxes, and carotenoids. The main triglyceride fatty acids are about 60% linoleic acid, 25% oleic acid, and 13.5% palmitic and stearic acid. The iodine value of refined maize oil produced in the US shows little variation (125.4 to 127.6). Oils from African maize may have iodine values as low as 110 and correspondingly reduced ratios of polyunsaturates to saturates (Reiners, 1978). The production and properties of maize oil are further discussed in Chapter 18.

IV RICE

Brown rice contains about 1.5 to 2.5% total fat (Houston, 1972; Chang *et al*., 1980). About 80% of the lipids of brown rice is found in the bran and polish fractions after milling; about one-third of the grain lipids is located in the embryo. In degermed brown rice about 70% of the total lipids are in the outer 8% milling fraction. The outermost 1.4% milling fraction contains 40% of fat, which constitutes about 40% of the total lipids of degermed brown rice.

The lipids of brown rice, bran and embryo have similar constants: relative density 0.91–0.92, ηD^{40}_{25} 1.465–1.470, iodine value 95–106 saponification value 177–196 (Juliano, 1972; 1980).

The crude fat content of brown rice and its fractions is as follows (dry matter basis):

Brown rice 1.8–4.0%

Milled rice 0.2–1.1%

Bran 14.6–21.7%

Embryo 15.2–23.8%

Polish 8.8–15.3%

Most rice lipids are removed during milling with the bran (which contains the germ) and polish. Commerical bran contains 10.1 to 23.5% oil and polish 9.1 to 11.5% oil; the milled rice contains 0.3 to 0.7% oil (Houston and Kohler, 1970).

Commercial bran oil contains germ oil. The unsaponifiable matter of bran oil consists of about 40% sterols, 25% higher alcohols, 20% ferulic acid esters, 10% hydrocarbons, 2% cholesteryl esters and small amounts of unidentified compounds. The hydrocarbon fraction is mainly squalene or its isomer (Houston, 1972; Juliano, 1972, 1980; Chang *et al.*, 1980; Cornelius, 1980).

Oryzanol, a mixture of ferulate esters of unsaturated triterpenoid alcohols, is a potent antioxidant present at a concentration of 1.0 to 3.0% in bran oil. Other antioxidants include the tocols which have vitamin E activity. Rice-bran oil also contains, depending on the nature and temperature of extractant, various amounts of waxes (up to about 10%).

Fat contents of 241 milled rice samples ranged from 0.19 to 2.73% (average 0.65%) determined as petroleum-extractable material. About 20% of the fatty acids are polyunsaturated, linoleic and linolenic acids. The oleic to linoleic acid ratio is about 1 : 1 (Kennedy, 1980).

Fat is unevenly distributed within the endosperm; the highest concentration is present in the outer layer and the lowest in the central portion. In a milling study of 12 rice lots, fat in the whole kernel ranged from 0.20 to 0.92%; in the flour passing through a 40-mesh screen the fat content was 4.1 to 11.6%, 17 times as much as in the whole kernel. The flour fraction retained on the 40-mesh sieve contained four times as much fat as did the whole kernel, and the residual kernel contained 0.12% fat, about one-quarter that of the original rice.

The composition of fat differs within the endosperm. Neutral fats account for 85 to 90% of total lipids in the outer layers and for only 60% in the centre. Generally, unsaturated fatty acids, oleic and linoleic show an inverse pattern of distribution. Oleic acid decreases and linoleic acid increases from the outer to the inner layers of the kernel in both the free fatty acids and the neutral fat fraction (Kennedy, 1980). Further details of rice lipids will be found in Chapter 15.

V SORGHUM

Sorghum grain has a general composition resembling that of maize. Wet milling processes for maize and grain sorghum are basically similar (Watson, 1970).

The object of dry milling sorghum is to separate the endosperm, germ and bran while recovering a maximum amount of endosperm (Hahn, 1969). Sorghum germ can be separated from other products and oil extracted from it. High yields of clean grits and minimum flour formation are desired. By-products of sorghum dry milling (bran, germ, and shorts) are used in the production of hominy feed. A decrease in the flour extraction in roller milling from 90 to 70% decreased the oil content of the groats from 2.8 to 2.0%. The oil content decreased from 3.4% in whole sorghum grain to 0.6% in decorticated grain where 39.0% of the kernel was abraded; the decrease was accompanied by decreases in fibre content (2.2 to 0.7%), mineral components (1.5 to 0.4%), and protein content (9.6 to 6.9%) and by an increase in brewers' extract (from 85.7 to 97.0%).

Impact or attrition degerminators (used after grain tempering and dehulling) are effective in germ separation and in the production (after sieving) of low oil content products. Relative density separators (used after dehulling, impaction, and size classification) can be used to separate the germ and endosperm during dry milling. Both waxy and non-waxy sorghum grain can be milled and fractionated. The germ of waxy grain, however, is more difficult to separate, since waxy grits contain more oil than non-waxy grits. Table 14.11 summarizes the composition of dry-milled sorghum grain products.

Table 14.11 Composition of dry-milled sorghum grain products†

Product	Protein (%)‡	Oil (%)	Crude fibre (%)	Ash (%)
Whole grain	9.6	3.4	2.2	1.5
Pearled	9.5	3.0	1.3	1.2
Flour-crude	9.5	2.5	1.2	1.0
Flour-refined	9.5	1.0	1.0	0.8
Brewers' grits	9.5	0.7	0.8	0.4
Bran	8.9	5.5	8.6	2.4
Gern	15.1	20.0	2.6	8.2
Hominy feed	11.2	6.5	3.8	2.7

†Hahn (1969).
‡Wet basis.

VI OATS

Oat hulls contain very little lipid; practically all the lipid is in the oat groats (dehulled grain). Oat groats contain higher concentrations of lipids than other cereal grains. In a comprehensive study of 4,000 entries from the world collection, the free lipid content of oat groats ranged from 3.1 to 11.6%; 90% of the entries contained from 5 to 9% lipid (Brown and Craddock, 1972). About 20% of the fatty acids in oat groats are palmitic; about 35%, oleic and more than 35%, linoleic acid (Youngs *et al.*, 1982). Triglycerides are the major component of oat lipids; digalactosyldiglycerides are the major glycolipid component and phosphatidylcholine is the major phospholipid component. Lysophosphatidylcholine comprises 51.6% of oat starch lipids (Acker and Becker, 1971). The gross composition of oat products is summarized in Table 14.12 (adapted from Caldwell and Pomeranz, 1973).

The composition of North American commercial oat products is listed in Table 14.13. The *steel oat* product is an oat that has been dehulled and cut

Table 14.12 Composition (%) of oats and oat products†

Component	Oats	Finished groats	Hulls	Oat shorts	Oat flour, chips, and meal
Protein (N × 6.25)	12.1	15.8	4.2	9.5	15.5
Crude fat	5.1	7.2	1.7	3.2	6.2
Crude fibre	11.0	1.5	32.9	22.0	3.6
Ash	3.4	1.9	6.0	6.7	2.1

†Wet basis, about 7% moisture; Caldwell and Pomeranz (1973).

Table 14.13 Composition of typical oat products†

Oat product	Moisture (%)	Protein (%)	Fat (%)	Crude fibre (%)	Ash (%)
		(Dry matter basis)			
Steel oat	9.0–12.0	16.5–18.5	7.0–8.0	1.4–1.8	2.0–2.5
Steam table	9.0–12.0	16.5–18.5	6.0–9.0	1.4–1.8	2.0–2.5
Regular buckeye rolled	9.0–12.0	16.0–18.5	6.0–9.0	1.2–1.8	2.0–2.5
Quick buckeye rolled	9.0–12.0	16.0–18.5	7.0–8.0	1.2–1.8	2.0–2.5
Flour	8.0–12.0	17.0–19.0	7.0–8.0	0.7–1.6	1.8–2.2

†Courtesy of the Quaker Oats Co., Chicago, Illinois, U.S.

to a smaller size. The *steam table* product is a dehulled oat that has been steamed and rolled into a thick flake. The *regular Buckeye rolled* product is a dehulled oat that has been steamed and rolled into a large flake while the *Quick Buckeye Rolled* product is a dehulled oat that has been steamed and rolled into a thin flake for quick cooking. The flour consists of dehulled oat that has been steamed and ground to produce a stable flour in which enzyme activity has been minimized. (Oat lipids are discussed in detail in Chapter 16.)

VII BARLEY

Most barley grain that goes into human food is consumed as pot or pearl barley. Both are manufactured by gradually removing the hull and outer portions of the barley kernel by abrasive action. The pearling, or decortication, process used to produce pot barley is merely carried out further to

Table 14.14 Composition of milled barley and the products of barley milling†

Product	Moisture (%)	Protein (%)	Fat (%)	*N*-free extract (%)	Crude fibre (%)	Ash (%)
Dehulled barley	12.5	10.6	1.7	72.1‡	1.6	1.5
Pearls	12.5	7.8	1.0	76.2	1.4	1.1
Pearling dust	12.5	9.5	1.4	74.3	0.8	1.5
Feedmeal	12.0	12.5	3.0	64.0	5.0	3.5
Bran	10.5	14.0	3.5	57.1	10.0	4.9
Hulls	10.4	3.6	1.0	49.2	28.6	7.2

†Adapted from Rohrlich and Bruckner (1966); wet basis.
‡In hulled barley.

Table 14.15 Composition of typical barley products†

Oat product	Moisture (%)	Protein (%)	Fat (%)	Fibre (%)	Ash (%)
		(Dry matter basis)			
Chester	9.0–10.0	11.0–12.0	2.0–2.5	1.0–1.5	0.7–1.3
Portage	9.0–10.0	11.0–12.0	1.0–1.5	1.0–1.5	0.7–1.3
Quick Cooking	9.0–10.0	10.0–12.0	1.0–1.5	0.5–1.0	0.7–1.3
Flakes	10.0–12.9	10.0–12.0	1.0–1.5	0.5–1.0	0.7–1.3
Flour	–	13.0–15.0	1.9–2.9	< 1.2	1.0–2.0

†Courtesy of the Quaker Oats Co., Chicago, Illinois, U.S.

Table 14.16 Average composition of raw materials and by-products of the brewing industry†

	Moisture (%)	Protein (%)	Fat (%)	Crude fibre (%)	Ash (%)	*N*-free extract (%)
Malted barley	7.7	12.4	2.1	6.0	2.9	68.9
Malted sprouts	7.6	27.2	1.6	13.1	5.9	44.6
Brewers' dried grains	7.2–7.7	21.1–27.5	6.4–6.9	15.3–17.6	3.9–4.2	39.4–42.9
Hops‡	12.5	17.5	18.7§	13.2	7.5	27.5
Spent hops	6.2	23.0	3.6	24.5	5.3	37.4
Yeast	4.3	50.0	0.5	0.5	10.0	34.7

†From G. Leavell, U.S. Department of Agriculture Bulletin No. 58 (1942); wet basis.
‡From Luers (1950).
§Total diethyl ether extract (includes non-lipid components; additional component, 3.0% tannins).

produce pearl barley; 100 kg of barley normally yields 65 kg of pot barley or 35 kg of pearl barley. Barley flour is a secondary product and polishings are a by-product of the pearling process. On the basis of decreases in weight and changes in chemical composition it has been calculated that six pearlings remove 74% of the protein, 85% of the fat, 97% of the fibre and 88% of the mineral ingredients contained in the original barley grain. Table 14.14 compares the gross composition of barley products obtained during the pearling of barley (Rohrlich and Bruckner, 1966). The composition of commercial barley products is listed in Table 14.15. *Chester* and *Portage* are creamy white, pearled barley products and both are used in food and pet food applications as thickeners and fillers. About 90 to 95% and 80 to 90% is retained on U.S. sieve no. 8 for Chester and Portage barley products, respectively. The *quick cooking* product is a creamy white, pearled barley that has been steamed and rolled into a flake to facilitate quick cooking (100% of the thick flakes is retained on U.S. sieve no. 8). The quick-cooking barley is used as a major ingredient in dry soups and as a thickener when a quick cooking product is required. The flakes are creamy white, pearled barley that has been steamed and rolled into thin flakes. Barley flakes provide a less chewy texture than oat flakes and can be used as an ingredient in granola products and speciality breads to provide texture. Barley flour is milled from barley grain that has been pearled, steamed and ground to produce a stable product in which enzyme activity has been minimized. No more than 2% is retained on U.S. sieve no. 2, and 70 to 80% pass U.S. sieve no. 100. Barley flour can be used as a thickener, stabilizer, binder or protein source for baby foods, malt beverages, prepared meats and pet foods.

The main products of malting are malted barley and sprouts (rootlets). The main by-products of brewing are brewers' spent grains, spent hops, and yeast (Pomeranz, 1973); Table 14.16 summarizes their composition.

REFERENCES

Acker, L. (1974). *Getreide, Mehl, Brot*. **7**, 181–187

Acker, L. and Becker, G. (1971). *Staerke*. **23**, 419–424.

Acker, L., Schmitz, H. J., and Hamza, Y. (1967). *Ber. Getreidechemiker Tagung; Arbeitsgemein. Getreideforschung*, pp. 30–34, Granum Verlag, Detmold.

Bartolome, L. G. (1981). Private communication, Henkel Corporation, Minneapolis, Minnesota.

Brekke, O. L. (1970). In *Corn: Culture, Processing, Products* (G. E. Inglett, ed.), Chapter 14, AVI Pub. Co., Westport, Conn.

Brockington, S. F. (1970). In *Corn: Culture, Processing, Products* (G. E. Inglett, ed.), Chapter 15, AVI Pub. Co., Westport, Conn.

Brown, C. M. and Craddock, J. C. (1972). *Crop Sci.*, **12**, 514–515.
Caldwell, E. F. and Pomeranz, Y. (1973). In *Industrial Uses of Cereals* (Y. Pomeranz, ed.), Chapter 12, American Association of Cereal Chemists, St. Paul, Minnesota.
Chang, S. C., Saunders, R. M. and Luh, B. S. (1980). In *Rice: Production and Utilization* (B. S. Luh, ed.) Chapter 23, AVI Pub. Co., Westport, Conn.
Cornelius, J. A. (1980). *Trop. Sci.*, **22**, 1–26.
Hahn, R. R. (1969). *Cereal Science Today*, **14**, 234–237.
Harness, J. (1978). In Seminar Proc. Products Corn Refining Industry in Food, 7–10, Corn Refiners Assoc. Inc., Washington, D.C.
Harris, D. W. (1981). Private Communication, Clinton Corn Processing Co.
Houston, D. F. (1972). In *Rice: Chemistry and Technology* (D. F. Houston, ed.), Chapter 11. American Association of Cereal Chemists, St. Paul, Minnesota.
Houston, D. F. and Kohler, G. O. (1970). *Nutritional Properties of Rice*. U.S. National Academy of Sciences, Washington, D.C.
Juliano, B. O. (1972). In *Rice: Chemistry and Technology* (D. F. Houston, ed.), Chapter 2, American Association of Cereal Chemists, St. Paul, Minnesota.
Juliano, B. O. (1980). In *Rice: Production and Utilization* (B. S. Luh, ed.), Chapter 10, AVI Pub. Co., Westport, Conn.
Kennedy, B. M. (1980). In *Rice: Production and Utilization* (B. S. Luh, ed.), Chapter 11, AVI Pub. Co., Westport, Conn.
Leavell, G. (1942). Brewers and distillers by-products and yeast in livestock feeding. *U.S. Dept. Agr. Bureau Animal Ind. Bull.* No. 58, Washington, D.C.
Luers, H. (1950). *Die Wissenschaftlichen Grundlagen von Malzerei und Brauerei*. Verlag Hans Huber, Nurnberg, W. Germany.
Morrison, W. R. (1978). In *Advances in Cereal Science and Technology* (Y. Pomeranz, ed.), Vol. 2, pp. 221–348, American Assocation of Cereal Chemists, St. Paul, Minnesota.
Pomeranz, Y. (1973). In *Industrial Uses of Cereals* (Y. Pomeranz, ed.) American Association of Cereal Chemists, St. Paul, Minnesota.
Rapp, W. (1978). In Seminar Proc. Products Corn Refining Industry in Food, 11–17, Corn Refiners Assoc., Inc., Washington, D.C.
Reiners, R. A. (1978). In Seminar Proc. Products Corn Refining Industry in Food, 18–21, Corn Refiners Assoc., Inc., Washington D.C.
Rohrlich, M. and Bruckner, G. (1966). *Das Getreide und seine Verarbeitung*, 2nd ed. Paul Parey Verlag, Berlin.
Watson, S. A. (1970). In *Soghum: Production and Utilization* (J. S. Wall and W. M. Ross, eds), Chapter 17, AVI Pub. Co., Westport, Conn.
Youngs, V. L., Peterson, D. M. and Brown, C. M. (1982). In *Advances in Cereal Science and Technology* (Y. Pomeranz, ed.), American Association of Cereal Chemists, St. Paul, Minnesota.

15 Lipids in Rice and Rice Processing

BIENVENIDO O. JULIANO

The International Rice Research Institute, Los Baños, Laguna, Philippines

I LIPIDS IN RICE

Several reviews have been published recently on aspects of rice grain lipids (Juliano, 1977; Fujino, 1978; Morrison, 1978a). Therefore this chapter will emphasize data since 1976 and properties specific to rice and rice processing. Acyl lipids are discussed by Morrison (Chapter 2) and non-saponifiable lipids by Barnes (Chapter 3).

"Lipids in Cereal Technology"
ISBN 0-12-079020-3

A Gross Structure of the Rice Grain

The structure of the cereal grain and location of lipid reserves are discussed by Angold (Chapter 1). The gross structure and properties of the rice grain have been reviewed recently (Juliano, 1980). The grain consists of the edible portion (the caryopsis or fruit) called brown rice and the covering hull composed mainly of two modified glumes: the lemma and the palea (Juliano, 1983) (Fig. 15.1). The pericarp, testa, nucellus and aleurone layer cover the starchy endosperm and the germ. The pericarp and aleurone layers are thicker in the dorsal region than in the lateral and ventral regions. The starchy endosperm consists of the subaleurone layer and the inner endosperm. The subaleurone layer is 2 cells in thickness with fewer starch granules and more protein bodies than the inner endosperm (Harris and Juliano, 1977). The hull constitutes 18–28% of the grain weight. The embryo accounts for 1 to 2% of the weight of brown rice. Commercial germ, which includes the outer layers enclosing the embryo, represents up to 3% of brown rice weight.

A cuticular layer envelops the nucellus and the rest of the grain under the testa except on the dorsal region at the pigment strand connecting the vascular bundles with the nucellus (Oparka and Gates, 1981). This cuticular layer is probably a major source of bran wax.

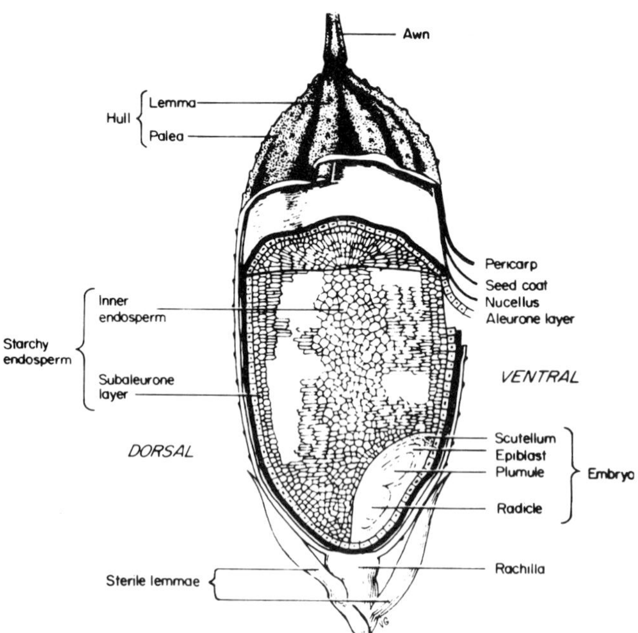

Fig. 15.1 Longitudinal section of the rice grain (Juliano, 1983).

Lipids are present in the aleurone layer, germ and subaleurone layer of the starchy endosperm in the form of lipid bodies or spherosomes (Buttrose and Soeffky, 1973; Bechtel and Pomeranz, 1977, 1978; Bechtel and Juliano, 1980). Lipid bodies have an electron-dense boundary but the presence of a membrane around them is questionable (Bechtel and Pomeranz, 1977). The lipid bodies are largest (≤1.5 μm) in the aleurone layer, followed by the subaleurone layer (≤1 μm) (Bechtel and Juliano, 1980) and small (≤ 0.7 μm) in all tissues of the embryo.

Unlike aleurone and germ protein bodies, those of the starchy endosperm, particularly the core portion of the large spherical protein bodies, are rich in lipids (Tanaka *et al.,* 1978; Resurreccion and Juliano, 1981). Starch granules also contain lipids from the enclosing plastid membrane and bound within the granules. Light-microscope examination of the inner cells of the starchy endosperm showed Sudan IV staining along the cell walls (which suggests lipids) probably derived from the cell membrane (Yoshizawa *et al.,* 1980).

B Composition of Lipids of Brown Rice and Hull

Recent reviews have been published on the chemistry of rice lipids (Juliano, 1977; Fujino, 1978; Morrison, 1978a). Most of the studies on lipid distribution of the rice grain used non-polar solvents, but even polar solvents do not completely extract starch lipids (SL) unless the granules are swollen by absorption of water (Choudhury and Juliano, 1980a; Maniñgat and Juliano, 1980; Azudin and Morrison, 1982).

The hulls of IR42 rice had about 0.5% lipids of which 64% was non-polar lipid (NL), 25% glycolipid (GL), and 11% phospholipid (PL) (Choudhury, 1979). In terms of whole (rough) rice lipids, hull contributed 0.4% to total non-starch lipids (NSL), 0.3% of the NL fraction, 17% of the GL fraction, and 7% of the PL fraction (Choudhury and Juliano, 1980a). The hull has no starch granules and, therefore, no SL. Hexane-extracted hull lipids have a fatty acid (FA) composition of 18% 16 : 0, 42% 18 : 1, and 28% 18 : 2 (Hartman and Lago, 1976).

More studies have been done on the distribution and composition of lipids in brown rice than in hull (Choudhury and Juliano, 1980a, b). Hexane and diethyl ether extract mainly NSL* which consist primarily of free fatty acids (FFA) and triglycerides (TG). Extraction was more efficient with chloroform-methanol (2 : 1) followed by water-saturated butan-1-ol (WSB). The NSL content of three brown rices ranged from 2.9–3.4% (dry basis) and consisted mainly of NL (Choudhury and Juliano, 1980b) (Table

* Non-starch lipids (NSL) and starch lipids (SL) are described in Chapter 2 and references to preferred methods of extraction are given in Appendix 1.

Table 15.1 The NSL composition of three brown-rice samples, differing in amylose content, and of their milling fractions†

Property	Brown rice	Bran	Embryo	Polish	Subaleurone layer	Inner endosperm
Wt. % of brown rice	100	5.90–6.4	1.3–1.5	4.1–4.4	4.9–5.2	82.5–83.8
NSL content (% dry basis)	2.9–3.4	19.4–25.5	34.1–36.5	10.2–15.0	5.6–8.5	0.41–0.81
NSL composition (%)						
NL	85–87	88–90	91–92	86–88	82–86	66–76
GL	4–6	4–5	2–3	4–5	5–7	12–18
PL	8–9	7–8	6–7	8–9	8–12	12–17
Lipid classes (% of total)						
TG	69–71	75–76	77–79	70–74	58–62	30–37
FFA	6–7	4–5	4	5–8	13–17	27–29
ASG	2–3	2	1	2–3	2–4	5–6
SG	< 1	< 1	< 1	1	1–2	2–3
DGDG	< 1	< 1	< 1	< 1	1	1–2
PE	3–4	3	3–4	3	3–4	3–5
PC	4	3–4	3–4	3–4	4	3–5
LPE	< 1	< 1	< 1	< 1	1–2	2–4
Others	8–11	8–9	8–9	9–12	11–12	15–17
FA composition‡ (wt. %)						
16 : 0	22–24	22–25	23–25	23–25	26–28	32–34
18 : 1	33–37	36–37	36–38	36–39	26–28	20–21
18 : 2	36–40	36–37	35–38	37–39	40–41	38–41

†Choudhury and Juliano (1980b).
‡Includes <1–3% 14 : 0, 2–4% 18 : 0, and 1–2% 18 : 3.

15.1). In samples extracted immediately after grinding, the major NSL classes were TG and FFA, acyl steryl glycoside (ASG), phosphatidylethanolamine (PE), and phosphatidylcholine (PC) (Table 15.1). The FA of brown rice NSL were mainly 16 : 0, 18 : 1, and 18 : 2. The NL fraction had identical FA composition to that of the total NSL but the GL and PL fractions had more 16 : 0 and less 18 : 1 (Table 15.2).

The lipid distribution in fractions from the three brown rices expressed as percentage of total NSL was 39–41% in bran, 14–18% in germ, 15–21% in polish, 12–14% in outer endosperm or subaleurone layer, and 12–19% in inner endosperm (Choudhury and Juliano, 1980b). Content of lipids (dry basis) was highest for germ, followed by bran, polish, subaleurone layer, and, lastly, inner endosperm (Table 15.1).

SL are mainly confined to the starchy endosperm where starch granules are present in the mature rice grain (Juliano, 1980). Choudhury and Juliano (1980a, b) used WSB (3 × 8 h, 25 °C, 5 : 1 solvent to sample ratio) to extract SL from the residue after NSL extraction (Morrison, 1978a). However, WSB at 90°–100 °C extracts SL more effectively than does WSB at ambient temperature, probably due to simultaneous starch granule swelling.

Although the cold WSB extraction used by Choudhury and Juliano did not extract 100% of the SL, the FA composition of SL as determined by fat-by-hydrolysis was identical to that of SL extracted with cold WSB: 42% 16 : 0, 15% 18 : 1, 38% 18 : 2 and 5% other FA (Choudhury and Juliano, 1980a). For IR42 brown rice, after removal of NSL, fat-by-hydrolysis was 0.41% and residual fat-by-hydrolysis, after extraction with cold WSB, was 0.09%. This is equivalent to a yield of 78% of SL using this cold solvent. In

Table 15.2 The range of FA composition of total NSL and SL and of the NL, GL and PL fractions of three brown rices†

		FA composition‡ (wt. %)		
Lipid fraction	% composition	16 : 0	18 : 1	18 : 2
Total NSL	100	23–24	32–37	36–40
NL	86	22–24	35–37	35–39
GL	5	30–33	24–26	36–38
PL	9	25–29	30–34	34–38
Total SL	100	43–48	12–17	29–40
NL	33	24–38	20–23	30–53
GL	20	53–55	7–12	26–32
PL	47	45–54	9–14	30–39

†Choudhury and Juliano (1980b).
‡Includes <1–3% 14 : 0, 2–4% 18 : 0, and 1–2% 18 : 3.

this example, 0.66% SL extracted by the cold WSB was equivalent to 0.32% fat-by-hydrolysis (SL was measured gravimetrically, whereas fat-by-hydrolysis represents the amount of FA derived from SL released by hydrolysis).

The SL of brown rice extracted by cold WSB were lower in a waxy rice (≤2% amylose) than in two non-waxy rices (Choudhury and Juliano, 1980b) (Table 15.3). The SL of waxy rice had more NL, but less PL than that of non-waxy rices. The principal SL classes were FFA, lysophosphatidylethanolamine (LPE) and lysophosphatidylcholine (LPC) for both waxy and non-waxy rices. The FA composition of SL showed more variation among the three brown rices (Table 15.3) than that of the NSL (Table 15.1); non-waxy rice SL had more 18 : 2 and less 18 : 1 than did waxy rice SL. Analysis of the lipids showed that the NL fraction had the highest 18 : 1 content and the lowest 16 : 0 content but that the opposite was true for the GL fraction (Table 15.2). The SL fractions also had a wider range of FA compositions than did the NSL fractions.

Table 15.3 The composition of SL in three brown rices differing in amylose content†

Property‡	IR4445-63-1	IR480-5-9	IR42
SL content (%)	0.21	0.76	0.66
SL composition (%)			
NL	41	28	28
GL	21	20	18
PL	37	52	54
Lipid classes (% of total)			
TG	5	4	4
FFA	28	20	21
ASG	3	2	2
SG	1	1	1
MGMG	3	3	2
PE	3	4	4
PC	5	5	5
LPE	15	21	21
LPC	18	21	23
Others	18	19	17
FA composition (wt. %)			
16 : 0	45	43	48
18 : 1	17	12	13
18 : 2	29	40	35
Others	9	5	4

†Choudhury (1979); Choudhury and Juliano (1980b).
‡Amylose content: 2% for IR4445-63-1 (waxy), 24% for IR480-5-9, and 29% for IR42.

C Lipids of Rice Bran and Milled Rice

1 *Rice Bran Lipids*

Brown rice is processed by abrasive milling to produce about 5% by weight of bran, 5% polish and 90% well-milled rice. Commercial bran usually includes the germ. Bran lipids represent the source of commercial bran oil, and have a NL : GL : PL ratio close to that of brown rice NSL (Choudhury and Juliano, 1980b) (Table 15.1); a similar ratio is obtained for germ lipids. The lipid classes of bran and germ are mainly TG, plus FFA, ASG, PE, and PC and the FA compositions of bran and germ lipids are similar (Table 15.1). The NL has a similar FA composition to total NSL, but GL have more 16 : 0 and less 18 : 1 particularly in the germ (Choudhury and Juliano, 1980b).

Miyazawa *et al.* (1977) reported the major bran PL of waxy and non-waxy rice bran to be PC, PE, and phosphatidylinositol (PI). Total lipids of rice bran contain 90.6% NL (Miyazawa *et al.*, 1978).

A detailed study of the GL fraction of rice bran oil has been made at Obihiro University, Hokkaido, Japan (Fujino, 1978), including oligogalactosyl glycerides (Fujino and Miyazawa, 1979) and cellooligosyl sitosterol (Ohnishi and Fujino, 1980).

2 *Milled Rice Lipids*

The lipid compositions of 85% milled rice from the brown rice samples described in Table 15.1 are given in Table 15.4. Waxy rice has more NSL and less SL than non-waxy rice, a greater proportion of NL but less GL and PL in the NSL and greater proportions of NL and GL but less PL in the SL. Major NSL classes are TG and FFA for all three samples and major SL classes are FFA, LPC, and LPE. The FA composition of the NSL was similar among the samples but exhibited less 16 : 0 and more 18 : 1 than the FA of SL. NSL showed a closer range of composition among the three samples than did SL.

3 *Lipids of Destarched Milled Rice*

The residue from cooked milled rice (IR480-5-9) consisted mainly of protein bodies after destarching with α-amylase (*Aspergillus oryzae*), particularly after sieving (Tanaka *et al.*, 1978). It contained 7.3% total lipids and 84% crude protein. The lipids consisted of 76–79% NL, 10–12% GL, and 12–13% PL (Tanaka *et al.*, 1978). A similar destarched preparation from IR32 milled rice contained 12–13% lipids.

Another destarched preparation from IR480-5-9 milled rice contained 9.5% lipids with a NL : GL : PL ratio of 92 : 5 : 3 (Resurreccion and

Table 15.4 The lipid composition of three 82–84% milled rices differing in amylose content†

Property‡	NSL			SL		
	IR4445-63-1	IR480-5-9	IR42	IR4445-63-1	IR480-5-9	IR42
Content (%)	0.81	0.41	0.45	0.12	0.57	0.55
Lipid composition (%)						
NL	76	66	66	47	26	27
GL	12	17	18	29	15	17
PL	12	17	16	24	59	56
Lipid classes (% of total)						
TG	36	27	35	5	2	1
FFA	29	28	26	37	21	22
ASG	4	6	6	3	2	1
SG	2	2	3	2	2	1
DGDG	3	4	4	tr	tr	tr
MGMG	tr	tr	tr	3	2	3
PE	3	4	4	3	4	4
PC	3	5	5	3	4	5
LPE	3	3	3	12	23	22
LPC	3	4	4	13	24	24
Others	14	17	10	16	17	17
FA composition (wt. %)						
16 : 0	32	34	33	44	44	46
18 : 1	21	21	20	18	10	13
18 : 2	40	38	41	30	42	37
Others	7	7	6	8	4	4

†Choudhury (1979); Choudhury and Juliano (1980b).
‡Amylose content: 2% for IR 4445-63-1 (waxy), 24% for IR480-5-9, and 29% for IR42.

Juliano, 1981). The NL fraction was mainly FFA plus TG, diglycerides (DG) and monoglycerides (MG), and the major PL were LPC and LPE plus traces of PC and PE. Major GL were SG, ASG, DGDG, MGDG, and ceramide monohexoside (Mano, 1982). The FA composition of total lipids of two samples was 1% 14 : 0, 30–35% 16 : 0, 2% 18 : 0, 20–26% 18 : 1, 38–40% 18 : 2, and 2% 18 : 3 (Mano, 1982). Protein body preparations derived from treatment with pancreatic α-amylase had a NL : GL : PL ratio of 65 : 7 : 39. The ratio for milled rice destarched with *A. oryzae* α-amylase was 92 : 6 : 2 with a waxy cooked rice (IR29).

Pepsin-treated protein bodies and fecal protein particles are derived from the core of the large spherical protein bodies of milled rice and have a higher lipid content (18–22%) than whole protein bodies (Tanaka *et al*.,

1978; Resurreccion and Juliano, 1981). The core must be rich in lipid since protein bodies from pepsin-treated cooked IR480-5-9 milled rice, representing 35% of protein-body weight, accounted for 80% of the lipids and 27% of the protein of the whole preparation (Resurreccion and Juliano, 1981). The core lipids had similar composition to whole protein-body lipids in terms of lipid classes and FA composition.

4 *Lipids of Starch Granules*

The level and composition of SL depend to a large measure on the method used for preparing rice starch granules (Maniñgat and Juliano, 1980; Azudin and Morrison, 1982). Starch may be prepared either from brown or milled rice, but the recovery on a starch basis is higher for brown rice due to less damaged starch.

The highest recovery of starch lipids (~ 1%) was obtained using alkaline protease to remove protein (Maniñgat and Juliano, 1980). The extraction of protein from IR480-5-9 milled rice with sodium dodecyl benzene sulphonate (DoBS) reduced the lipids extractable in cold WSB from 0.8 to 0.44% (Maniñgat and Juliano, 1980), probably due to part extraction and simultaneous exchange with lipids (Fujii, 1972). DoBS further contaminates the GL fraction on lipid fractionation (Maniñgat and Juliano, 1980). The alkali-treatment to remove protein reduces further the lipids extractable in cold WSB (0.28% for 0.1% NaOH, and 0.09% for 0.2% NaOH) probably by saponifying the PL and glycerides to FFA (Maniñgat and Juliano, 1980).

Complete extraction of SL is possible only after starch swelling as occurs with hot WSB; the lipid content of a Japanese rice starch prepared with 0.1% NaOH was 0.62% by hot methanol, 0.83% by acid hydrolysis for 30 min, and 0.82% by two hot WSB extractions (Ohashi *et al.,* 1980). Another Japanese rice starch sample prepared by sonication and DoBS treatment had 0.69% hot WSB lipids (Ito *et al.,* 1979). Eight DoBS-prepared rice starches had 0.03–0.44% lipids extractable with cold WSB (Maniñgat and Juliano, 1980) and 0.10-0.82% lipids extractable with hot WSB (IRRI, 1983). The SL content followed the amylose content closely but tended to be highest for intermediate-amylose content rices (Fig. 15.2).

Analysis by thin-layer chromatography (tlc) of lipid fractions of eight DoBS-prepared rice starches showed that the main NL class was FFA (88–97%) with minor amounts of MG, 1,2- and 1,3-DG, and sterols (Maniñgat and Juliano, 1980). Major GL were digalactosylmonoglyceride (DGMG) and monogalactosylmonoglyceride (MGMG) together with smaller amounts of SG and contaminant DoBS. Major PL were LPC (61–91%) followed by LPE with smaller amounts of PC and PE. Ito *et al.* (1979) reported the SL (0.69%) of a DoBS-prepared starch to consist of

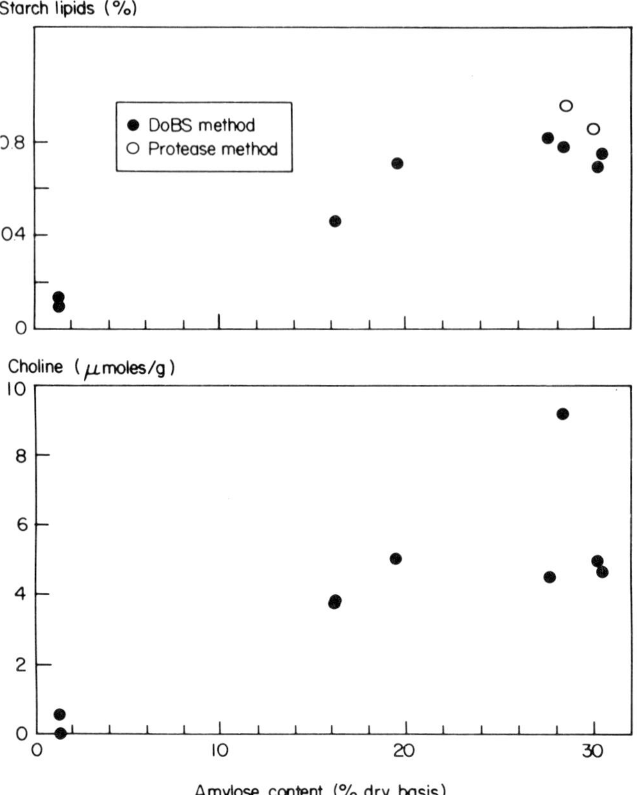

Fig. 15.2 Relationship between amylose content of starch and the content of lipids and of choline of rice starches prepared from milled rice by DoBS extraction of rice proteins.

32.2% FFA, 30.0% LPC, 11.4% MGMG, 7.5% DGMG, 6.9% LPE, 3.4% MG, and 8.6% other lipids.

A sample of alkaline protease-prepared rice starch had 0.17% NSL or surface SL (30 min extraction with cold WSB) and 0.66% SL (Azudin and Morrison, 1982). The NSL was composed of 5% DG plus FFA, 29% lysoPL, 6% DGXG, 3% each of MG plus ASG and TG, and 2% each of SE and MGXG. The SL fraction had 44% DG plus FFA, 28% LPC, 6% LPE, 5% MG plus ASG, 4% each of DGXG and lysophosphatidylglycerol (LPG), 3% TG, and 3% each of SE and MGXG. FFA, LPC and LPE were the major SL of two non-waxy starches (Sawada and Mano, 1980).

Rice starch prepared using NaOH (0.1%) yielded 0.15% NSL or surface SL and 0.55% SL (Azudin and Morrison, 1982). Major NSL were 91% DG plus FFA, and 6% TG, and major SL were 70% DG plus FFA, 22% lysoPL (mainly LPC), 4% DGXG, 3% MG plus ASG, 2% each of MGXG, and SE, and <1% TG. Alkali-treatment was confirmed to lower the proportion of phospholipids in the SL as reported by Maniñgat and Juliano (1980).

Choline content also followed closely the amylose content of rice starch (Fig. 15.2) and the choline : phosphorus ratio was 0.75–0.89 in six DoBS-prepared non-waxy rice starches, and 0 and 0.03 for two waxy rice starches (Hizukuri and Juliano, 1982). The greatest choline content found was for IR480-5-9, an intermediate-amylose content rice. The ratio supports the presence of 73.5% LPC in the lysoPL of SL of non-waxy rice starch as obtained by Azudin and Morrison (1982).

The FA composition of SL from DoBS-prepared rice starches consisted of 47% 16 : 0 and 40% 18 : 2 (Ito *et al.*, 1979). The SL of a waxy rice starch had a FA composition of 58% 16 : 0, 19% 18 : 1, and 14% 18 : 2 (Maniñgat and Juliano, 1980). The SL of two protease-prepared starches had FA composition of 37 and 40% 16 : 0, 11% 18 : 1, and 37 and 46% 18 : 2 (Sawada and Mano, 1980).

D Factors Affecting Lipid Composition

An earlier cropping season (daily mean ripening temperature 27.1–27.8 °C) in Japanese rice resulted in a higher crude fat content of the brown rice grain (2.83%) of six non-waxy varieties than did normal-season culture (2.65%, ripening temperature 24.5–26.5 °C) and late-season culture (2.37%, ripening temperature 17.0–21.3 °C) (Taira *et al.*, 1979). With decreasing ripening temperature, 18 : 1 content decreased progressively (41.0, 40.0 and 37.3%) and 18 : 2 content increased progressively (36.6, 37.3, and 40.4%). The content of 18 : 3 also increased progressively (1.4, 1.7, and 2.0%). The 18 : 1 and 18 : 2 contents of the oil showed the highest correlation ($r = -0.98^{**}$). The changes in lipid content were mainly in crude fat since mean fat-by-hydrolysis (after diethyl ether extraction) remained constant at 0.96–0.98% for the six milled rices (Taira *et al.*, 1980).

Lugay and Juliano (1964) found two japonica (= sinica) rices to have a higher 18 : 2 content in hexane-extracted bran lipids than two indica rices. Wang *et al.* (1980) suggested that 18 : 2 and 18 : 3 content of membrane lipids of the rice embryo might be correlated with adaptability to low temperature. Recent studies on japonica and indica rices grown at the IRRI indicated no major distinction in FA composition among the hexane-extracted brown rice lipids suggesting that environmental factors are probably involved (IRRI, 1981).

Changes in lipid composition during grain development followed closely the early development of the aleurone and embryo which are richer in NSL than the endosperm (Choudhury and Juliano, 1980a).

The amount of SL in milled rice and starch correlates positively with the amylose content of the starch or milled rice (Maniñgat and Juliano, 1980; Choudhury and Juliano, 1980b) (Table 15.4). Differences in lipid composition would be expected among samples differing in amylose content, particularly in the NL : GL : PL ratio because of the differences in lipid classes between NSL and SL. The SL have a higher PL content and relatively higher 16 : 0 and lower 18 : 1 content than do the NSL (see Section IC). For example, Kawashima and Kiribuchi (1980) reported 1.16% lipids in waxy rice flour (0.78% NSL and 0.38% SL) as compared with 1.21% lipids for a non-waxy rice flour (0.59% NSL and 0.62% SL). Thus, non-polar solvents, such as diethyl ether or hexane, extract mainly NSL and give essentially the same FA composition for all brown rice milling fractions. Different FA compositions are obtained with the use of more polar solvents, such as chloroform-methanol (2 : 1), since the SL are also partly extracted (Resurreccion and Juliano, 1975).

The degree of milling has a greater effect on NSL than on SL, the latter, as measured by fat-by-hydrolysis, remains constant at 0.6-0.7%, even on milling to 60% by weight of brown rice, whereas the NSL content dropped from 3% to 0.07% (Yoshizawa *et al.*, 1973). Similar changes were noted in recent studies on three brown rices by Choudhury and Juliano (1980b) (Table 15.1).

The NSL contain more GL and PL and less NL in three 82–84% milled rice than do the lipids of the subaleurone layer, polish, bran, and germ (Choudhury and Juliano, 1980b) (Table 15.1). The SL of the three overmilled rices (Table 15.4) had ratios of NL : GL : PL similar to those of the SL of brown rice (Table 15.3). A recent study of the milling fractions of three brown rices showed that the NSL of the subaleurone layer and the inner endosperm are richer in 16 : 0 and 18 : 2, and contain less 18 : 1 than the lipids of the bran-embryo (Choudhury and Juliano, 1980b). (See Table 15.1.) The FA composition changed from 18% 16 : 0, 31% 18 : 1 and 48% 18 : 2 in 80% milled rice to 26% 16 : 0, 10% 18 : 1, and 59% 18 : 2 in the 60% overmilled rice (Yoshizawa *et al.,* 1973) with overmilling between 80% and 60% (milled rice yield from brown rice). The SL of overmilled rice and brown rice have similar FA compositions (Tables 15.3 and 15.4). With overmilling from 80% to 60% yield, the SL (fat-by-hydrolysis) increased in 16 : 0 content but decreased in 18 : 2 content (Yoshizawa *et al.,* 1973).

II RICE BRAN OIL PRODUCTION

A Heat Stabilization of Bran

A major problem in the widespread extraction of oil from rice bran is the high lipase activity which results in FFA formation within a few days of milling, particularly at high temperature and humidity (Desikachar, 1977; Yokochi, 1977; Barber and Benedito de Barber, 1980). The freshly recovered bran of stored brown rice in Japan increases in FFA content of the oil during storage, from 2–4% in a fresh crop to 5–8% in a 1-year-old crop and >10% for a 2-year-old crop (Yokochi, 1977). The FFA are mainly derived from NL glycerides (Ryu and Cheigh, 1980).

The lipase is readily inactivated by heating bran, for example 3 min at 95–100 °C, with a decrease in moisture content to 3–4% (Yokochi, 1977). Lower temperatures are less effective and increase in FFA may occur on storage in humid conditions. This second FFA formation, even in heat-stabilized bran (Viraktamath and Desikachar, 1971), is probably due to microbial lipase. The lipolytic bacteria in rice are mainly *Xanthomonas* species (DeLucca II *et al.,* 1978).

Heat-stabilized bran may be stored for about 3 months, but should be used preferably for oil extraction within 1 month. Stabilized bran stored for 2–3 months yields 1-2% less oil and the oil is darker (Yokochi, 1977).

Various continuous heating processes have been proposed for bran stabilization at the level of the village mill (Desikachar, 1977; ASRCT, 1977). Low-cost extrusion cookers, developed for American farmers to cook soyabeans for animal feed, have recently been proposed for bran stabilization (Cheigh *et al*., 1980; Sayre *et al*., 1982; Enochian *et al*., 1981). Rice bran can be stabilized at throughputs up to 500 kg h^{-1} at 130 °C at a cost of $10 or less per ton of rice bran, depending on the price of electricity.

B Oil Extraction

Details of the solvent extraction of oil from rice bran are given in monographs on rice and rice processing (Yokochi, 1977; Chang *et al.,* 1980; Houston, 1972). Extraction by hydraulic pressing yields high-quality crude oil and oil cake, but the yield of oil is less (10–12%) than by solvent extraction (16–18%) (Yokochi, 1977).

The typical steps in the Japanese pressure expelling system are:

(1) purification of raw bran by sifting and air to remove whole and broken grains;

(2) steam cooking of the purified bran at 4–5 kg cm^{-2} steam pressure followed by drying;

(3) prepressing at 70 kg cm^{-2} and oil cake discharging;

(4) pressing with ring or cage hydraulic-type presses at 105–316 kg cm^{-2} (Yokochi, 1977).

Hexane extraction may either be batch, battery or continuous type (Yokochi, 1977). In Japan, the batch and battery types are considered adequate, but continuous extraction systems are operating in Burma, Egypt, India, Mexico, Taiwan, and Thailand. Bran is also prepurified, steamed, pelletized if necessary, and dried. Batch extraction is the oldest system. The pretreated bran is placed in one or more extractors, hexane is pumped in, and the solvent level is maintained to percolate the bran and extract the oil. The miscella is passed through the filter onto the evaporator for desolventizing. In battery extraction the fresh solvent is applied only to one batch, whereas the miscella obtained is used to treat the contents of all the other extraction vessels consecutively in a countercurrent system.

Continuous extraction uses the countercurrent principle; it may be used for other oil seeds (Yokochi, 1977) and is either the immersion or percolation type (Takeshita, 1972). Some mills use an expeller prior to continuous solvent extraction.

Bran that is contaminated with hull, such as that produced by Engelberg-type dehullers which dehull and mill in one step, cannot be economically used for oil extraction because of its lower oil content of 4–9% (Yokochi, 1977).

The solvent-extractive rice milling process of Riviana Foods, Inc. (X-M process) does not have the problem of FFA formation in the bran (Hunnel and Nowlin, 1972). The bran-removal step is done in the presence of an alkane solvent to increase milled rice yield, reduce the amount of broken grains and extract fat from the entire grain, resulting in edible defatted bran and crude dewaxed rice oil. Brown rice is pretreated with warm rice oil (0.5%; 2–4 h) and milled in the presence of a rice oil-hexane miscella. The milled rice is desolventized and the slurry of bran, hexane and rice oil is used for oil extraction. The bran product contains a residual small proportion of oil and some lipase activity (Lynn, 1969).

Takeshita (1982) reported a new Japanese extraction method for rice bran oil involving premoulding of rice bran at 14% moisture and ≤40 °C and extraction with hexane at ≤15 °C to obtain light-coloured crude oil free of wax. The wax may be subsequently extracted from the defatted bran.

C Refining Bran Oil

Refining crude rice bran oil usually involves dewaxing to remove the relatively high melting temperature wax, deacidification to remove FFA and removal of gums, bleaching to improve colour, and steam deodorizing (Cornelius, 1980).

The simplest way to recover rice wax is by using settling tanks, cooling crude oil gradually and filtering or centrifuging the sludge at low temperature (Chang *et al.*, 1980). The wax is then removed from the settlings by various processes (Chang *et al.*, 1980).

Deacidification with alkali and bleaching with clay are used in Japan for refining crude rice bran oil (Yokochi, 1977). However, some refiners have adopted molecular distillation of FFA at high temperature and low pressures. Miscella alkali refining is also practised to reduce the entrainment of TG by the soap stock which results in large edible oil losses in the conventional alkali-refining processes. Phospholipids also act as detergents during the deacidifying process and are removed as part of the gums. Details of these processes are given by Chang *et al.* (1980). FFA are converted to soap and settle down as "soap stock" or "foots" together with gums when the emulsion is disrupted by heating. Degumming involves the use of water or a suitable organic or inorganic chemical, and may be done before deacidification, the swelled gum and water being usually removed by centrifugation (Takeshita, 1972). The degummed oil (with 0.10–0.14% water) is preferably deacidified immediately.

Bleaching is done to lighten the colour of the deacidified oil, using either activated carbon (Chang *et al.,* 1980) or bleaching earth (Yokochi, 1977).

Deodorizing the neutralized, bleached oil involves steam treatment to remove objectionable odours such as peroxides, aldehydes, and ketones, as well as the pleasant characteristic odours of rice bran oil (Chang *et al.,* 1980). The more modern units use steam-vacuum deodorization.

D Bran Oil and Wax Properties

Yokochi (1977) reports the properties of Japanese crude rice bran oil as follows: iodine value 92-115, saponification value 175–192, energy value 9.438 kcal g^{-1}, and 3–8% unsaponifiable matter. Japanese standards for refined rice bran oil are quoted as cloud point ≤15 °C, ≤5% unsaponifiable matter, saponification value 180–195, iodine value 92–115, relative density (25°/25 °C) 0.913–0.919, and refractive index 1.470–1.473 (Takeshita, 1972). The winterized oil passes the cold test at 0 °C for 1 h. The NSL from other brown-rice fractions have similar properties to bran oil (Choudhury and Juliano, 1980b). Thus crude rice oil obtained from the X-M process, which represents not only bran oil but also oil from the surface of the milled rice, had 2.5–3.5% wax, a relative density of 0.917–0.920 at 25 °C, iodine value of 100–105, saponofication value of 188–192 and 2% unsaponifiable matter (Hunnel and Nowlin, 1972).

The chemical composition of bran oil has been discussed in Section IC and minor constituents of rice oils are described in recent reviews (Juliano, 1977; Fujino, 1978; Morrison, 1978a; Yokochi, 1977; Chang *et al.,* 1980).

Crude wax is contaminated by oil, but the purified wax has a melting point of 75–76 °C, a saponification value 73, iodine value 13, and 49% unsaponifiable matter (Tsuchiya, 1948). Ito *et al* (1981) recently characterized the lipids of rice bran wax and found the major lipids to be SE, long-chain alkyl ester, short-chain alkyl ester and hydrocarbon. Th major FA are 52% 18 : 2, 33% 18 : 1, and 11% 16 : 0 for SE; 33% 22 : 0, 31%

24 : 0, and 24% 16 : 0 for long-chain alkyl ester; and 60% 18 : 1 and 36% 16 : 0 for short-chain alkyl esters. Sitosterol was 57% of the free sterols and 64% of SE. Major alcohols of long-chain alkyl esters were 19% 30 : 0, 18% 34 : 0 (branched), 12% 28 : 0, 11% 24 : 0, and 10% 32 : 0. Representative molecular species of the short-chain alkyl ester were methyl 18 : 1, methyl 16 : 0, and ethyl 18 : 1. The hydrocarbon fraction consisted of alkanes (46% 29 : 0, 24% 31 : 0, and 6% 33 : 0), alkenes (39% 29 : 1, 21% 31 : 1, 19% 33 : 1), and branched-chain alkene (squalene). Bran wax has a composition similar to that of epicuticular wax from the rice plant (Bianchi *et al.*, 1979).

E Parboiling and its Effects on Lipids

Parboiling is a process of steeping freshly-harvested grain either in cold or hot water below the starch gelatinization temperature, followed by heating to gelatinize the starch without much brown rice expansion (with or without steam pressure) and slow drying. Mahadevappa and Desikachar (1968) noted that parboiling results in the rupture of lipid bodies in the bran layers, a phenomenon which makes the resulting bran more gummy, resulting in clogging of the sieves during milling. Parboiling has been consistently reported to result in a bran with higher oil content than the bran from raw rice, the increase being greater for medium and coarse grains than for slender grains (Mukherjee and Bhattacharjee, 1978). Coarse-grain varieties have more and thicker aleurone cells, particularly in the dorsal region, than fine-grain varieties (Juliano, 1980). Padua and Juliano (1974) explained the higher oil content of bran from parboiled rice as due to the harder endosperm with greater resistance to milling, resulting in a bran fraction with lower endosperm contamination.

Recent studies by Bhat Sondi *et al.* (1980) showed that the oil content of residual milled grain was consistently lower and that of bran higher, in parboiled compared with raw rice of one variety at all degrees of milling. The total oil content was unchanged by parboiling. Bhat Sondi *et al.* (1980) considered endosperm contamination of bran as most unlikely at 2% bran removal but did not present confirmatory histochemical and starch analysis. In addition, bran from parboiled rice is richer in protein than is bran from raw rice (Padua and Juliano, 1974) and it is unlikely that proteins also migrate outward into the bran layers during starch gelatinization. The outward diffusion of thiamine during parboiling is more readily explained as due to its diffusibility through cell membranes (Padua and Juliano, 1974).

Parboiling inactivates rice bran lipase (Viraktamath and Desikachar, 1971) and FFA production during storage of bran obtained from parboiled rice is slower than that for bran from raw rice. However, in addition to some decomposition of thiamine (Padua and Juliano, 1974), the vitamin E (tocol) content of the grain is also reduced by parboiling, as evidenced by the more rapid lipid autoxidation in milled parboiled rice than in milled raw rice (Sowbhagya and Bhattacharya, 1976; Desikachar, 1977). Lipid degradation in parboiled rice is primarily an oxidative rather than hydrolytic process.

III LIPIDS AND MILLED RICE PROPERTIES

A Fat Deterioration during Storage

The storage of rice grain, particularly as milled rice, is accompanied by both hydrolysis and oxidation of lipids (Villareal *et al.*, 1976). Yasumatsu and Moritaka (1964) showed that the contents of NSL and SL (fat-by-hydrolysis) of two Japanese milled rices remained the same during ambient temperature storage for 6 months. However, the FFA increased in the NSL fraction at the expense of glycerides. Carbonyl compounds which increased in the volatiles of cooked rice during storage were propanal/acetone, pentan-1-al and hexan-1-al (Endo *et al.*, 1978; Yasumatsu and Moritaka, 1979) derived mainly from the oxidation of 18 : 2 and 18 : 3 in the NSL (Yasumatsu *et al.*, 1966).

Waxy rices are more prone to lipid degradation than non-waxy rices (Villareal *et al.*, 1976; Perez and Juliano, 1981), probably because of the higher content of NSL in the waxy rices (Choudhury and Juliano, 1980b). Stored waxy rices have higher contents of FFA and carbonyl compounds than stored non-waxy rices. The NSL are probably involved in these reactions since Morrison (1978b) reported the SL in wheat flour to be resistant to oxidation by chlorine.

The storage of milled rice results in an increase in the Amylograph peak viscosity of the flour (Yasumatsu *et al.*, 1964). The difference in flour paste viscosity between milled rices stored at room temperature (15–20 °C) and at 9 °C disappeared when the flours were defatted with chloroform-methanol (2 : 1). In addition to decreasing peak viscosity, defatting also reduced the temperature of initial increase in viscosity of the rice flours (Yasumatsu *et al.*, 1964; Kawashima and Kiribuchi, 1980) and reduced the gelatinization temperature of the starch (Maniñgat and Juliano, 1980; Ohashi, *et al.*, 1980). Yasumatsu *et al.* (1964) showed that only the FFA fraction of the NSL caused the increase in Amylogram viscosity, with 0.7% giving the maximum effect.

Storage often results in a hardening of gel consistency of milled rice (Juliano *et al.*, 1980; Perez and Juliano, 1981). However, even after the

3–4 months period of rapid change, some samples during storage continue to harden in gel consistency or increase in gel viscosity in 0.2 N KOH. The addition of formaldehyde drastically changed the gel consistency of IR32 flour suggesting that carbonyl compounds from oxidation of polyunsaturated FA are responsible for this change in gel consistency in 0.2 N KOH, but not in Amylograph consistency (Juliano *et al.*, 1980; Perez and Juliano, 1981).

Surface-defatting of milled rice with light petroleum had little effect on the storage changes in cooked-rice hardness and stickiness, but reduced the changes in gel consistency and Amylograph setback and consistency (Perez and Juliano, 1981).

B Lipid Extraction during Water-Washing and Cooking

Hayakawa and Igaue (1979) reported that four washings of milled rice with water reduced crude fat content from 0.55% to 0.19%. Cold-water extraction of 8 milled rices (including 2 waxy rices) removed 0.10–0.29% (mean of 0.18%) of light petroleum soluble fraction (IRRI, 1983). Boiling the washed rice in excess water to the optimum cooking time required for the centre of the grain to gelatinize further extracted 0.03–0.18% (mean of 0.09%) of light petroleum-soluble material. The total extracted petroleum-soluble lipids during washing and cooking ranged from 0.19 to 0.34% of the milled rice weight (mean of 0.27%). The hot 80% ethanol-soluble fraction of the boiling water extract ranged from 0.07 to 0.30% of the milled rice weight (mean of 0.20%).

Cooking results in a reduction of the true digestibility, in monogastric animals, of milled-rice protein bodies, e.g. from 100% to 85–90% in rats (Eggum *et al.*, 1977). Similar values of 82–93% are reported for man for the protein of cooked milled rice (Hopkins, 1981). Recent studies suggest that the lower digestibility and solubility of the low-lysine, cysteine-rich core proteins of the large spherical protein bodies are due to increased disulphide cross-linking, brought about during cooking by disruption of hydrophobic bonding between protein and lipids (Tanaka *et al.*, 1978; Resurreccion and Juliano, 1981).

C Lipids and Texture of Cooked Milled Rice and Starch

Cooking involves the gelatinization of starch and melting of the amylose-lipid complex (Eberstein *et al.*, 1980). Thus, although waxy rice starch has a high enthalpy of gelatinization, the total cooking enthalpy is less than for a high-amylose non-waxy rice starch due to this latter process. Kugimiya and Donovan (1981) utilized the reversible nature of the transition of the amylose-lipid complex for the calorimetric determination of amylose content of starches based on formation and melting of the amylose-LPC complex. Waxy starch does not have this enthalpy because of the minimal amount of amylose and SL.

Rice starch has higher Amylograph viscosity than milled-rice flour, particularly waxy starch samples (Merca and Juliano, 1981; Shoji and Kurasawa, 1981). A plausible explanation for this phenomenon seems to be the removal of protein during starch preparation. However, the removal of lipids due to cold-water washing increases the Amylograph viscosity of rice flours, particularly waxy rice (Merca and Juliano, 1981). A cold-water extract of waxy rice flour, when boiled, cooled and re-added to the washed flour lowers paste viscosity, thus favouring lipids as the active fraction rather than α-amylase which would have denatured at 100 °C. Washing waxy rice flour with cold water also results in harder gel consistency of the washed flour and this supports the Amylograph data (Merca and Juliano, 1981).

Defatting waxy flour and starch also effects a greater increase in Amylograph viscosity than defatting non-waxy flour and starch, particularly for removal of NSL (Kawashima and Kiribuchi, 1980; Shoji and Kurasawa, 1981). Defatting rice starch with light petroleum followed by cold WSB results in low gel viscosity in 0.2 N KOH using a Wells-Brookfield cone-plate microviscometer, and in a lower gelatinization temperature of the defatted starch (Maniñgat and Juliano, 1980; Ohashi *et al.*, 1980). Adding palmitic acid, but not rice bran oil, to waxy milled rice further suppressed Amylograph peak viscosity as occurred with non-waxy milled rice (Merca and Juliano, 1981).

An investigation of the nature of FFA-amylopectin complex formation suggested that the outer chain segments of amylopectin were involved (about 12 glucose units). The β-amylolysis limit dextrin from amylopectin does not complex as well with palmitic acid as the native IR29 waxy starch does (IRRI, 1981) (see Table 15.5). LPC also complexes with aqueous IR29 waxy-starch gels resulting in a harder, more viscous gel at 1–2% of the starch. Thus, complexing of lipids with amylopectin probably involves – at the most – two helices of 6 glucose units each, based on the V-complex of amylose with relatively hydrophobic compounds such as butan-1-ol.

With increasing severity of parboiling, the resultant parboiled rice takes a longer period to cook to acceptable texture. Even rice parboiled at 100 °C, or steamed without pressure, cooks longer than raw rice. The extent to which starch retrogradation is influenced by the recrystallization of the amylose-lipid complex (Kugimiya and Donovan, 1981) remains to be determined. Parboiled rice definitely has a slightly more rancid taste than raw rice as do all pregelatinized starchy foods when compared with raw food.

Table 15.5 Effect of ß-amylase treatment on the complexing capacity of defatted waxy (IR29) rice starch as indexed by a gel consistency test†

Treatment	Gel consistency‡ (mm)	
	Starch	ß-limit dextrin
None	91	97
16 : 0 (1 mg)	47	98
Carboxymethyl cellulose (2 mg)	85	98

†IRRI (1981).
‡100 mg IR29 starch or ß-limit dextrin in 1.6 ml water and 0.7 ml 95% ethanol (wetting agent).

Lipids are added in some of the cooked-rice products. Examples are coconut milk added to waxy rice cake (Antonio *et al.,* 1975), ghee or oil to boiled rice, olive oil to *arroz Valenciana,* and oil to canned, precooked rice. Unless the oil is rich in FFA as in ghee, the major effect is to coat the milled rice surface and reduce the stickiness of the cooked rice. FFA can complex with starch to increase starch viscosity as discussed above.

Extrusion cooking is becoming popular for the preparation of precooked weaning foods (Wilson, 1979). SL protect the amylose of cereal starches from heat degradation during extrusion cooking at high temperatures (Mercier, 1981). Corn, wheat, and rice starches differ in the properties of their products extruded at 135 °C and 225 °C (Mercier and Feillet, 1975). With the wide range in amylose content of starch from 1 to 33% among rice varieties (Juliano, 1983), rice would be an ideal material with which to study the effect of starch amylose content and gelatinization temperature on properties of extrusion-cooked cereal grains.

During the cooling of the cooked rice, recrystallization of the amylose-lipid complex probably occurs and MG, FFA, and lysoPL may also complex with amylose, as demonstrated by Kugimiya and Donovan (1981) by adding LPC to gelatinized starches. Such recrystallization probably contributes to starch retrogradation (cooked rice hardness).

The lipids of milled rice reduce the apparent amylose content of the rice, as determined by iodine colorimetry, by forming a complex with amylose and making part of the amylose (2 percentage points of milled rice) unavailable for iodine complexing (Perez and Juliano, 1978). Bolling and El Bayâ (1975) reported that petroleum-extracted lipids did not interfere with the amylose assay but those extracted with 85% methanol did interfere (mainly LPC, with some LPE).

D Sake Brewing and Flavour

Sake, or Japanese rice wine, is made with rice and water and the most characteristic features are the use of rice-*koji* and of parallel fermentation (Nunokawa, 1972). *Koji* is a culture of *Aspergillus oryzae* on steamed rice which results in degradation of starch by amylase, and protein by protease. The sugars liberated are successively fermented by yeast in a *moromi* mash.

White-core Japanese rice, which is usually larger grained than translucent rice, is preferred for sake. To reduce the fat and protein content of the grain, the rice is overmilled to at least 70–80% recovery of head milled rice from brown rice, or up to 50 to 60% for superior sake. Nunokawa (1972) found that brown rice had 0.1% crude fat, 0.2% ash, and 5–6% protein; and 50% milled rice had 0.04% crude fat, 0.2% ash, and 4–5% protein. However, fat-by-hydrolysis (SL) remained nearly constant at 0.6–0.7% of milled rice (Yoshizawa *et al.*, 1973). The FA composition of the TG of overmilled (80%) rice NSL was 18% 16 : 0, 30% 18 : 1, and 48% 18 : 2, while that of 60% milled rice was 26% 16 : 0, 10% 18 : 1, and 59% 18 : 2 (Yoshizawa *et al.*, 1973). By contrast, the FA composition of fat-by-hydrolysis was 44% 16 : 0, 13% 18 : 1, and 38% 18 : 2 in 80% milled rice, and of 60% milled rice was 46% 16 : 0, 12% 18 : 1, and 36% 18 : 2, indicating less difference in FA composition than found for the NSL.

During the water-washing of 75% milled rice (25% bran-polish removal from brown rice) to remove adhering bran and subsequent steeping, only small amounts of lipids were removed from the grain (Ishikawa and Yoshizawa, 1974a). The lower loss of lipids during washing of overmilled rice as compared with ordinary milled rice (See Section IIIB) is probably due to the lower NSL content. During the steaming process, 31–59% of the NSL were removed, with the glyceride fraction reduced to 60% and little increase in FFA. The FFA, mainly polyunsaturated FA produced by hydrolysis of glycerides, volatilize during steaming, resulting in residual NSL with a higher proportion of 16 : 0 in the FA.

The SL (residual fat-by-hydrolysis) decreased progressively from 0.68% to 0.15% of the rice and the NSL content increased from 0.05% to 0.47% with complete disappearance of TG and an increase in FFA during digestion of steamed rice with α-amylase and glucoamylase, (Ishikawa and Yoshizawa, 1974b). The SL also changed drastically in FA composition from 41% 16 : 0, 13% 18 : 1, and 41% 18 : 2 to 72% 16 : 0, 10% 18 : 1, and 12% 18 : 2. Only a small amount of lipids, mainly FFA, were liberated in the aqueous saccharified liquor, but the amount of dissolved lipids increased exponentially with ethanol concentration. In the preparation of destarched milled rice by treatment of gelatinized milled rice with food-

Table 15.6 FA composition of the lipid fractions of 22–24 day sake mash fermentation and of sake yeast lipids†

Lipid fraction	Amount (% of milled rice)	FA composition (wt. %)					
		14 : 0	16 : 0	18 : 0	18 : 1	18 : 2	18 : 3
Solid mash fraction							
Crude fat FFA	0.168	3	74	4	5	14	tr
Crude fat TG	0.130	3	26	4	11	55	1
Crude fat FA ethyl ester	0.083	3	42	5	11	39	tr
Fat-by-hydrolysis	0.237	3	46	7	15	29	tr
Extracted lipids	0.116	1	55	19	10	14	1
Sake yeast lipids							
Diethyl ether soluble	–	1	28	21	50	–	–
$CHCl_3$–MeOH (2 : 1) soluble‡	–	1	29	18	52	–	–

†Ishikawa and Yoshizawa (1976).
‡After diethyl ether extraction.

grade *Aspergillus oryzae* ∝-amylase, the loss of SL in the residue was greater than the loss of NSL (Resurreccion and Juliano, 1981).

During the actual fermentation of steamed rice with rice *koji,* about 58% of the SL (residual fat-by-hydrolysis) became extractable with diethyl ether (Ishikawa and Yoshizawa, 1976). The major proportion of lipids remained in the residual solid fraction because of the low solubility of lipids in water. The extracted lipids increased with increasing alcohol concentration which reached 0.05% of the liquor fraction at the final stage of fermentation (24 days). The crude fat and liberated lipids consisted mainly of FFA, TG and ethyl esters of FA. Lipids which were analysed as SL were found mainly in the solid fraction as FFA. Since ethyl esters of FA were not found in the raw materials, they are probably formed by yeast during fermentation.

The ratio of polyunsaturated FA to saturated FA decreased during sake mash fermentation (Ishikawa and Yoshizawa, 1976). The NSL in the solid fraction showed an increase in the proportion of 16 : 0 from 28% to 51% and a decrease in 18 : 2 from 53% to 34% between days 6 and 19 of sake mash fermentation. The FA compositions and amounts of the lipid fractions in sake mash after 22 days are shown in Table 15.6 together with the FA composition of the yeast lipids. The results suggest that yeast cultured under the alcoholic fermentation conditions synthesized a large quantity of 18 : 0. The increase in 18 : 0 in the liberated lipid must have been caused by the yeast.

The major flavour esters of sake produced by the sake yeast are ethyl acetate, ethyl caproate and isoamyl acetate (Ishikawa and Yoshizawa,

1979). Isoamyl acetate formation was shown to be linearly correlated to the melting point of the FA incorporated into yeast cellular lipids over the range −20 °C to + 40 °C. It was inversely related to the percentage of18 : 2 in yeast cellular FA, suggesting that flavour ester formation in the fermentation medium is affected by the permeability of cell membranes, which in turn depends largely on the FA chain length or FA composition of the membrane lipids. The higher 16 : 0 content of the SL relative to the NSL, and the loss of linoleic acid during steaming contribute to the better sake flavour from overmilled as compared to regular milled rice. The SL of the inner endosperm (60% milled) also has 2 percentage points more 16 : 0 and 2 percentage points less 18 : 2 than the SL of 80% milled rice (Yoshizawa *et al.,* 1973).

ACKNOWLEDGEMENT

The author wishes to thank Dr W. R. Morrison who allowed citation of the unpublished data of N. S. Azudin.

REFERENCES

Antonio, A. A., Juliano, B. O. and del Mundo, A. M. (1975). *Philippine Agriculturist* **58**, 351–355.
ASRCT (Applied Scientific Research Corporation of Thailand) (1977). Study on the verification and definition of the most suitable rice bran stabilizing technology and specification of its technical parameters. Report to UNIDO Dec. 1976. Bangkok. 52 pp.
Azudin, N. S. and Morrison, W. R. (1982). University of Strathclyde, Glasgow, Scotland. Unpublished data.
Barber, S. and Benedito de Barber, C. (1980). In *Rice: Production and Utilization* (B. S. Luh, ed.), pp. 790–862. Avi, Westport, Connecticut.
Bechtel, D. B. and Juliano, B. O. (1980). *Ann. Bot.* **45**, 503–509.
Bechtel, D. B. and Pomeranz, Y. (1977). *Am. J. Bot.* **64**, 966–973.
Bechtel, D. B. and Pomeranz, Y. (1978). *Am. J. Bot.* **65**, 75–85.
Bhat Sondi, A., Mohan Reddy, I. and Bhattacharya, K. R. (1980). *Food Chem.* **5**, 277–282.
Bianchi, G., Lupotto, E. and Russo, S. (1979). *Experientia* **35**, 1417.
Bolling, H. and El Bayâ, A. W. (1975). *Chem. Mikrobiol. Technol. Lebensm.* **3**, 161–163.
Buttrose, M. S. and Soeffky, A. (1973). *Austr. J. Biol. Sci.* **26**, 357–364.
Chang, S.-C., Saunders, R. M. and Luh, B. S. (1980). In *Rice: Production and Utilization* (B. S. Luh, ed.), pp. 764–789. Avi, Westport, Connecticut.
Cheigh, H.-S., Kim, C.-J. and Kim, D.-C. (1980). Paper presented at International

Symposium on Recent Advances in Food Science and Technology, Taipei, Jan. 9–11.
Choudhury, N. H. (1979). Ph.D Thesis. University of the Philippines at Los Baños.
Choudhury, N. H. and Juliano, B. O. (1980a). *Phytochemistry* **19**, 1063–1069.
Choudhury, N. H. and Juliano, B. O. (1980b). *Phytochemistry* **19**, 1385–1389.
Cornelius, J. A. (1980). *Trop. Sci.* **22**, 1–26.
DeLucca II, A. J., Plating, S. J. and Ory, R. L. (1978). *J. Food Protection* **41**, 28–30.
Desikachar, H. S. R (1977). In Proceedings of the Rice By-Products Utilization, International Conference, Valencia, Spain, 1974, Vol. II Rice By-Products Preservation, pp. 1–32. Instituto Agroquímica y Tecnología de Alimentos, Valencia, Spain.
Eberstein, K., Höpke, R., Konieczny-Janda, G. and Stute, R. (1980). *Stärke* **32**, 265–270 (in German).
Eggum, B. O., Resurreccion, A. P. and Juliano, B. O. (1977). *Nutr. Rep. Int.* **16**, 649–655.
Endo, I., Chikubu, S. and Tani, T. (1978). *Rep. Nat. Food Res. Inst. Tokyo* **33**, 1–6 (in Japanese).
Enochian, R. V., Saunders, R. M., Schultz, W. G., Beagle, E. C. and Crowley, P. R. (1981). *U.S. Dept. Agric. Market Res. Rep.* 1120.
Fujii, T. (1972). *Denpun Kogyo Gakkaishi* **19**, 159–168.
Fujino, Y. (1978). *Cereal Chem.* **55**, 559–571.
Fujino, Y. and Miyazawa, T. (1979). *Biochim. Biophys. Acta* **572**, 442–451.
Hayakawa, T. and Igaue, I. (1979). *Nippon Nogeikagaku Kaishi* **53**, 321–327.
Harris, N. and Juliano, B. O. (1977). *Ann. Bot.* **41**, 1–5.
Hartman, L., and Lago, R. C. A. (1976). *J. Sci. Food Agric.* **27**, 939–942.
Hizukuri, S. and Juliano, B. O. (1982). Unpublished data [typescript].
Hopkins, D. T. (1981). In *Protein Quality in Humans: Assessment and in Vitro Estimation* (C. E. Bodwell, J. S. Adkins and D. T. Hopkins, eds), pp. 169–193. Avi, Westport, Connecticut.
Houston, D. F. (1972). In *Rice Chemistry and Technology* (D. F. Houston, ed.), pp. 272–300. American Association of Cereal Chemists, St. Paul, Minnesota.
Hunnell, J. W. and Nowlin, J. F. (1972). In *Rice Chemistry and Technology* (D. F. Houston, ed.), pp. 201–214. American Association of Cereal Chemists, St. Paul, Minnesota.
IRRI (International Rice Research Institute) (1981). Annual Report for 1980, pp. 67–71. Los Baños, Laguna, Philippines.
IRRI (International Rice Research Institute) (1983). Annual Report for 1981, pp. 21–34. Los Baños, Laguna, Philippines.
Ishikawa, T. and Yoshizawa, K. (1974a). *Nippon Nogeikagaku Kaishi* **48**, 337–341.
Ishikawa, T. and Yoshizawa, K. (1974b). *Nippon Nogeikagaku Kaishi* **48**, 657–662.
Ishikawa, T. and Yoshizawa, K. (1976). *Nippon Nogeikagaku Kaishi* **50**, 131–136.
Ishikawa, T. and Yoshizawa, K. (1979). *Agric. Biol. Chem.* **43**, 45–53.
Ito, S., Sato, S. and Fujino, Y. (1979). *Stärke* **31**, 217–221.
Ito, S., Suzuki, T. and Fujino, Y. (1981). *Nippon Nogeikagaku Kaishi* **55**, 247–253.
Juliano, B. O. (1977). *Riso* **26**, 3–21.
Juliano, B. O. (1980). In *Rice Production and Utilization* (B. S. Luh, ed.), pp. 404–438. Avi, Westport, Connecticut.
Juliano, B. O. (1983). In *Starch Chemistry and Industry* (R. L. Whistler, E. F. Paschall and J. N. BeMiller, eds), Academic Press, London and New York (in press).

Juliano, B. O., Perez, C. M., Blakeney, A. B., Breckenridge, C., Castillo Toro, D., Choudhury, N. H., Kongseree, N., Laignelet, B., Merca, F. E., Paule, C. M. and Webb, B. D. (1980). *Riso* **29**, 233–237.

Kawashima, K. and Kiribuchi, T. (1980). *Kaseigaku Zasshi* **31**, 625–628.

Kugimiya, M. and Donovan, J. W. (1981). *J. Food Sci.* **46**, 765–770, 777.

Lugay, J. C. and Juliano, B. O. (1964). *J. Am. Oil Chemists' Soc.* **41**, 273–275.

Lynn, L. (1969). In *Protein-Enriched Cereal Foods for World Needs* (M. Milner, ed.), pp. 154–172. American Association of Cereal Chemists, St. Paul, Minnesota.

Mahadevappa, M. and Desikachar, H. S. R. (1968). *J. Food Sci. Technol.* **5**, 72–73.

Maniñgat, C. C. and Juliano, B. O. (1980). *Stärke* **32**, 76–82.

Mano, Y. (1982). *Bull. Obihiro Ohtani Junior College* **19**, 37–44 (in Japanese).

Merca, F. E. and Juliano, B. O. (1981). *Stärke* **33**, 253–260.

Mercier, C. (1981). *Cereal Foods World* **26**, 504 (abstr).

Mercier, C. and Feillet, P. (1975). *Cereal Chem.* **52**, 283–297.

Miyazawa, T., Yoshino, Y. and Fujino, Y. (1977). *J. Sci. Food Agric.* **28**, 889–894.

Miyazawa, T., Tazawa, H. and Fujino, Y. (1978). *Cereal Chem.* **55**, 138–145.

Morrison, W. R. (1978a). In *Advances in Cereal Science and Technology* (Y. Pomeranz, ed.), Vol. 2, pp. 221–348, American Association of Cereal Chemists, St. Paul, Minnesota.

Morrison, W. R. (1978b). *J. Sci. Food Agric.* **29**, 365–371.

Mukherjee, R. K. and Bhattacharjee, M. (1978). *J. Am. Oil Chemists' Soc.* **55**, 463–464.

Nunokawa, Y. (1972). In *Rice Chemistry and Technology* (D. F. Houston, ed.), pp. 449–487. American Association of Cereal Chemists, St. Paul, Minnesota.

Ohashi, K., Goshima, G., Kusuda, H. and Tsuge, H. (1980). *Stärke* **32**, 54–58.

Ohnishi, M. and Fujino, Y. (1980). *Agric. Biol. Chem.* **44**, 333–338.

Oparka, K. J. and Gates, P. (1981). *Planta* **151**, 561–573.

Padua, A. B. and Juliano, B. O. (1974). *J. Sci. Food Agric.* **25**, 697–701.

Perez, C. M. and Juliano, B. O. (1978). *Stärke* **30**, 424–426.

Perez, C. M. and Juliano, B. O. (1981). *J. Texture Studies* **12**, 321–333.

Resurreccion, A. P. and Juliano, B. O. (1975). *J. Sci. Food Agric.* **26**, 437–439.

Resurreccion, A. P. and Juliano, B. O. (1981). *Qual. Plant. Plant Foods Hum. Nutr.* **31**, 119–128.

Ryu, C.-H. and Cheigh, H.-S. (1980). *Korean J. Food Sci. Technol.* **12**, 278–284.

Sawada, T. and Mano, Y. (1980). *Bull. Obihiro Ohtani Junior College* **17**, 31–38 (in Japanese).

Sayre, R. N., Saunders, R. M., Enochian, R. V., Schultz, W. G. and Beagle, E. C. (1982). *Cereal Foods World* **27**, 317–322.

Shoji, I. and Kurasawa, H. (1981). *Kaseigaku Zasshi* **32**, 167–171.

Sowbhagya, C. M. and Bhattacharya, K. R. (1976). *J. Food Sci.* **41**, 1018–1023.

Taira, H., Taira, H. and Fujii, K. (1979). *Japan. J. Crop Sci.* **48**, 371–377.

Taira, H., Taira, H. and Fujii, K. (1980). *Japan. J. Crop Sci.* **49**, 559–568.

Takeshita, Y. (1972). *Trans. Kokushikan Univ. Dept. Engineering* **5**, 1–22.

Takeshita, Y. (1982). Kokushikan University, Tokyo, Japan. Personal communication.

Tanaka, Y., Resurreccion, A. P., Juliano, B. O. and Bechtel, D. B. (1978). *Agric. Biol. Chem.* **42**, 2015–2023.

Tsuchiya, T. (1948). *J. Nippon Oil Technol. Soc.* **1** (1), 1–6.

Villareal, R. M., Resurreccion, A. P., Suzuki, L. B. and Juliano, B. O. (1976). *Stärke* **28**, 88–94.

Viraktamath, C. S. and Desikachar, H. S. R. (1971). *J. Food Sci. Technol.* **8**, 70–74.
Wang, H.-C., Tang, Z.-C., Su, W.-A., Wang, W.-Y. and Li, J.-S. (1980). *Acta Phytophysiol. Sinica* **6**, 227–236.
Wilson, D. E., ed. (1979). Proceedings Low-Cost Extrusion Cookers, Second International Workshop, Dar es Salaam, Tanzania, 1979. Colorado State University, Fort Collins.
Yasumatsu, K. and Moritaka, S. (1964). *Agric. Biol. Chem.* **28**, 257–264.
Yasumatsu, K. and Moritaka, S. (1979). *J. Takeda Res. Lab.* **38**, 62–72.
Yasumatsu, K., Moritaka, S., and Kakinuma, T. (1964). *Agric. Biol. Chem.* **28**, 265–272.
Yasumatsu, K., Moritaka, S. and Wada, S. (1966). *Agric. Biol. Chem.* **30**, 483–486.
Yokochi, K. (1977). In Proceedings of the Rice By-Products Utilization, International Conference, Valencia, Spain, 1974, Vol. III Rice Bran Oil Utilization, pp. 1–38. Instituto Agroquímica y Tecnologia de Alimentos, Valencia, Spain.
Yoshizawa, K., Ishikawa, T. and Noshiro, K. (1973). *Nippon Nogeikagaku Kaishi* **47**, 713–717.
Yoshizawa, K., Hasuo, T. and Ishikawa, T. (1980). *Nippon Jozo Kyokai Zasshi* **75**, 432–435.

16 Oat Lipids

E. G. HAMMOND

Department of Food Technology, Iowa State University, Ames, Iowa 50011, U.S.A.

I INTRODUCTION

Špinka (1939a,b), Hutchinson (1953) and Shukla (1975) reviewed the literature on oat composition and the food uses of oat, and Youngs (1978) reviewed the recent literature on oat lipids.

"Lipids in Cereal Technology"
ISBN 0-12-079020-3

II LIPID CONTENT

A Range of Oil Content

The earliest examination of oat grain for oil content was by König (1871) who found an oil content of 4.45%. Table 16.1 summarizes the investigations of variation in oil content of oats. In general, the lipid extractable with non-polar solvents ranges from 2.0–11.0%; the more extreme values are associated with *Avena* species other than *Avena sativa*, the cultivated

Table 16.1 Oil content of oat grain (*Avena sativa* L.)

Reference	No. of samples	Range of oil content	Mean	Place of growth
Krarup (1904)	–	4.95– 7.13%	–	Germany
Berry (1920)	117	4.85–10.35%†	8.0	Britain
Gardner and Hutchinson (1951)	9	6.0 – 9.8%	–	Britain
Hutchinson and Martin (1955)	54	4.5 –11.1%	6.4	Britain
Hoffman (1951)	10	3.9 – 6.2%	5.0	Germany
Hübner (1951)	8	4.5 – 6.5%	–	Germany
Ackerman (1954)	7	4.73– 6.37%	–	Sweden
Persson and Bingefors (1956)	9	5.0 – 6.7%	–	Sweden
Frimmel (1958)	17	2.25– 5.2%	–	Germany
Stuke (1960, 1961a)	396‡	3.5 – 9.5%	6.0	Germany
Pokorný *et al.* (1963)	5	5.3 –6.3%	5.8	Czech
Brown *et al.* (1966)	169	3.8 – 9.8%	5.6§	U.S.A.
Ganssmann (1967)	14	5.6 – 7.4%	5.8	Germany
Novozhilova *et al.* (1967)	3	4.35– 5.39%	–	U.S.S.R.
Fritz (1970)	5	5.9 –10.4%	–	Germany
Brown and Craddock (1972)	4533¶	3.1 –11.6%	7.0	U.S.A.
Forsberg *et al.* (1974)	10	5.0 – 9.7%	6.7	U.S.A.
Frey and Hammond (1975)	445§§	2.0 –11.0%	–	U.S.A.

†12.87 in a "wild oat".
‡Includes 54 *Avena* species other than *Avena sativa*.
§Summer oats 5.6, winter oats 8.2.
¶Includes an undesignated number of *Avena sp.* other than *A. sativa*.
§§Including 330 strains of *Avena sterilis*.

oat. Brown and Craddock (1972) reported that oil contents were normally distributed with a mean of 7% and that 90% of the samples fell between 5 and 9%. The 8% mean of Berry's (1920) data is significantly higher than later reports. Recent European mean values range from 5.8–6.4%.

B Lipid Extraction

Non-polar solvents do not extract all the lipids from oats. Reiser (1947) reported that ethanol-diethyl ether (3 : 1) extracted more lipid and much more lipid phosphorous than diethyl ether alone, but that the diethyl ether extracted the unsaponifiables more thoroughly. Hutchinson and Martin (1955), Janíĉek and Pokorný (1959), and Pokorný *et al.* (1961a,b) compared the lipids extracted from oats by solvents of different polarities in combination with acid hydrolysis. Solvents of greater polarity extracted additional material that contained a smaller proportion of fatty acids, and hydrolysis released additional lipid material. Salun and Kalugina (1974) and Dehghan (1979) found that water-saturated butan-1-ol (WSB) extracted more lipids from oats than hexane, and Acker and Becker (1971) showed that WSB extracted much more lysophosphatidylcholine from oat starch than did diethyl ether. Youngs *et al.* (1977) also obtained higher yields of oat lipid with WSB. Sahasrabudhe (1979) compared the total lipid and amounts of several complex lipids extracted from oats by several solvents. Ethanol gave the greatest yield of total lipid.

Physical interactions between oat lipids and other constituents, especially proteins, may make extraction of the lipid difficult. Janíĉek and Pokorný (1960) reported that heat treatment of oats at 130–150 °C, decreased the extractable lipid by 25–60%, and this decrease was a linear function of temperature. Rohrlich and Niederauer (1967) reported isolating a lipoprotein from oat and other cereals by extraction with chloroform-methanol. Youngs (1974) isolated a protein-rich fraction from oat-water slurries by centrifugation. This fraction, which comprised 17–19% of the oat weight, contained 18–21% "free" lipid, extractable with petroleum ether, and 4% "bound" lipid, extractable with butanol-water. It is not clear whether the reported lipoproteins exist in the oats or are artefacts of the extractions.

Oats are normally ground to extract the lipids, although Clements (1977) was able to extract about 57% of the lipid from whole oats that could be extracted with hexane from ground oats. This was a much greater portion of the total lipid than could be extracted from other whole cereals, and Clements believed that this was because the oat grain was cracked during dehulling, thus allowing solvent penetration. The methodology for lipid extraction is discussed further in Appendix 1.

C Effect of Agronomic Variables on Lipid Content

It is generally agreed that the location and weather can alter the total lipid content of oats (Berry, 1920; Mix, 1931; Hübner, 1951; Hutchinson and Martin, 1955; Frimmel, 1958; Stuke, 1961a,b; Kurten and Ganssmann,

1966; Beringer, 1967, 1971b; Fritz, 1970; Frey and Hammond, 1975) although the magnitude of the environmental effect is small compared with that of the varietal effect. Hutchinson and Martin (1955) suggested that environmental effects were ±1% from the mean value, which agrees with Stuke (1961b) who observed a maximum change in oil content of 1.93% caused by environment. Beringer (1967, 1971b) has established that the temperature during the first three weeks after flowering is particularly important and that a low temperature, 12 °C, during this period favours greater lipid content at maturity compared with 30 °C. Large plant spacings and wet conditions at harvest reduce the total lipid content of the grain (Stuke, 1961b; Hübner, 1951). Fertilizer treatment is usually considered to have little effect on lipid content in oat grain (Berry, 1920; Mix, 1931; Beringer, 1966; Kurten and Ganssmann, 1966; Jahn-Deesbach *et al.*, 1971), but Mix (1931) claimed that lipid content was increased slightly by potassium, and Stuke (1961b) found a decrease in oil content due to nitrogen treatment.

Correlations have been found between oat lipid content and a number of agronomic variables although there are several conflicting reports. Thus, Hüber (1951) found lipid negatively correlated with grain size, but Frimmel (1958) found a weak positive correlation. Frimmel (1958) reported a negative correlation of lipid with yield and proportion of husk. Lipid content and yield were positively correlated in one year but were not significant for others in a study by Forsberg *et al.* (1974). The authors reported that lipid and kernel weight were positively correlated for most strains but Brown *et al.* (1966) found a slight negative correlation. Lipid and protein were claimed to be negatively correlated (Brown *et al.*, 1966). Thro (1982) found oil content positively correlated or independent of the grain yield.

III LIPASE AND OTHER ENZYMES

Early investigations of oat lipids revealed high acid values (Stellwaag, 1890; Dubovitz, 1918; Paul, 1921; Amberger and Wheeler-Hill, 1927; and Munro and Binnington, 1928) that indicated as much as 30% or more of the lipid was free fatty acids (FFA). Berry (1920) and Hutchinson *et al.* (1951) noticed that the acid value was low if the oats were extracted immediately after grinding, but that the value increased if the ground grain was stored prior to extraction.

Bamann and Ullmann (1942) demonstrated the presence of a tributyrinase in oats and Frey and Hammond (1975) have used the clearing of a tributyrin emulsion in agar to screen grain for lipase. Other investigators have measured oat lipase by titrating the FFA released in oat doughs (Hutchinson and Martin, 1952; Templeton and Carpenter, 1953; Pokorny *et al.*, 1962; Acker and Beutler, 1965). A rapid colorimetric test for residual lipase in oats has been proposed and was designed to test for adequate heat stabilization during processing (Kazi and Cahill, 1969). It was based on the change of pH as FFA was released by lipase. Frey and Hammond (1975) measured lipase by a colorimetric determination of FFA released from a dough made from five grains, and Sahasrabudhe (1982) has used a similar method to measure lipase in individual grains. Matlashewski *et al.* (1982) measured lipase by the release of ^{14}C-oleic acid from triolein. The methods of measuring lipase activity are discussed in Appendix 2.

Pokorný *et al.* (1962) measured the lipase activity of five Czech oat varieties grown at different locations and found small variety and variety x location effects, but a strong location x year effect. Frey and Hammond (1975) screened 352 oat varieties by the tributyrin method and found 58 samples with no apparent lipase, but their colorimetric test of the release of long-chain fatty acids confirmed that all varieties had lipase activity. The activity ranged from 10–210 μmols FFA g^{-1} hr^{-1}. Kalbasi-Ashtari (1976) examined 27 U.S. oat cultivars by Frey and Hammond's method and found values of 13.8 to 69 μmols FFA g^{-1} hr^{-1}.

The properties of oat lipase, its distribution in the oat grain, the effect of water on its activity and the specificity of its attack on triglycerides have been studied (Hutchinson *et al.*, 1951; Peers, 1953; Martin and Peers, 1953; Acker and Beutler, 1965; Berner and Hammond, 1970; and Sahasrabudhe, 1982). See Chapter 6 for a discussion of these topics.

Oat lipase may be destroyed by heat but the temperature required varies with the moisture content (Hutchinson *et al.*, 1951; Moran, 1952; Pokorný *et al.*, 1968). Pokorný *et al.* (1968), in agreement with Janecke (1951), found that heat treatments below 100 °C could activate oat lipase. The reason for this is not clear, but the activation may be caused by destruction of a lipase inhibitor that is less stable than the lipase itself. Other methods of destroying or removing oat lipase include soaking in hydrochloric acid, abrasion of the surface of the caryopses and treatment with solvents, but there is poor agreement about the effectiveness of these methods (Hutchinson *et al.*, 1951; Moran, 1952; Frey and Hammond, 1975; and Youngs, 1978).

The release of FFA in ground oats can be delayed significantly by storage at 4 °C compared with 20 or 30 °C (Thomke, 1970). Salun and Kalugina (1974) studied the change in various lipid fractions of oats and oat products stored for 6 months at water activity of 0.55 and 0.75; FFA were released primarily from triglycerides. Proportionately more FFA is released from oats stored at high water activity (Frey and Hammond, 1975) and oats that have been damaged (Welch, 1977). It has been reported that naked oats are more susceptible to damage than glumed oats (Welch, 1977).

Martin (1956) reported that oatmeals that have received inadequate heat stabilization gave dark oatcakes. Martin believed that FFA released by lipase played a role in the darkening, but it is possible that lipoxygenase was also involved. During the storage of oats, the concentration of unsaturated fatty acids, tocols and carotenoids decrease (Nechaev *et al.,* 1972a; Sosedov *et al.,* 1971), especially at relatively high moisture contents. Nechaev *et al.* (1972b) found that heat stabilization decreased the formation of carbonyl compounds and polar products of lipid oxidation in oats during one year of storage. Heydanek and McGorrin (1981) found that the volatiles isolated from dry oat groats were distinct from those isolated from hydrated oat groats. Hydrated oats yielded the alcohols and aldehydes typical of the products of fat oxidation. Oxidation and lipoxygenase activity in oats is further discussed in Chapter 6.

IV FATTY ACID COMPOSITION

Stellwaag (1890), Dubovitz (1918), Van Kampen (1920), Paul (1921), Amberger and Wheeler-Hill (1927), Janíĉek and Pokorný (1959), Pokorný *et al.* (1960), Pokorný *et al.* (1961a), and Novozhilova *et al.* (1967) reported classical analytical constants for oat oil such as iodine, thiocyanogen, saponification, and acid values. The acid values were usually high because of lipase action. A number of early studies were made of the fatty acid (FA) composition of oat lipids using classical methods (Konig, 1874; Amberger and Wheeler-Hill, 1927; Takahashi *et al*., 1935; Janíĉek and Pokorný, 1959) but these have been superseded by instrumental methods of analysis.

Pokorný *et al.* (1961b) first applied gas-liquid chromatography (glc) to the analysis of oat lipid, and Lindberg *et al.* (1964a) reported the first glc results that gave resolution of all the major oat fatty acids. Since then glc has been used in many studies of the fatty acid composition of oat lipid and these are summarized in Table 16.2. Many of the earlier analyses exceed the ranges of the data in Table 16.2 and it is possible that the techniques used by early investigators were subject to large errors. The data of Welch (1975) clearly differ from other reports and the author attributed this to the fact that the oats had been grown in a greenhouse. Berrie *et al.* (1975) have reported that there are traces of 6 : 0 – 10 : 0 in oats and that these

Table 16.2 Fatty acid composition of oats

Reference	No. of varieties	Percentage composition						Location
		14 : 0	16 : 0	18 : 0	18 : 1	18 : 2	18 : 3	
Lindberg *et al.* (1964a)	2	0.1–0.2	14.6–17.8	0.6–1.3	37.1–41.4	39.1–43.1	1.4–2.3	Sweden
Forsberg *et al.* (1974)	24	0.1–1.3	20.0–28.0	0.5–2.1	24.5–32.0	38.4–47.7	0.7–2.0	U.S.A.
Frey and Hammond (1975)	64	–	14.0–23.0	0.5–4.0	29.5–51.0	26.0–47.0	0.5–5.0	U.S.A.
Welch (1975)	6‡	–	15.4–23.9	0.9–2.4	18.7–35.0	43.5–53.0	1.8–3.6	Britain
Youngs and Püskülcü (1976)	15	0.4–0.8	16.2–21.8	1.2–2.0	28.4–40.3	36.6–45.8	1.5–2.5	U.S.A.
de la Roche *et al.* (1977)	9	–	17.2–23.6	0.8–1.8	26.5–47.5	33.2–46.2	0.9–2.4	U.S.A.
Sahasrabudhe (1979)	12	0.5–4.9	14.9–25.8	1.6–3.9	25.8–41.3	31.3–41.0	1.7–3.6	Canada
Dehghan (1979)	6	0.2–1.2	17.5–21.6	1.0–1.9	30.4–35.8	40.0–46.3	0.7–1.0	U.S.A.
Pan (1981)	103§	–	13.0–21.0	1.0–4.0	33.0–48.0	28.0–44.0	1.2–4.0	U.S.A.

†12 : 0, 14 : 1, 16 : 1, 17 : 0, 17 : 1, 20 : 0, 20 : 1, and 22 : 0 reported in trace amounts.
‡Includes both winter and spring sowings.
§Includes 43 *A. sterilis* strains.

inhibit germination. Table 16.2 concludes that 18 : 1 and 18 : 2 are the major FA of oat lipids together with a large contribution from 16 : 0, whereas 18 : 0 and 18 : 3 are minor components.

Forsberg *et al.* (1974), Frey and Hammond (1975), Welch (1975) Youngs and Püskülcü (1976), and de la Roche *et al.* (1977) reported that lipid content is positively correlated with the percentage of 18 : 1 and negatively with the percentage of 18 : 2 in the FA. All but Frey and Hammond found lipid content negatively correlated with the percentage of 16 : 0.

Lindberg *et al.* (1964b) reported that the growth environment, particularly the temperature, could affect the polyunsaturated FA (PUFA) content of oats. Beringer (1967, 1971a) and Beringer and Saxena (1968) reported that oats held at 12 °C after flowering had more unsaturated FA than those held at 30 °C. When developing oats were exposed to $^{14}CO_2$, the same effect of temperature influenced the incorporation of the ^{14}C into saturated and unsaturated FA. When Australian oat varieties were grown in Australia and Germany, the German-grown oats had FA that were much richer in 18 : 2 and lower in 18 : 1 than those of the Australian-grown oats (Marquard, 1973). Welch (1975) found significant differences in the FA composition of grain from oats sown in spring and winter. Youngs and Püskülcü (1976) and de la Roche *et al.* (1977) found only slight environmental influence on FA composition in their studies.

V LIPID CLASS COMPOSITION

A Acyl Lipids

Brown *et al.* (1970) applied thin-layer chromatography (tlc) to the separation of oat lipids and reported the presence of triglycerides (TG), diglycerides (DG), FFA, sterols and lipids of greater polarity. Subsequently, there have been a number of studies of the composition of lipid classes in oat grain. The results of these studies are summarized in Table 16.3 (for general information about the acyl lipids found in cereals see Chapter 2). Oat lipids contain from 70 to 76% non-polar lipids (NL), from 3 to 10% phospholipids (PL) and from 7 to 17% glycolipids (GL) (Price and Parsons, 1975; Dehghan, 1979). There is considerable variation in the reported proportions of individual lipid classes but as shown in Table 16.3 the major component of the NL is TG. The fatty acid composition of the polar lipids differs from that of the NL; the NL contain a greater proportion of 18 : 1 and less 18 : 2 and 16 : 0 (Beringer, 1971b; de la Roche *et al.*,

Table 16.3 Percentage composition of the lipid classes in oat lipids

Lipid class	Reference			
	Salun and Kalugina (1974)	de la Roche *et al.* (1977)	Youngs *et al.* (1977)	Sahasrabudhe (1979)
Triglycerides	43.3–45.2	64.2–85.0	37.5–50.6	50.3–56.4
Free fatty acids	11.9–12.0	2.8– 7.2	2.0–11.0	4.0–10.5
Steryl esters	7.8– 9.4	0.2– 0.6		2.8– 4.0
Sterols	7.6– 8.6		0.5– 1.7	1.8– 4.1
Steryl glycosides			0.9– 1.4	
Phospholipids	18.0–19.4†	5.0–17.2	7.1–12.1	11.6–26.0
Glycolipids		6.9–11.2‡	10.9–11.9	5.8– 8.6
1,2-Diglyceride	5.6– 5.9 §		0.6– 1.3	4.3– 8.9†
1,3-Diglyceride	3.5– 4.8 §		1.2– 1.7	

†Includes monoglycerides.
‡Includes monoglycerides and diglycerides.
§Isomers not designated.

1977; Dehghan, 1979; Sahasrabudhe, 1979). In addition to 18 : 0 and 18 : 3, minor quantities of 12 : 0, 14 : 0, 14 : 1, 20 : 1, 20 : 2, 20 : 3, 20 : 4, 20 : 5, 22 : 0, 22 : 1 and 24 : 0 have also been reported for oat lipids. Beringer (1971b) found that grain from oats which developed at 12 °C immediately after flowering, rather than at 28 °C, contained NL and polar lipids with a greater proportion of 18 : 2.

Many early observers determined FFA in oat lipid, but few took enough precaution against lipase activity to obtain accurate figures. In addition to the results in Table 16.3, Hutchinson and Martin (1955) published valuable data. Fourteen varieties grown at a number of locations exhibited FFA contents from 2.7 to 10.2% of the total lipid. Variety and location caused significant differences. Low FFA content is an important quality characteristic of grain used for oatmeal milling.

Early investigators estimated PL by the phosphorous extractable into various solvents and reported for oats 0.76% "lecithin" (Stellwaag, 1890), and 0.037% lipid phosphorous (Guerrant, 1927). Acetone insolubility was used to separate PL from other lipids, and Schulze and Pfenniger (1911) claimed to have isolated both choline and a small amount of betaine from oat PL. Trier (1913) found no evidence for betaine in oat PL, but detected choline and galactose. Diemer *et al.* (1937) found 1.25–1.8% acetone-insoluble PL consisting of 8.5–11.4% choline and 62.5% FA in oats. Aylward and Showler (1962b) separated the acetone-insoluble oat lipids into "lecithin" and "cephalin" fractions and demonstrated the presence of phosphatidylserine (PS) and phosphatidylethanolamine (PE) in the cephalin fraction. The same authors (1962a) reported the isolation of phosphatidic acid (PA) and phosphatidylinositol (PI). More recent results for the PL and GL composition of oats have been published by Tevekelev (1970), Price and Parsons (1975), Youngs *et al.* (1977), Dehghan (1979) and Sahasrabudhe (1979); the data of Sahasrabudhe (1979) for the percentage composition of polar lipids are presented in Table 16.4. Digalactosyldiglyceride and phosphatidylcholine were the major GL and PL respectively. In addition to the lipids listed in Table 16.4 other polar components have been found in oat lipids, including sulpholipid (Price and Parsons, 1975; Dehghan, 1979), acyl steryl glycoside (Dehghan, 1979) and diphosphatidylglycerol (Price and Parsons, 1975). Carter *et al.* (1961) demonstrated the presence of mono- and digalactosylglycerol in saponified oat oil.

Acker and Becker (1971) reported that the starch lipid of oats contained lysophosphatidylcholine (LPC) (51.6%), lysophosphatidylethanolamine (5.1%), lysophosphatidylinositol (7.0%) and FFA (7.7%). The FA composition of the LPC was 1% 14 : 0, 46% 16 : 0, 1% 18 : 0, 10% 18 : 1 and 42% 18 : 2.

Table 16.4 Composition of oat grain polar lipids†

Glycolipids	
Monogalactosylmonoglyceride	6.2
Monogalactosyldiglyceride	18.5
Digalactosylmonoglycerides	4.8
Digalactosyldiglyceride	41.5
Sterylglucoside	7.0
Others	22.0
	100.0%
Phospholipids	
Phosphatidylcholine }	29.9
Phosphatidic acid }	3.9
Phosphatidylinositol	
Phosphatidylglycerol	9.5
Phosphatidylethanolamine	14.8
Phosphatidylserine	3.2
Lysophosphatidylethanolamine, Lysophosphatidylserine, Lysophosphatidylcholine	20.4
Others	19.0
	100.0%

†From Sahasrabudhe (1979).

B Sterols

A number of early observations were made on the unsaponifiable fraction of oat oil and the data are summarized in Table 16.5.

Idler *et al.* (1953), in the first published investigation of oat sterols, reported that β-sitosterol made up 53–56% of the total sterols and that Δ^5- and Δ^7-stigmastadiene-3β-ols were present. Knights (1965) produced evidence that the latter two sterols were $\Delta^{5,24}$- and $\Delta^{7,24}$-stigmastadien-3β-ol, and also reported that cholesterol, brassicasterol, campesterol, stigmasterol and Δ^7-stigmastenol were also present. Knights and Laurie (1967) reported that detailed identification of oat sterols and quantitative analyses of the major sterols were given by Knights (1968) and Youngs *et al.* (1977). These are summarized in Table 16.6 and further details of cereal grain sterols are given in Chapter 3. Novozhilova *et al.* (1968) reported 0.19–0.32% sterols in oat oil.

C Antioxidants

In the 1930s and 1940s Musher obtained a number of patents which claimed that the addition to lard of flours and extracts from various veg-

Table 16.5 Content of unsaponifiables in oat oil

Reference	% Unsaponifiables in lipid
Stellwaag (1980)	2.65
Dubovitz (1918)	1.61
Paul (1921)	1.30
Munro and Binnington (1928)	2.26
Reiser (1947)	0.20–0.46†
Janícék and Pokorný (1959)	1.5–2.7†
Pokorný *et al.* (1961b)	1.8–4.3†
Novozhilova *et al.* (1967)	1.8–3.5

†Variation with extraction method.

etable products was effective in prolonging stability in the active oxygen test. Although Musher's (1935a) data show that soya flour was a more effective antioxidant than oat flour, the latter was preferred for its blandness and lack of colour (Musher, 1935a, 1935b). Peters and Musher (1937, 1938) and Supova *et al.* (1959) reviewed the claims that oat flour increased fat stability when added to fats, margarine and mayonnaise, dusted over bacon, potato chips and nuts, or added to wrapping papers. Supova *et al.* (1959) also confirmed that various extracts of oats had antioxidant properties for lard in the active oxygen test. Extraction solvents of higher polarity yielded extracts with greater antioxidant activity and a methanol extract of hexane-defatted oat flour was the most active. However, the extracts were less effective than propyl gallate and showed no synergism with it.

Table 16.6 Sterol composition of oat grain

	Composition as percentage of total sterols		
	Knights (1968)†		Youngs *et al.* (1977)
Sterol	*A. sativa*	*A. fatua*‡	*A. sativa*
Sitosterol	39.0	57.3	69
Campesterol	6.4	7.5	10
Stigmasterol	5.0	3.8	8
Δ^5-avenasterol	21.2	19.2	7
Δ^7-avenasterol	13.5	5.7	–
Cholesterol	5.8	3.7	2
Δ^7-stigmasten-3ß-ol	6.4	2.0	–
Δ^7-cholesten-3ß-ol	2.8	0.8	–

†Six other sterols were detected in trace amounts (see Chapter 3).
‡Wild oat.

1 *Tocols*

Oat grain and oat oil contain both tocopherols and tocotrienols. Table 16.7 summarizes the published compositional data. The major tocols appear to be α-tocotrienol and α-tocopherol but there is disagreement concerning the presence of the β- and γ-isomers. A further discussion of tocol composition is in Chapter 3. In addition to the data in Table 16.7, Beringer and Saxena (1968) detected the four tocopherols plus α-tocotrienol in oat oil and found a total tocol content ranging from 430 to 620 mg kg^{-1}. Kalbasi-Ashtari and Hammond (1977) found 105 mg kg^{-1} of α-tocopherol in oat oil. Values quoted by Sosedov *et al.* (1971) of 380 to 900 mg kg^{-1} for total tocols in oat grain may be incorrectly reported; these values are far too high and probably relate to oat oil rather than oat grain.

Low temperatures after flowering (12 °C as compared with 28 °C) were claimed to result in a lower tocol content of the oat grain at maturity (Beringer and Saxena, 1968); α-tocopherol and α-tocotrienol were the major tocols detected. It is possible that significant losses of tocols occur during the processing of oats to produce granular and shredded products, and in the preparation of oatmeal (Herting and Drury, 1969). Although tocols are important contributors of vitamin E activity and antioxidant activity in cereal grains, it is probable that the relatively high antioxidant activity of oat flours and oat extracts is due to the phenolic acid compounds described below rather than to the tocols.

2 *Caffeic and Ferulic Acid Antioxidants*

In 1961 Daniels and Martin reported the isolation of an antioxidant from oats that contained caffeic and ferulic acids. The purified antioxidant was as effective as propyl gallate and butylated hydroxytoluene in an automated version of the active oxygen method (Martin, 1961). The antioxidant was extracted from oats by diethyl ether, but not light petroleum, and the extract was fractionated by chromatography, first into 6 (Daniels *et al.*, 1963) and then into 24 active fractions (Daniels and Martin, 1967). The oat antioxidants were identified as caffeic and ferulic acid diesters and monoesters of C_{26} and C_{28} α, ω-diols (Daniels and Martin, 1964 and 1965; Daniels, 1966) and monoesters of hexacosan-1-ol, 26-hydroxyhexacosanoic acid and 28-hydroxyoctacosanoic acid (Daniels and Martin, 1967). Also, glycerol monoesters of 26-hydroxyhexacosanoic acid and 28-hydroxyoctacosanoic acid were identified in which the ω-hydroxyl group of the acid and the two remaining hydroxyl groups of the glycerol were esterified with caffeic and ferulic acids (Daniels and Martin, 1968). In general, the caffeic esters were more effective antioxidants than ferulic esters. Daniels *et al.* (1963) suggested that the phenolic acid esters were

Table 16.7 Tocol composition of oat grain and oat oil†

Reference	Total tocol content (mg kg^{-1})	Composition as percentage of total tocols							
		α-T	β-T	γ-T	δ-T	α-T-3	β-T-3	γ-T-3	δ-T-3
Oat grain									
Mason and Jones (1958)	13	13	–	61	–	20	6	–	–
Lindberg (1966)	29.7	39.7	11.3		–	41.3	4.5	–	–
Slover *et al.* (1969)	18.9	26.5	4.8	–	–	58.2	10.6	–	–
Lásztity *et al.* (1980)‡	29.8	18.3	3.1	7.4	1.4	43.1	3.4	9.1	3.2
Barnes (1982)	18.6	23	3	–	–	63	11	–	–
Barnes (1982)	18.8	23	5	–	–	63	9	–	–
Oat oil									
Green *et al.* (1955)	610	28	–	36	10	22	4	–	–
Novozhilova *et al.* (1968)	98–635	57.6–82.9	–	11.9–30.1	5.2–12.3	–	–	–	–
Chow *et al.* (1969)	175	19	6	0	0	51	12	7	5
Herting and Drury (1969)	300	81.1	–	2.7	–	16.2	–	–	–

†Further information on the composition of oat tocols is given in Chapter 3.
‡Includes 7.8% α-tocoquinone and 3.2% γ-tocoquinone.

responsible for the antioxidant properties noted by Musher, and King (1962) synthesized glycerol esters of caffeic acid and reported that they had antioxidant activities comparable to propyl gallate and butylated hydroxytoluene.

3 *Sterols*

Sims *et al.* (1972) and Boskou and Morton (1976) have shown that certain sterols, such as the avenasterols that are present in oats (see Section VB), exert an antioxidant effect in oils at frying temperatures. The mechanism of this effect is unknown.

D Pigments

Novozhilova *et al.* (1968) reported the presence of taraxanthin and xanthophyll epoxide and 9.8–38 mg kg^{-1} total carotenoids in oats. Sosedov *et al.* (1971) reported 104–113 mg kg^{-1} of carotenoids and 77–95 mg kg^{-1} of chlorophyll in oats.

E Triglyceride Structure

Amberger and Wheeler-Hill (1927) examined oat triglyceride structure by elaidinization and hydrogenation of oat oil and concluded that triolein and palmitodioleins were present. Pan (1981) developed a rapid method for the stereospecific analysis of triglycerides, applied the method to 60 *A. sativa* and 43 *A. sterilis* strains and concluded that the glyceride structure was similar to other vegetable fats. The 16 : 0 and 18 : 0 occurred primarily at the *sn*-1 and *sn*-3-positions while the *sn*-2-position was enriched with 18 : 1 and 18 : 2. The 18 : 3 occurred slightly more frequently at the *sn*-3-position. For varieties with different fatty acid compositions, the amount of any fatty acid on a particular position was linearly related to the amount of that acid in the whole fat. Pan used this relationship to look for varieties whose glyceride structure differed radically from the general pattern. About 12% of the varieties deviated from the linear regression by 3 standard deviations, but no variety had a radically different glyceride structure.

VI DISTRIBUTION OF LIPIDS IN OATS

Janíĉek and Pokorný (1959) and Pokorný *et al*. (1961b) extracted oat hulls with various solvents and characterized the lipid. Pokorný *et al.* (1960) found that the hull surface wax was rich in hydroxy fatty acids and

Table 16.8 Concentration of lipids in oat grain tissue fractions†

Tissue	Lipid content (% dry weight of tissue)	
	Free lipid‡	Bound lipid§
Caryopsis	5.5–8.0	1.4–1.6
Hull	2.0–2.3	0.6
Bran	6.4–9.5	1.2–1.3
Endosperm	5.2–6.8	1.0
Scutellum	20.4–20.6	2.8–4.2
Embryo axis	10.6–12.6	3.3–4.1

†Youngs *et al.* (1977).
‡Diethyl ether extract.
§Water-saturated butan-1-ol (WSB) subsequent to diethyl ether.

unsaponifiable matter. Stuke (1961a) showed that oat embryo was more rich in lipid than was the endosperm and Beringer (1966) found 30.7% lipid in the embryo axis. Recent analyses of the lipid content and composition of oat grain tissue fractions are given in Tables 16.8 and 16.9. Oat hulls have a relatively low content of lipid but the proportion of saturated fatty acids is greater than in the lipid from other oat grain tissues. The scutellum and embryo axis are rich in lipids compared with the bran-endosperm fraction but the difference is primarily due to the greater proportion of non-polar lipids in the germ tissues (Price and Parsons, 1979). Youngs *et al.* (1977) found a much higher ratio of 16 : 0 to 18 : 1 in the bound lipids compared with the free lipids of the bran and endosperm fractions. Similarly, the ratio was higher in polar compared with non-polar lipids in the bran-endosperm fraction (Price and Parsons, 1979).

Table 16.9 Composition of lipid classes in oat grain tissue fractions†

Tissue	Total grain lipid (%)	Composition of lipid classes (% of total lipid)		
		Non-polar lipid	Glycolipid	Phospholipid
Hull	4.4	66.9	27.6	5.5
Bran-endosperm	7.1	56.9	21.4	21.7
Embryo axis	21.2	87.4	3.8	8.8

†Price and Parsons (1979).

VII CHANGE IN LIPID CONTENT DURING GRAIN DEVELOPMENT

The lipid content of the oat grain is at a maximum during the early stages of development when it is expressed as a percentage of dry matter (Beringer, 1966; 1971b; Brown *et al.*, 1970). Also during the early stages, the percentage fatty acid composition changes to show a decreased proportion of 18 : 1 and 18 : 3 and an increased proportion of 18 : 2 (Lindberg *et al.*, 1964b; Brown *et al.*, 1970; Beringer, 1971b). Brown *et al.* (1970) analysed the grain lipid classes qualitatively by tlc during development, and noted the presence, in the early stage of ripening, of FFA and diglycerides which, it was concluded, could not be caused by lipase activity. However, Urquhart *et al.* (1982) found that oat lipase activity reached a maximum 24–30 days after anthesis, and had declined at maturity to about half the maximum value.

VIII GENETIC CONTROL OF OAT GRAIN OIL CONTENT

Stuke (1960, 1961a) estimated that 6–7 genes influenced the formation of oil in the oat grain and that heritability was 82.5–97.5% for crude oil content. Baker and McKenzie (1972) also concluded that oil content was highly heritable and not correlated with caryopsis weight, density, or hull content. Brown *et al.* (1974) concluded that oil heritability was generally >70% and under polygenic control; there was no evidence of cytoplasmic control. Frey *et al.* (1975) studied *A. sativa* and *A. sterilis* crosses and concluded that oil content was polygenically inherited and that high oil content was partly dominant. Oil content was not correlated with caryopsis weight, heading date or plant height. Thro (1982) found that oil content of oats was positively correlated with, or independent of, grain weight and concluded, from crosses of *A. sativa* and *A. sterilis,* that oil content was controlled by additive gene action and the interaction between the additive effects of two or more genes. *A. sterilis* appeared to contribute genes responsible for high oil content to the crosses, and the crosses yielded a number of transgressive segregates with both higher and lower oil content than their parents. Heritabilities were estimated for individual fatty acids and, when the experiment was replicated in three different environments, these values were 16 : 0, 0.68; 18 : 1, 0.72; 18 : 2, 0.64; and 18 : 3, 0.27.

IX OAT GRAIN AS A SOURCE OF EDIBLE OIL

As early as 1918, Dubovitz suggested that oats might be extracted to produce an edible oil and calculated that from an oat containing 7.17% oil in the dehulled grain, the value of the oil alone would equal the value of the oats together with the costs of solvent extraction. Munro and Binnington (1927) extracted oats with hexane on a pilot-plant scale to use the oil in dietary experiments rather than to study the feasibility of commercial production. Frey and Hammond (1975) considered the possibility of using the oats as an oil source and calculated that, to return as much to farms in the mid-western United States as corn and soyabeans do, oats would need to have an oil content of 17% (wet weight) while maintaining current grain yields and protein content. Kalbasi-Ashtari and Hammond (1977) showed that oat oil could be refined by conventional techniques with a 15% loss by degumming and an alkali-refining loss of 25–30%. The oil was more stable than soyabean oil in storage tests at 55 °C, using both flavour and peroxide values as criteria although the refined oil contained no caffeic or ferulic acid antioxidants.

Although oat yields a good quality oil, there are several disadvantages to its use as an oil source. The oil content is rather low for economic extraction, the oil-rich tissues of the grain are not easily separated (*cf.* maize), oat lipid is rich in phospholipids and glycolipids which result in high degumming losses and the high FFA content leads to high alkali refining losses. All these disadvantages could probably be overcome by plant breeding. Oat grain with 10–11% oil content seems to be within easy reach, apparently with little loss in yield (Baker and McKenzie, 1972; Frey *et al.* 1975), but it is not clear to what extent the oil content might be increased above this level. In the oats with high oil content the increases are mainly in the TG fraction (de la Roche *et al.* 1977) and therefore degumming losses should decrease as oil content rises. It may be possible to select for low FFA content (Hutchinson and Martin, 1955; de la Roche, 1977) and for low lipase activity, and perhaps for strains rich in avenasterol for high-temperature stability to oxidation. The inclusion of oat caffeic acid antioxidants in oat oil also might increase its stability. A method has been patented for the extraction of oil from oats using propan-2-ol (Boocock and Oughton, 1977).

Even if the oil content of oat grain cannot be raised beyond 10%, it could still be profitable to extract oat oil and the author has calculated that extraction might raise the value of oats by as much as $0.34 per bushel in the U.S.A. Oat grain is an under-utilized cereal and commercial oil extraction would be a more attractive prospect if food uses were developed for the other components of the grain, especially the starch, protein and hemicellulose gums.

REFERENCES

Acker, L. and Becker, G. (1971). *Staerke* **23**, 419–424.
Acker, L. and Beutler, H. O. (1965). *Fette. Seifen. Anstrichm.* **67**, 430–433.
Akerman, A. (1954). *Sveriges Utsadesforen. Tidskr*, **64**, 261–277.
Amberger, K. and Wheeler-Hill, E. (1927). *Z. Unters- Lebensm.* **54**, 413–431.
Aylward, F. and Showler A. J. (1962a). *J. Sci. Food Agric.* **13**, 92–95.
Aylward, F. and Showler, A. J. (1962b). *J. Sci. Food Agric.* **13**, 492–496.
Baker, R. J. and McKenzie, R. I. H. (1972). *Crop Sci.* **12**, 201–202.
Bamann, E. and Ullman, E. (1942). *Biochem. Z.* **312**, 10–40.
Barnes, P. J. (1982). In *Proceedings of the 7th World Cereal and Bread Congress, Prague, 1982* (J. Holas, ed.), Elsevier, Amsterdam, in press.
Beringer, H. (1966). *Z. Pflanzenernaehr. Dueng. Bodenkd.* **114**, 117–128.
Beringer, H. (1967). *Z. Pflanzenernaehr. Bodenkd.* **116**, 45–53.
Beringer, H. (1971a). *Plant Physiol.* **48**, 433–436.
Beringer, H. (1971b). *Z. Pflanzenernaehr. Bodenkd.* **128**, 115–122.
Beringer, H. and Saxena, N. P. (1968). *Z. Pflanzenernaehr. Bodenkd.* **120**, 71–78.
Berner, D. L. and Hammond, E. G. (1970). *Lipids* **5**, 572–573.
Berrie, A. M. M., Don, R., Buller, D., Mahbood, A. and Parker, W. (1975). *Plant Sci. Lett.* **6**, 163–173.
Berry, R. A. (1920). *J. Agric. Sci.* **10**, 359–414.
Boocock, J. R. B. and Oughton, R. W. (1977). U.S. Patent No. 4,053,492.
Boskou, D. and Morton, I. D. (1976). *J. Sci. Food Agric.* **27**, 928–932.
Brown, C. M., Alexander, D. E. and Carmer, S. G. (1966). *Crop Sci.* **6**, 190–191.
Brown, C. M., Aryeetey, A. N. and Dubey, S. N. (1974). *Crop Sci.* **14**, 67–69.
Brown, C. M. and Craddock, J. C. (1972). *Crop Sci.* **12**, 514–515.
Brown, C. M., Weber, E. J. and Wilson, C. M. (1970). *Crop Sci.* **10**, 488–491.
Carter, H. E., Ohno, K., Nojima, S., Tipton, C. L. and Stanacev, N. Z. (1961). *J. Lipid Res.* **2**, 215–222.
Chow, C. K., Draper, H. H. and Csallany, A. S. (1969). *Anal. Biochem.* **32**, 81–90.
Clements, R. L. (1977). *Cereal Chem.* **54**, 865–874.
Daniels, D. G. H. (1966). *J. Chromatog.* **21**, 305–306.
Daniels, D. G. H. and Martin, H. F. (1961). *Nature.* **191**, 1302.
Daniels, D. G. H. and Martin, H. F. (1964). *Chem. Ind. London.* 2058.
Daniels, D. G. H. and Martin, H. F. (1965). *Chem. Ind. London.* 1763.
Daniels, D. G. H. and Martin, H. F. (1967). *J. Sci. Food Agric.* **18**, 589–595.
Daniels, D. G. H. and Martin, H. F. (1968). *J. Sci. Food Agric.* **19**, 710–712.
Daniels, D. G. H., King, H. G. C. and Martin, H. F. (1963). *J. Sci. Food Agric.* **14**, 385–390.
Dehghan, M. (1979). The relative composition and distribution of the lipids in six common oat varieties and commercial flour. M.S. Thesis, Iowa State University, Ames.
de la Roche, I. A. Burrows, V. D. and McKenzie, R. I. H. (1977). *Crop Sci.* **17**, 145–148.
Diemar, W., Bleyer, B. and Schmidt, W. (1937). *Biochem. Z.* **294**, 353–364.
Dubovitz, H. (1918). *Chem. Ztg.* **42**, 13–14.
Forsberg, R. A., Youngs, V. L. and Shands, H. L. (1974). *Crop Sci.* **14**, 221–224.
Frey, K. J. and Hammond, E. G. (1975). *J. Am. Oil Chem. Soc.* **52**, 358–362.
Frey, K. J., Hammond, E. G. and Lawrence, P. K. (1975). *Crop Sci.* **15**, 94–95.
Frimmel, G. (1958). *Bodenkultur* **10**, 41–47.
Fritz, A. (1970). *Muehle* **107**, 130–132.

Ganssman, W. (1967). *Getreide und Mehl* **17**, 37–41.

Gardner, H. W. and Hutchinson, J. B. (1951). *Agriculture London* **58**, 208–216.

Green, J., Marcinkiewicz, S. and Watt, P. R. (1955). *J. Sci. Food Agric.* **6**, 274–282.

Guerrant, N. B. (1927). *J. Agric. Res. Washington, D.C.* **35**, 1001–1019.

Herting, D. C. and Drury, E. J. E. (1969). *J. Agric. Food Chem.* **17**, 785–790.

Heydanek, M. G. and McGorrin, R. J. (1981). *J. Agric. Food Chem.* **29**, 950–954.

Hoffman, W. (1951). *Z. Pflanzenzacht.* **29**, 318–345.

Hübner, R. (1951). *Z. Acker Pflanzenbau* **93**, 169–197.

Hutchinson, J. B. (1953). *Chem. Ind. London* 578–581.

Hutchinson, J. B. and Martin, H. F. (1952). *J. Sci. Food Agric.* **3**, 312–315.

Hutchinson, J. B. and Martin, H. F. (1955). *J. Agric. Sci.* **45**, 411–418.

Hutchinson, J. B., Martin, H. F. and Moran, T. (1951). *Nature* **167**, 758–759.

Idler, D. R., Nicksic, S. W., Johnson, D. R., Meloche, V. W., Schuette, H. A. and Baumann, C. A. (1953). *J. Am. Chem. Soc.* **75**, 1712–1715.

Jahn-Deesbach, W., Marquard, R. and Schipper, A. (1971). *Z. Acker Pflanzenbau* **133**, 36–46.

Janíĉek, G. and Pokorný, J. (1959). *Sb. Vy. Sk. Chem.-Technol. Praze, Oddil Fak. Technol. Potravin. Technol.* **3**, 503–520.

Janíĉek, G. and Pokorný, J (1960). *Sb. Vy. Sk. Chem.-Technol. Praze* **4**(2), 327–329.

Kalbasi-Ashtari, A. (1976). Oat oil – its refining and oxidative stability. M. S. Thesis, lowa State University, Ames.

Kalbasi-Ashtari, A. and Hammond, E. G. (1977). *J. Am. Oil Chem. Soc.* **54**, 305–307.

Kazi, T. and Cahill, T. J. (1969). *Analyst* **94**, 417.

King, H. G. C. (1962). *Chem. Ind. London* 1468.

Knights, B. A. (1965). *Phytochemistry* **4**, 857–862.

Knights, B. A. and Laurie, W. (1967). *Phytochemistry* **6**, 407–416.

Knights, B. A. (1968). *Phytochemistry* **7**, 2067–2068.

König, J. (1871). *Landwirtsch. Vers. Stn.* **13**, 241–255.

Krarup, A. V. (1904). *Biedermans Zentr.* **33**, 94–106.

Kürten P. W. and Ganssmann, W. (1966). *Z. Acker Pflanzenbau* **123**, 121–144.

Lásztity, R., Berndorfer-Kraszner, E. and Huszar, M. (1980). In *Cereals for Food and Beverages* (G. E. Inglett and L. Munck, eds), pp. 429–445 Academic Press, New York and London.

Lindberg, P. (1966). *Acta Agric. Scand.* **16**, 217–220.

Lindberg, P., Bingefors, S., Lannek, N. and Tanhuanpää, E. (1964a). *Acta Agric. Scand.* **14**, 3–11.

Lindberg, P., Tanhuanpää, E., Nilsson, G. and Wass, L. (1964b). *Acta Agric. Scand.* **14**, 297–306.

Marquard, R. (1973). *Z. Lebensm. Unters. Forsch.* **153**, 354–362.

Martin, H. F. (1956). *Food Manuf.* **31**, 463–465.

Martin, H. F. (1961). *Chem. Ind. London* 364–367.

Martin, H. F. and Peers, F. G. (1953). *Biochem. J.* **55**, 523–529.

Mason, E. L. and Jones, W. L. (1958). *J. Sci. Food Agric.* **9**, 524–527.

Matlashewski, G. J., Urquhart, A. A., Sahasrabudhe, M. R. and Altosaar, I. (1982). *Cereal Chem.* **57**, 418–422.

Mix, A. (1931). *Landwirtsch. Jahrb. Berlin* **73**, 796–840.

Moran, T. (1952). *Food Manuf.* **27**, 73–74.

Munro, L. A. and Binnington, D. S. (1928). *Ind. Eng. Chem.* **20**, 425–427.

Musher, S. (1935a). *Food Inds.* **7**, 167–168.
Musher, S. (1935b). *Food Inds.* **7**, 329–330.
Nechaev, A. P., Denisenko, Ya. I., Shvartsman, M. I., Terent'eva, G. N. and Baikov, V. G. (1972a). *Iz. Vyssh. Ucheb. Zaved. Pisch. Tekhnol.* 145–146.
Nechaev, A. P., Shvartsman, M. I., Seit-Ablaeva, S. K., Rogover, V. S. and Lapshina, G. E. (1972b). *Iz. Vyssh. Ucheb. Zaved. Pisch. Tekhnol.* 54–56.
Novozhilova, G. N., Denisenko, Ya. I. and Nechaev, A. P. (1967). *Izv. Vyssh. Ucheb Zaved. Pisch. Technol.* 28–29.
Novozhilova, G. N., Denisenko, Ya. I., Nechaev, A. P. and Shnaidman, L. O. (1968). *Prikl. Biokhim. Mikrobiol.* **4**, 333–336.
Pan, W. P. (1981). A rapid method for stereospecific glyceride analysis and its application to soyabean and oat varieties. Ph.D. Thesis, Iowa State University, Ames.
Paul, E. (1921). *Analyst* **46**, 238–239.
Peers, F. G. (1953). *Nature* **171**, 981–982.
Persson, P. J. and Bingefors, S. (1956). *Sverigis Utsadesforen. Tidskr.* **66**, 174–181.
Peters, F. N. and Musher, S. (1937). *Ind. Eng. Chem.* **29**, 146–151.
Peters, F. and Musher, S. (1938). *Food Inds.* **10**, 129–175.
Pokorný, J. Zeman, I. and Janíĉek, G (1960). *Sb. Vy. Sk. Chem. Technol. Praze Potraviny* **4**(2), 343–354.
Pokorný, J., Janíĉek, G., Žák, L. and Kubátová, J. (1961a). *Sb. Cesk. Akad. Zemed. Ved, Rostl. Vyroba* **7**, 299–310.
Pokorný, J., Zeman, I. and Janíĉek, G. (1961b). *Sb. Vy. Sk. Chem. Technol. Praze Potraviny* **5**(1), 351–364.
Pokorný, J.,Pliška, V., Pokorná, V., Karvánková-Kubátová, J. and Janíĉek, G. (1962). *Sb. Vy. Sk. Chem. Technol. Praze Potraviny* **6**(3), 189–197.
Pokorný, J., Hrdlička, J., Pliska, V. and Janíĉek, G. (1963). *Sb. Vy. Sk. Chem. Technol. Praze Potraviny* **7**(1), 223–230.
Pokorný, J., Pokorná, V., Zwain, H. and Janíĉek, G. (1968). *Sb. Vy. Sk. Chem. Technol. Praze Potraviny* E20, 85–88.
Popov, M. P. and Cheleev, D. A. (1959). *Biokhim. Zerna Sbronik* **5**, 263–279.
Price, P. B. and Parsons, J. G. (1975). *J. Am. Oil Chem. Soc.* **52**, 490–493.
Price, P. B. and Parsons, J. (1979). *J. Agric. Food Chem.* **27**, 813–815.
Reiser, R. (1947). *J. Am. Oil Chem. Soc.* **24**, 199–204.
Rohrlich, M. and Niederauer, T. (1967). *Fette Seifen Anstrichm.* **69**, 63–67 and 226–230.
Sahasrabudhe, M. R. (1979). *J. Am. Oil Chem. Soc.* **56**, 80–84.
Sahasrabudhe, M. R (1982). *J. Am. Oil Chem. Soc.* **59**, 354–355.
Salun, I. P. and Kalugina, S. A. (1974). *Vopr. Pitan.* 64–67.
Schulze, E. and Pfenninger, U. (1911). *Hoppe Seyler's Z. Physiol. Chem.* **71**, 174–185.
Shukla, T. P. (1975). *Crit. Rev. Food Sci. Nutr.* **6**, 383–431.
Sims, R. J., Fioriti, J. A. and Kanub, M. J. (1972). *J. Am. Oil Chem. Soc.* **49**, 298–301.
Slover, H. T., Lehmann, J. and Valis, R. J. (1969). *J. Am. Oil Chem. Soc.* **46**, 417–420.
Sosedov, N., Nechaev, A., Shvartsman, M. and Terent'eva, G. (1971). *Mukomol. Elevator. Prom.* **11**, 39–40.
Špinka, J. (1939a). *Chem. Listy.* **33**, 60–69.
Špinka, J. (1939b). *Chem. Listy.* **33**, 38–45.
Stellwaag, A. (1890). *Landwirtsch. Ver. stn.* **37**, 135–154.

Stuke, H. (1960). *Z. Pflanzenzuch.* **42**, 147–189.
Stuke, H. (1961a). *Getreide und Mehl* **11**, 123–127.
Stuke, H. (1961b). *Z. Pflanzenzuch.* **45**, 143–177.
Šupová, J., Pokorný, J. and Janíĉek, G. (1959). *Sb. Vy. Sk. Chem. Technol. Praze. Oddil Fak. Potravin. Technol.* **3**, 525–544.
Takahashi, E., Tase, S. and Saeki, Y. (1935). *J. Ag. Chem. Soc. Japan* **11**, 199–205, Abstr. 44–45.
Templeton, W. H. and Carpenter, B. R. (1953). *Analyst* **78**, 726–727.
Tevekelev, D. (1970). *Izv. Bulg. Akad. Nauk. Sofia Inst. Khranene* **9**, 5–19.
Thomke, S. (1970). *Z. Tierphysiol. Tiernaehr. Futtermittelkd.* **27**, 31–35.
Thro, A. M. (1982). Feasibility of oats as an oil seed crop. Ph. D. Thesis, lowa State University, Ames.
Trier, G. (1913). *Hoppe Seyler's Z. Physiol Chem.* **86**, 153–173.
Urquhart, A. A., Brumell, C. A., Altosaar, I. and Matlashewski, G. J. (1982). *Plant Physiol.* (submitted).
Van Kampen, G. B. (1919). *Ollen en Vetten* **3**, 203. [Janíĉek and Pokorný, 1959].
Welch, R. W. (1977). *J. Sci. Food Agric.* **28**, 269–274.
Welch, R. W. (1975). *J. Sci. Food Agric.* **26**, 429–435.
Youngs, V. L. (1974). *J. Food Sci.* **39**, 1045–1046.
Youngs, V. L. (1978). *Cereal Chem.* **55**, 591–597.
Youngs, V. L. and Püskülcü, H. (1976). *Crop Sci.* **16**, 881–883.
Youngs, V. L., Püskülcü, H. and Smith, R. R. (1977). *Cereal Chem.* **54**, 803–812.

17 Lipids in Maize Technology

E. J. WEBER

U.S. Department of Agriculture, University of Illinois, Urbana, Illinois 61801 U.S.A.

I INTRODUCTION

Maize (*Zea mays* L.) is the only major cereal grain that originated in the Americas; Columbus introduced it to Europe in 1492. Today it is grown on all continents in many countries. The countries that are the major produc-

The mention of a trademark, proprietary product, or vendor does not constitute a guarantee or warranty of the product by the U.S. Department of Agriculture and does not imply its approval to the exclusion of other products or vendors that may also be suitable.

"Lipids in Cereal Technology"
ISBN 0-12-079020-3

Table 17.1 Major producers and exporters of maize in 1981†

Country	Production	Export
	Thousand metric tons	
United States	208,314	53,975
Peoples' Republic of China	59,000	–‡
Brazil	23,600	–
Mexico	12,000	–
Argentina	–	6,500
South Africa	–	4,800
World Total	437,658	75,301

†USDA Foreign Agricultural Service (1982).
‡No value given.

ers and exporters of maize are shown in Table 17.1. In 1981, the U.S. produced 48% of the world maize crop (USDA, 1982) and was by far the largest producer. The Peoples' Republic of China was second, producing 14% of the world crop. The U.S. exported more than one-fourth of their maize, providing 72% of the world export market, making it the world's largest exporter. Argentina was a distant second in exports with 9%.

The importance of maize in the U.S. is illustrated by the fact that maize production exceeds the combined totals of all the other major grains – wheat, oats, soyabeans, barley and rye – grown in the U.S. (USDA, 1981). Domestically about 88% of the U.S. maize crop is used for livestock feed; approximately 40% is fed to hogs, 29% to cattle and 19% to poultry (Jugenheimer, 1976). Maize also yields more industrial products than any other grain. The industrial users of maize in the U.S. are the mixed feed manufacturers, the wet millers, dry millers and the distilling and fermentation industries.

The types and proportions of the chemical components in the grain affect both the feeding efficiency and the industrial usage of maize. The average values for the major components of dry, dent-maize grain are 80% carbohydrate, 10% protein, 4.5% oil, 3.5% fibre and 2% minerals (Jugenheimer, 1976). The percentages of these components vary slightly according to genotype and environmental conditions, but the values listed are typical for commercial hybrids. Carbohydrates make up the major fraction of the maize grain. Protein and oil content are lower in maize than, for example, in soyabean which has about 40% protein and 22% oil, but the average yield of maize is about three times that of soyabean. Among the cereal grains, only pearl millet with 5.4% oil and oat groats with 7% oil have higher oil contents than maize. Brown rice (2.3% oil), wheat (1.9%), barley (2.1%) and sorghum (3.4%) all have lower oil contents.

Morrison (1978) and Weber (1973, 1978, 1980) have written reviews about maize lipids. This review will highlight recent literature in the area that may have important impacts on lipids in maize technology. These areas include agricultural practices, plant breeding and biochemical research on maize lipids.

II OIL CONTENT

A Effect of Agricultural Practices on Oil Content

Earle (1977) obtained data from three commercial processors on the variation in oil content of maize by crop years from 1917 to 1972. The range was only 4.0 to 4.9%. No correlations between year-to-year variations in temperature, rainfall or fertilization rates were found. One company had consistently lower oil recovery than the other two because a different solvent, light petroleum rather than hexane, was used to extract the oil.

Welch (1969) found that the addition of the fertilizer elements N, P and K increased the oil content of the grain slightly, but the most important effect was the increased grain yield which produced more oil per unit of land area. During the past 25 years, the average increase in yield per year for maize in the U.S. has been about 132 kg ha^{-1} (2.1 bu/a*) (Johnson, 1982). In 1931, before the use of hybrids became widespread, the U.S. average yield was 1,540 kg ha^{-1} (24.5 bu/a) (Jugenheimer, 1976). In 1981, the average was a record of 6,899 kg ha^{-1} (109.9 bu/a). Yields exceeding 18,800 kg ha^{-1} (300 bu/a) have been reported (Marten and Nelson, 1982). The potential for continuing increases in yield certainly exists. The major factors contributing to the higher yields have been the use of improved hybrids, increased fertilization, denser plant populations and effective weed and pest control.

Freeman (1973) discussed several agricultural practices that may affect the milling quality of maize. Until recently, whole ears of maize were harvested, allowed to dry in an open crib and then shelled. Now almost all U.S. maize is shelled in the field. Maize that is field shelled at 20 to 25% moisture or higher may be chipped, cracked and bruised during harvesting. For example, when maize at 28% moisture was shelled mechanically, 25 to 30% of the grains had visible damage. The germs are particularly susceptible to abrasions and cuts. Some of the oil may be lost or migrate to the endosperm. The oil in mechanically damaged grain is much more susceptible to deterioration by autolytic enzymes which catalyse fatty acid (FA) oxidation and triglyceride (TG) hydrolysis. Fungi also attack damaged

*U.S. bushels per acre; 1 U.S. bushel = 0.352 hectolitres; 1 acre = 0.405 hectares (ha).

Table 17.2 Oil recovered after wet milling of artificially dried maize†

Drying temperature (°C)	Extent of combine harvester damage	Oil recovery‡ (%)
Ambient	Low	75.7
49	Low	70.3
82	Low	68.2
149	Low	61.0
82	High	55.0

†Vojnovich *et al.* (1975).
‡Moisture-free basis.

grain more easily. Pericarp damage over the germ makes the grain especially susceptible to fungal invasion and the fungi grow preferentially on the germ, depleting the oil. Fungi apparently are the primary factors responsible for the production of free fatty acids (FFA) in stored grain. Free fatty acids are undesirable because they must be removed during oil refining and the yield of oil is reduced.

Sprouting, weather and disease may also damage the grain. When maize was subjected to drought during grain fill, the grain weight was reduced to 52% of the control (Jurgens *et al.*, 1978). The drought increased the protein content of the grain from 8.3 to 11.0% but decreased the oil content from 3.8 to 3.1%. In 1970, when an epidemic of southern maize leaf blight (*Helminthosporium maydis*) occurred, no effect was noted on the starch and protein content of the grain, but oil content was reduced essentially in proportion to the amount of blight damage (Freeman, 1973).

Maize should have a moisture content of not more than 13.5% to be safely stored (Jugenheimer, 1976). Above this moisture level, fungi and endogenous degradative enzymes exhibit a sharp increase in activity. The quality of maize in storage is influenced by the fungal and insect populations, moisture, temperature and handling. The critical factors in determining the degree of damage that may result from artificial drying are the initial moisture content of the grain, drying temperature and time. Brown *et al.* (1979) found that if maize with 20–30% initial moisture was batch dried at 80 °C, 63.5% of the grains had stress-cracks, and of these damaged grains, 25.5% had single cracks and 38.0% multiple cracks. Grain with stress-cracks is susceptible to additional breakage during handling. Vojnovich *et al.* (1975) observed a significant decrease in oil recovery as drying temperatures were increased (Table 17.2). The loss of oil was still greater when damage from the combine harvester was high. If maize is harvested at moisture contents above 25%, Brown *et al.* (1981) recommended drying temperatures of 60 °C or less to prepare grain that would be of good quality for milling.

B Oil Content of Maize Grain Fractions

The distribution of lipids in component parts of maize grain is shown in Table 17.3. The lipid distribution is similar despite the differences in the four maize strains. H51 is an inbred; LG-11 is a three-way cross hybrid forage maize. Both the waxy maize and the amylomaize are endosperm mutants, and the amylomaize is also a higher oil strain. The maize strains came from different parts of the world – H51 and amylomaize from the U.S.; LG-11 from France; and waxy maize from Italy. The mean dry weights of the grains were: H51 = 233 mg; LG-11 = 250 mg; waxy maize = 284 mg; and amylomaize = 249 mg. The conditions for extraction of the lipids differed slightly (Weber, 1979; Tan and Morrison, 1979), but hot, water-saturated butan-1-ol was used as the final extraction solvent for all the grain parts.

The proportions of the maize grain represented by the various fractions are typical of the values reported by other investigators and summarized in reviews of cereal lipids by Weber (1973) and Morrison (1978). In Table 17.3, amylomaize illustrates the fact that higher-oil strains have larger germs. Endosperm makes up 81–86% of grain with normal oil content, but the percentage of lipid in endosperm is low, about 1%. The aleurone, non-starch and starch regions of the endosperm each have distinctive lipid class distributions. The quantities of all the lipid classes found in these regions of the endosperm and also in germ, pericarp and tip cap are discussed by Morrison (see Chapter 2). The lipid classes were isolated fom the same maize strains shown in Table 17.3.

Table 17.3 Distribution of lipids in maize grain

	H51†		LG–11‡		Waxy maize‡		Amylomaize‡	
	Wt. (%)	Lipid§ (%)	Wt. (%)	Lipid (%)	Wt. (%)	Lipid (%)	Wt. (%)	Lipid (%)
Whole grain	100.0	4.9	100.0	4.9	100.0	5.1	100.0	9.3
Germ	10.9	35.5	8.4	38.7	10.8	35.6	15.0	33.7
Endosperm	82.6	1.0	86.0	1.0	81.0	0.8	74.7	2.4
Pericarp	5.8	0.2	4.2	0.3	6.6	0.2	7.8	0.2
Tip cap	0.7	0.3	1.4	1.6	1.6	1.7	2.5	1.8

†Weber (1979).
‡Tan and Morrison (1979).
§Lipid % of dry wt. of respective part quantified as fatty acid methyl esters.

In maize with normal oil content, 78 to 85% of the acyl lipid of the grain is found in the germ (Table 17.3). Approximately 90% of this lipid is TG. In the milling industries, germ is separated from the grain and extracted with hexane. This solvent extracts mainly non-polar lipids, and the TG content of crude maize oil is about 95%. Mounts and Anderson give the details of maize oil production, processing and use in Chapter 18. Maize oil has the highest value per unit weight of any of the major products of the milling industries, and millers have expressed some interest in higher oil maizes. However, the greatest advantage for higher oil maizes would be an increase in caloric value for livestock feeding. Oil has 2.25 times the energy values of carbohydrate or protein per unit weight.

C Breeding Maize for Higher Oil Content

The classic experiment in breeding maize for high and low oil contents was started at the University of Illinois in 1896 and is still being conducted today (Dudley, 1974, 1977). The original Burr's White corn had 4.7% oil. In 1981 after 82 generations of mass selection among ears for high and low oil contents, the Illinois High Oil (IHO) strain had 19% oil and the Illinois Low Oil strain 0.3% oil (Dudley, personal communication). Significant genetic variability still exists in the IHO strain (Dudley, 1977) and further increases in oil should occur as selection is continued. Both germ size and oil percentage in the germ have increased in IHO, but endosperm and total grain weight have decreased (Curtis *et al.*, 1968). With selection only for oil, the yield of IHO has fallen to about 30% of that of commerical hybrids.

With more effective breeding schemes and attention to yield, hybrids have been produced that have 6–8% oil content and yields equivalent to those of commercial varieties (Watson, 1975; Creech and Alexander, 1978). The most significant development in breeding for higher oil maizes has been the adaptation of wide-line nuclear magnetic resonance (NMR) spectroscopy to nondestructive analysis of oil content in maize (Bauman *et al.*, 1963; Alexander *et al.*, 1967). Large numbers of samples may be screened because the scan time may be as brief as 2 s. Alexander (1982) has, with NMR selection of single grains, achieved an increase in oil content in Alexho Synthetic from 4.4 to 15.8% in only 22 cycles, a gain of 0.52% per year. IHO, in the long-term experiment originally based on mass selection among ears and destructive chemical analysis for oil, has shown an increase of only 0.17% per year.

Miller *et al.* (1981), using high intensity selection by NMR, were able to increase the oil content of "Reid Yellow Dent" maize from 4.0 to 9.1% in only seven cycles with no reduction in yield. Trifunovíc and others (1975) at the Maize Research Institute in Yugoslavia are also using NMR analysis to develop lines that vary in oil content. Among 490 inbred lines from their breeding programme, the oil content ranged from 2.7 to 12.5% with a mean value of 6.14%.

In a combined electron microscopy and nuclear magnetic resonance study, Ratkovic *et al*. (1978) found that the size of the oil bodies in maize germ increases as the oil content increases and that the oil bodies are ellipsoidal rather than spherical.

Alexander (1982) has released three high oil inbreds to the seed trade (R802A–7% oil, R855–9%, R806–9%). A hybrid with 6.5 to 7.0% oil is now being commercially marketed to farmers in the U.S. Until recently, contracting was the only inducement that farmers could be offered to produce higher oil maize for milling. Today low cost, infra-red grain analysers are available which may be used to identify maize with higher oil contents (Hymowitz *et al*., 1974) and a premium price may be paid for grain with higher than normal oil levels.

III FATTY ACID COMPOSITIONS AND LIPID CLASSES

A Fatty Acid Compositions of Maize Grain Fractions

The FA compositions of the total lipids from the various fractions of the grain of hybrid LG-11 (Tan and Morrison, 1979) are shown in Table 17.4. The germ lipids have the lowest percentage of palmitic acid and the highest percentages of oleic and linoleic acids. This FA composition reflects the high proportion (89%) of TG in the germ acyl lipids. The FA patterns of the non-starch lipids (NSL) + aleurone lipids and NSL − aleurone lipids suggest that the aleurone has a high content of TG. The major lipid class in the pericarp and tip cap fractions is also TG. Pericarp has the highest level of steryl esters (SE). The starch lipids are the most saturated, and their high percentage of 16 : 0 is indicative of the FFA and lysophospholipids which constitute the major starch lipids. The 18 : 3 contents of all the grain fractions could be a problem for millers because the triunsaturated acid oxidizes readily to produce off-flavours (Ho *et al*., 1978; Smouse, 1979).

B Breeding for Modified Fatty Acid Composition in Maize

A representative of a maize cereal manufacturing company has stated that in the past five years the company has observed an increase in unsaturated FFA in endosperm fractions. These unsaturated FFA can cause increased rancidity formation in the cereal products (Watson, 1981). This problem is believed to be due to chipped grain where exposed starch adsorbs water

Table 17.4 Fatty acid compositions of total acyl lipids in grain fractions of maize hybrid LG-11†

	% of total acyl lipid					Fatty acid composition (wt. %)				
Grain fraction	SE	TG	FFA	GL	PL	16 : 0	18 : 0	18 : 1	18 : 2	18 : 3
Germ	2	89	1	2	4	11	1	24	63	1
Endosperm										
NSL‡ + aleurone lipids	8	54	21	7	5	17	2	19	58	4
NSL − aleurone lipids	4	14	53	12	11	24	2	12	59	4
SL‡	1	1	48	6	42	31	1	11	54	3
Pure starch	< 1	1	56	3	38	37	2	10	48	4
Pericarp	31	42	2	7	11	25	5	21	46	3
Tip cap	11	52	15	8	8	22	3	20	49	6

†Adapted from Tan and Morrison (1979).
‡NSL = non-starch lipids; SL = starch lipids.

when the grain is soaked for processing. Then the water is replaced by oil from broken germ during processing and produces a product which has a higher percentage of oil than is desirable. The oil has also increased in polyunsaturation by about 5–8% over the past 20 years (Weber, 1978; Watson, 1981). At the same time, the average oil content of maize received at U.S. Midwest milling plants has declined from about 4.8 to 4.2% (Freeman, 1973; Watson, 1981). All of these properties – quantity of oil, polyunsaturation of the oil, and even susceptibility to breakage – are influenced by the selection of maize genotypes. Producers of commercial maize seed usually do not monitor the oil content or FA composition of the hybrids that they sell. Yield potential, disease resistance and other desirable agronomic traits are the determining factors in the selection of parental lines. When superior lines are developed, the older lines are replaced.

Table 17.5 shows evidence that changes in parental lines over the years may have caused the indicated decrease in oil content and increase in polyunsaturation. The table also shows the use of public lines for the production of hybrid maize seed. These inbred lines have been developed by university and government maize breeders. In 1964, Oh43 and C103 were the most widely used lines; C103 had 4.6% oil. In 1970, the top two lines, B37 and W64A, had 4.9 and 4.6% oil, respectively. The latest survey in 1979 showed that the oil content of the three most widely used lines ranged from only 3.8 to 4.3%. Recently selected lines do have less oil. In 1964 the FA composition of C103 had only 38% 18 : 2. In 1979, Mo17 with 66.6% 18 : 2 and A632 with 67.9% 18 : 2 were involved in 21.9% of the total seed produced. The trend appears to be toward higher unsaturation.

Table 17.5 Oil and polyunsaturated fatty acid in public inbred lines used as parental sources for U.S. hybrid maize seed

Line	Usage of inbred line as % of total seed†				Content of oil‡ (%)	Proportion of 18 : 2‡ (%)
	1964	1970	1975	1979		
B73	–	–	3.1	16.1	4.3	59.0
Mo17	–	1.8	7.0	12.2	3.8	66.6
A632	–	7.4	15.2	9.7	4.0	67.9
B37	2.0	25.7	6.8	2.4	4.9	57.2
W64A	0.9	13.0	1.5	0.6	4.6	63.3
Oh43	15.7	11.7	0.9	0.1	4.0	68.3
C103	11.9	4.2	0.3	–	4.6	38.0

†Based on data from Sprague (1971), Zuber (1975), Zuber and Darrah (1980).
‡Weber (unpublished data); oil as % of grain, 18 : 2 as % of total FA.

An interesting aspect of these surveys is that they give some indication of the status of the U.S. germ plasm base. Three lines in the 1979 survey accounted for the same percentage (38%) of the total seed production as six lines in the 1975 survey. Adequate genetic variability is assumed to exist although these surveys do not indicate the possible relationships between public and commercial inbred lines. Most maize breeding programmes now include some exotic germ plasm in recognition of the dangers of reducing the genetic base for selection. However, germ plasm resources are fast disappearing, and there is an urgent need to collect, catalogue and preserve maize germ plasm (Vincent *et al*., 1978; Timothy and Goodman, 1979).

The known variability for fatty acid composition in maize covers a wide range. Jellum (1970) has made the most extensive search for unusual FA compositions. He found a range among 788 maize strains from 14 to 64% for oleic acid and 19 to 71% for linoleic acid (Table 17.6) and a strong negative correlation between these FA. A greater diversity for FA composition was present in maize of foreign origin than in U.S. maize. Among four strains from India, Sharma *et al*. (1975) found one strain with only 10% 18 : 2 and 28% 16 : 0. Italian maize oil was more saturated than commercial U.S. oil with higher levels of 16 : 0 and lower levels of 18 : 2 (Camussi *et al*., 1980). In the study of Italian maize, only weak correlations were observed between FA and morphological traits of maize. These traits included ear weight, grain shape, tassel flowering time and plant height.

Environmental factors such as temperature, year, location, planting date and fertility have little effect on FA composition as compared with genetic effects (Jellum and Marion, 1966; Poneleit and Bauman, 1970; Jahn-Deesbach *et al*., 1975). The 4.2% increase in 18 : 2 content of a commercial U.S. maize oil over a 10-year period (Table 17.6) probably reflects

Table 17.6 Range of fatty acid composition in maize

	Source	Fatty acid composition as % of total FA				
		16 : 0	18 : 0	18 : 1	18 : 2	18 : 3
Mazola oil (1972)†	U.S.	12.0	1.8	27.3	57.8	1.1
Mazola oil (1982)†	U.S.	10.9	1.6	24.4	62.0	1.0
788 strains‡	World	6–22	0.6–15	14–64	19–71	0.5–2.0
4 strains§	India	10–28	2–8	32–53	10–53	0.3–1.2
102 strains¶	Italy	12–18	0.6–4	22–42	39–54	–

†U.S. commercial maize oil analysed by Weber.
‡Jellum (1970).
§Sharma *et al.* (1975).
¶Camussi *et al.* (1980).

changes in genotypes of the hybrids, as suggested earlier, rather than the influence of environmental factors.

FA composition is highly heritable in maize (Poneleit and Alexander, 1965; de la Roche *et al*., 1971a; Poneleit, 1972; Sun *et al*., 1978). Factors have been located on chromosome 2 (Plewa and Weber, 1975) and on the long arm of chromosome 5 (Shadley and Weber, 1980) that are involved in the determination of 18 : 1 and 18 : 2 levels. The gene action for controlling the levels of 16 : 0, 18 : 1 and 18 : 2 appears to be mainly additive (Widstrom and Jellum, 1975). Breeding systems, such as mass selection or recurrent selection, would be effective for modification of FA composition. Fairly accurate predictions of the FA distribution in a hybrid can be made by averaging the values of the two inbreds used as parents.

Although the effect may be limited by the concentrations of FA available for esterification, the stereospecific distribution of FA in the TG molecule is under independent genetic control (de la Roche *et al*., 1971b).

Table 17.7 Fatty acid composition of classes of lipids from grain of four maize inbreds†

		Fatty acid composition (mol %)				
Lipid Class	Inbred	16 : 0	18 : 0	18 : 1	18 : 2	18 : 3
Triglyceride	H21	16.5	2.9	37.4	42.2	1.0
	IHO	12.9	2.1	35.4	48.8	0.8
	K6	11.3	1.1	22.1	64.1	1.3
	NY16	7.4	1.6	20.1	69.6	1.3
Phosphatidyl-choline (PC)	H21	20.0	1.5	42.1	36.0	0.5
	IHO	19.0	1.5	37.5	41.4	0.6
	K6	23.9	1.8	28.3	45.2	0.8
	NY16	18.6	2.2	25.8	52.3	1.1
Phosphatidyl-inositol (PI)	H21	42.1	2.6	18.7	35.6	1.0
	IHO	39.5	2.2	20.6	37.4	0.3
	K6	40.2	1.9	12.3	44.1	1.5
	NY16	38.3	2.2	12.8	45.8	1.0
Phosphatidyl-ethanolamine (PE)	H21	20.8	0.8	23.5	54.2	0.7
	IHO	25.1	1.0	18.9	54.3	0.7
	K6	24.0	1.0	15.3	59.0	0.7
	NY16	20.6	1.9	18.1	58.3	1.0
Phosphatidyl-glycerol (PG)	H21	36.3	2.0	19.7	40.3	1.7
	IHO	36.7	2.7	23.6	36.6	0.5
	K6	35.5	2.5	15.6	42.6	3.8
	NY16	34.2	3.6	16.6	44.3	1.4

†Weber (1978).

The placement of a specific FA at each position of the TG is important because the FA in the outer positions of the TG molecule are more susceptible to oxidation (Sahasrabudhe and Farn, 1964; Raghuveer and Hammond, 1967; Catalano *et al*., 1975) and hydrogenation (Drozdowski, 1977) than the fatty acid in the middle position.

Genetic modification of the FA of maize oil will affect not only the FA composition of the TG but also the FA compositions of other lipid classes. This fact may have particular significance for polar lipids which are essential components of all membranes. The FA compositions of the TG and phospholipids (PL) of four maize inbreds are shown in Table 17.7. Each lipid class has a characteristic FA pattern. Within each inbred, TG has a lower percentage of saturated 16 : 0 and a higher percentage of polyunsaturated 18 : 2 than the other lipid classes. PC has the highest level of 18 : 1. PE has the highest percentage of 18 : 2 among the PL. Both PI and PG have high percentages of saturated FA, but the PG tends to have more 18 : 1 and less 18 : 2 than PI. When the inbreds are ranked in the same order for each class of lipid, increasing levels of 18 : 2 generally are noted. The range is much larger for TG, from 42.2% for H21 to 69.6% for NY16, but differences also are apparent in each PL class.

Table 17.8 Fatty acid compositions of the lipids of maize inbreds C103D and B73 and their reciprocal crosses†

			Fatty acid composition (mol %)				
		n‡	16 : 0	18 : 0	18 : 1	18 : 2	18 : 3
Triglyceride	C103D	6	13.3	2.0	43.4	40.3	1.1
	C103D × B73	9	12.6	2.1	38.6	45.7	1.0
	B73 × C103D	9	12.2	2.2	33.7	50.8	1.0
	B73	3	11.4	1.9	29.6	55.9	1.3
Phosphatidyl-choline	C103D	6	20.8	2.3	49.5	26.6	0.7
	C103D × B73	9	19.1	1.8	44.1	34.2	0.8
	B73 × C103D	9	20.1	1.8	37.6	39.6	0.9
	B73	3	20.9	1.6	30.5	45.9	1.1
Phosphatidyl-inositol	C103D	6	38.9	4.0	29.8	26.8	0.5
	C103D × B73	9	36.3	2.8	26.8	33.5	0.7
	B73 × C103D	9	37.5	2.5	20.9	38.4	0.7
	B73	3	33.7	2.4	19.8	42.9	1.2
Phosphatidyl-ethanol-amine	C103D	6	23.0	3.9	38.4	34.2	0.5
	C103D × B73	6	20.7	3.3	34.0	41.6	0.4
	B73 × C103D	9	20.6	2.2	26.7	49.8	0.7
	B73	3	21.5	2.7	21.9	52.8	1.0

†Weber (1983).
‡Number of individual grains.

The heritability of FA composition in PL was tested by crossing two inbreds, C103D and B73, which differ widely in 18 : 2 content (Table 17.8). The lipids were isolated from the germ only. Inheritance is more easily determined in the diploid germ, which has equal inheritance from both parents, than in the triploid endosperm. The TG and PL of the crosses show intermediate 18 : 2 values, possibly with maternal effects. The characteristic FA patterns for each PL – high 18 : 1 for PC, high 16 : 0 for PI, and high 18 : 2 for PE – are present, but genotype superimposes variations in the FA composition within the lipid class patterns.

IV UNSAPONIFIABLE LIPIDS

A Sterols and Hydrocarbons

The unsaponifiable lipids of cereals have been discussed by Barnes in Chapter 3 and by Morrison (1978). Only a few very recent studies on maize will be mentioned here.

Worthington (1982) used a combination of preparative column chromatography, thin-layer chromatography (tlc) and gas-liquid chromatography (glc) to isolate and quantify hydrocarbons, sterols and SE without destroying the SE, In two commercial samples of maize oil, the average percentage values for hydrocarbons, free sterols and SE are: Mazola oil – 0.039, 1.42, 0.37; Kroger oil – 0.039, 0.95, 0.32. Squalene is the major hydrocarbon and makes up 59 and 34% of total hydrocarbons in Mazola and Kroger oils, respectively. The remaining hydrocarbons consist of a complex mixture of both even- and odd-carbon number hydrocarbons with the even-number compounds predominating. Maize oil SE contain a higher proportion of 18 : 2 than do the total oil lipids. Sitosterol is the major sterol of both free sterols and SE and together with campesterol, stigmasterol and Δ^5-avenasterol makes up 90–95% of all sterols.

Miric *et al.* (1981) analysed the sterols of maize oil from the grain of five dwarf Yugoslavian hybrids. The hybrids have normal oil content (4.2–5.1% of dry matter). The unsaponifiable fraction makes up 1.9 to 2.2% of the oils. The sterols of the oils consist of sitosterol (66.3–70%), campesterol (15.6–21.4%), stigmasterol (4.3–9.5%), Δ^5-avenasterol (1.6–8.7%), Δ^7-stigmasterol (0.2–1%), and cholesterol (0.2–0.4%).

A new sterol, 24-methyl-*E*-23-dehydrolophenol, has been identified in maize germ oil by Itoh *et al.* (1981). Capillary glc of the 4-methyl sterols has been suggested as a method of identifying various vegetable oils and estimating the degree of processing of an oil (Kornfeldt and Croon, 1981). Sterols may undergo interconversion in maize germ infested by fungi and ratios between different sterols might be used to characterize the various fungi (Farag *et al*., 1981). Each fungus also has an independent effect on the hydrocarbon constituents, producing an individual pattern of changes.

B Tocols

Maize oil is considered a premium oil in the world market because of its high level of PUFA and good stabilty for cooking or frying without hydrogenation (Huang *et al*., 1981). In an examination of ancient maize grain from caves in the arid southwestern U.S., Priestley *et al*. (1981) found that although most of the 18 : 2 is oxidized in the first 100–200 years, about 4% 18 : 2 by weight of FA found in modern maize is still detectable in grain more than 1500 year's old. This is an amazing example of the capacity of dry seeds to protect at least part of the PUFA from atmospheric oxidation.

Intact maize grain has been exposed to air, 15 p.p.m. NO_2 or 1.5 p.p.m. O_3, continuously for 100 h at room temperature without loss of PUFA (Brooks and Csallany, 1978). Only intact grains of maize were tested, but in the same study, when soyabeans were split into halves, air caused the destruction of tocols without detectable PUFA oxidation. Tocols are more oxidatively labile than PUFA and thus function as antioxidants in inhibiting lipid oxidation.

Traditionally γ-tocopherol (γ-T) has been considered the predominant tocol in maize oil. Screening maize inbreds, however, has revealed large genotypic variation in the tocol isomers. The tocols are extracted with hexane after direct saponification of the ground maize samples rather than first extracting the lipids (McMurray *et al*., 1980). The individual tocopherols and tocotrienols are determined by high-performance liquid chromatography (hplc). The conditions for hplc are similar to those used by Carpenter (1979) except that the mobile phase is 1.3% hexane in propan-2-ol. The predominant isomer in most of the maize inbreds is γ-T (Table 17.9), but in B37 the percentage of α-tocopherol (α-T) is higher

Table 17.9 Tocol isomers of maize inbreds†

	Content as % of total tocols‡				Total grain tocols
Inbred	α-T	α-T-3	γ-T	γ-T-3	(mg kg^{-1})
B37	44.8	13.2	36.1	5.9	52
C103D	25.3	14.3	43.3	17.1	26
B84	23.3	9.3	65.1	2.3	54
A619	8.1	4.0	69.7	18.2	102
W64A	2.1	5.1	87.8	5.0	67

†Weber (unpublished data).
‡T = tocopherol; T-3 = tocotrienol.

Table 17.10 Tocopherol composition of germ oil from maize inbreds and their reciprocal crosses†

	Tocopherols extracted in the germ oil ($mg\ kg^{-1}$ germ)	
	α-T	γ-T
Mo17	47	165
Mo17 × W64A	32	199
W64A × Mo17	28	232
W64A	10	420

†Weber (unpublished data).

than that of γ-T. The percentages of the tocotrienols also vary among the inbreds. Although all of these inbreds have normal oil content and nearly the same grain weight, the total weight of the tocols ranges from 26 to 102 $mg\ kg^{-1}$ of grain.

Maize oil is extracted from the germ only and injected directly into the hplc column for analysis to study the inheritance of the tocols. No tocotrienols (T-3) are found in maize germ. Intermediate weights for α-T and γ-T in the reciprocal crosses of Mo17 and W64A indicate that the proportions of vitamin E isomers can be modified by breeding (see Table 17.10). We do not yet know the exact functions of the various tocol isomers, but the proportions of the isomers present in the grain may be important. Although α-T is considered to have the most biological activity, γ-T may be a better antioxidant for 18 : 2 (Wu *et al*., 1979). Cereal tocols are further discussed in Chapter 3.

C Cutin

Espelie *et al*. (1979) have studied the composition of cutin from maize pericarp. In this lipid-derived polymer, ω-hydroxylated acids constitute the major class of monomers. Only minor amounts of alcohols and FA are present. In maize the major components are ω-OH-18 : 1 (51%), diOH-16 : 0 (20%), 9,10,18-triOH-18 : 0 (7%) and ω-OH-16 : 0 (6%).

V THE FUTURE?

A Maize Milling Industries

The wet milling industry in the U.S. is expected to continue to expand in the immediate future. Seventeen of the current 26 U.S. plants have been built since 1960 (USDA, 1982). Much of this growth can be attributed to the increasing demand for maize sweeteners. In 1970, U.S. per capita consumption of high-fructose syrup was only 0.3 kg $year^{-1}$; in 1980 it was 8.7 kg. Ten years ago maize sweeteners accounted for 16.7% of the total domestic sugar market, now it is 33%, and by 1985 perhaps 43% of the market. Increasing milling capacity also means increasing production of by-products such as maize oil, and perhaps new products such as maize lecithin.

Commercial maize lecithin was available from the mid-1930s through the 1950s, but in recent years the only source of commercial lecithin in the U.S. has been soyabean. A sample of commercially prepared maize lecithin (A. E. Staley Manufacturing Co, Decatur, IL) was recently analysed in our laboratory and compared with a sample of commercial soyabean lecithin (Weber, 1981). Some differences were noted in the distribution of polar lipids in two lecithins that may give maize lecithin slightly different properties (see Table 17.11). The ratio of GL to PL is 0.44 for maize lecithin and 0.17 for soyabean. The percentage of PE is much lower in maize than in soyabean. Alcohol fractionation which increases the ratio of PC to PE improves the emulsifying and antispattering properties of soyabean lecithin (van Nieuwenhuyzen, 1976). Corn lecithin naturally has a higher ratio of PC to PE. Some degradation may have occurred during processing; the

Table 17.11 Composition of polar lipids in commercial maize and soyabean lecithins†

	Maize	Soyabean
Lipids	% of total polar lipids	
Acyl sterylglycoside	15.0	4.3
Monogalactosyldiglyceride	1.8	0.8
Digalactosyldiglyceride	3.7	3.0
Other glycolipids	9.8	6.4
N-acyl phosphatidylethanolamine	2.6	2.2
N-acyl lysophosphatidylethanolamine	3.7	10.4
Phosphatidylethanolamine	3.2	14.1
Phosphatidylglycerol	1.4	1.0
Phosphatidylcholine	30.4	33.0
Phosphatidylinositol	16.3	16.8
Phosphatidic acid	9.4	6.4
Phosphatidylserine	1.0	0.4
Lysophosphatidylethanolamine	trace	0.2
Lysophosphatidylcholine	1.7	0.9

†Data calculated from fatty acid methyl esters in Weber (1981).

Table 17.12 Fatty acid compositions of lipids of commercial maize and soyabean lecithins†

	Fatty acid composition (mol %)				
Lipid	16 : 0	18 : 0	18 : 1	18 : 2	18 : 2
Total lecithin					
Maize	17.7	1.8	25.3	54.2	1.0
Soyabean	17.4	4.0	17.7	54.0	6.8
Triglyceride					
Maize	12.2	2.0	25.7	59.2	1.0
Soyabean	11.4	4.3	25.2	51.9	7.2
Monogalactosyldiglyceride					
Maize	20.5	3.3	23.1	46.6	6.5
Soyabean	23.8	5.4	16.9	46.0	7.9
Digalactosyldiglyceride					
Maize	21.0	2.5	15.8	55.5	5.2
Soyabean	15.7	4.8	10.2	47.7	21.6
Acyl sterylglycoside					
Maize	54.9	3.1	15.1	26.1	0.8
Soyabean	38.7	6.9	12.2	37.0	5.2
Phosphatidylethanolamine					
Maize	23.9	1.8	21.6	52.1	0.6
Soyabean	23.2	2.6	11.6	57.1	5.5
Phosphatidylcholine					
Maize	21.7	1.7	31.0	45.0	0.6
Soyabean	15.5	4.0	14.5	60.3	5.6
Phosphatidylinositol					
Maize	33.8	1.5	18.5	45.6	0.6
Soyabean	35.5	6.9	7.8	44.3	5.4

†Weber (1981).

levels of PE and PC are lower and that of phosphatidic acid (PA) is higher than those normally found in PL isolated directly from mature maize germ (Tan and Morrison, 1979; Weber, 1979).

The FA compositions of the crude maize and soyabean lecithins are shown in Table 17.12. Noticeable differences between the two lecithins are lower levels of 18 : 0 and 18 : 3 and a higher level of 18 : 1 in maize than in soyabean. The lower percentage of 18 : 3 should give maize lecithin greater resistance toward autoxidation and the development of off-flavours than soyabean lecithin has.

B Research and Development

Increasingly sophisticated instrumentation will enable us to make progress in research. For example, in 1977 Plattner *et al*. separated the TG of maize oil into five major peaks by hplc, but in 1981 El-Hamdy and Perkins obtained eight peaks from separation both by carbon number and unsaturation. Waltking and Zmachinski (1977) have correlated flavour evaluations of maize oil by glc with those of a taste panel.

C Maize Breeding

Selection for higher oil and better quality protein would improve maize both as a livestock feed and for milling. Endosperm mutants, such as opaque-2 (o2), floury-2 and sugary-1, also affect oil content and FA composition in the germ (Arnold *et al*., 1974; Sgarbieri *et al*., 1977; Gur'ev and Tymchuk, 1978), but Poneleit (1972) has determined that the o2 gene effect on 18 : 2 is independent of the 18 : 2 control loci in the germ.

Oil content, FA composition, FA placement in lipids, the proportions of the various lipid classes and tocols all appear to be heritable traits. Exciting possibilities exist for genetically altering the lipids of maize.

REFERENCES

Alexander, D. E. (1982). *J. Am. Oil Chem. Soc.* **59**, 284A.

Alexander, D. E., Silvela, S. L., Collins, F. I. and Rodgers, R. C. (1967). *J. Am. Oil Chem. Soc.* **44**, 555–558.

Arnold, J. M., Piovarci, A., Bauman, L. F. and Poneleit, C. G. (1974). *Crop Sci.* **14**, 598–599.

Bauman, L. F., Conway, T. F. and Watson, S. A. (1963). *Science* **139**, 498–499.

Brooks, R. I. and Csallany, A. S. (1978). *J. Agric. Food Chem.* **26**, 1203–1209.

Brown, R. B., Fulford, G. N., Daynard, T. B., Meiring, A. G. and Otten, L. (1979). *Cereal Chem.* **56**, 529–532.

Brown, R. B., Fulford, G. N., Otten, L. and Daynard, T. B. (1981). *Cereal Chem.* **58**, 75–76.

Camussi, A., Jellum, M. D. and Ottaviano, E. (1980). *Maydica* **25**, 149–165.

Carpenter, A. P., Jr. (1979). *J. Am. Oil Chem. Soc.* **56**, 668–671.

Catalano, M., de Felice, M. and Sciancalepore, V. (1975). *Ind. Aliment.* **14**, 89–92.

Creech, R. G. and Alexander, D. E. (1978). In *Maize Breeding and Genetics* (D. B. Walden, ed.), 249–264. John Wiley and Sons, New York.

Curtis, P. E., Leng, E. R. and Hageman, R. H. (1968). *Crop Sci.* **8**, 689–693.

de la Roche, I. A., Alexander, D. E. and Weber, E. J. (1971a). *Crop Sci.* **11**, 856–859.

de la Roche, I. A., Weber, E. J. and Alexander, D. E. (1971b). *Crop Sci.* **11**, 871–874.

Drozdowski, B. (1977). *J. Am. Oil Chem. Soc.* **54**, 600–603.

Dudley, J. W. (ed.). (1974). *Seventy Generations of Selection for Oil and Protein in Maize*. Crop Science Society of America, Madison, Wisconsin.

Dudley, J. W. (1977). In *Proceedings of the International Conference on Quantitative Genetics*, pp. 459–473. Iowa State University Press, Ames, Iowa.

Earle, F. R. (1977). *Cereal Chem.* **54**, 70–79.

El-Hamdy, A. H. and Perkins, E. G. (1981). *J. Am. Oil Chem. Soc.* **58**, 867–872.

Espelie, K. E., Dean, B. B. and Kolattukudy, P. E. (1979). *Plant Physiol.* **64**, 1089–1093.

Farag, R. S., Youssef, A. M., Sabet, K. A., Fahim, M. M. and Khalil, F. A. (1981). *J. Am. Oil Chem. Soc.* **58**, 722–727.

Freeman, J. E. (1973). *Trans. Am. Soc. Agric. Engr.* **16**, 671–679.

Gur'ev, B. P. and Tymchuk, S. Ya (1978). *Kukuruza* (#9), 21–23.

Ho, C.-T., Smagula, M. S. and Chang, S. S. (1978). *J. Am. Oil Chem. Soc.* **55**, 233–237.

Huang, A.-S., Hsieh, O. A.-L., Huang, C.-L. and Chang, S. S. (1981). *J. Am. Oil Chem. Soc.* **58**, 997–1001.

Hymowitz, T., Dudley, J. W., Collins, F. I. and Brown, C. M. (1974). *Crop Sci.* **14**, 713–715.

Itoh, T., Shimizu, N., Tamura, T. and Matsumoto, T. (1981). *Phytochemistry* **20**, 1353–1356.

Jahn-Deesbach, W., Marquard, R. and Heil, M. (1975). *Z. Lebensm. Unters-Forsch.* **159**, 271–278.

Jellum, M. D. (1970). *J. Agric. Food Chem.* **18**, 365–370.

Jellum, M. D. and Marion, J. E. (1966). *Crop Sci.* **6**, 41–42.

Johnson, R. R. (1982). *Better Crops* **66**, 3–7.

Jugenheimer, R. W. (1976). *Corn Improvement, Seed Production, and Uses.* John Wiley and Sons, Inc., New York, NY.

Jurgens, S. K., Johnson, R. R. and Boyer, J. S. (1978). *Agron. J.* **70**, 678–682.

Kornfeldt, A. and Croon, L.-B. (1981). *Lipids* **16**, 306–314.

Marten, J. F. and Nelson, W. L. (1982). *Better Crops* **66**, 10–13.

McMurray, C. H., Blanchflower, W. J. and Rice, D. A. (1980). *J. Assoc. Off. Anal. Chem.* **63**, 1258–1261.

Miller, R. I., Dudley, J. W. and Alexander, D. E. (1981). *Crop Sci.*, **21**, 433–437.

Miric, M., Lalic, Z. and Mihajlovic, M. (1981). *Hrana Ishrana* **22**, 125–128.

Morrison, W. R. (1978). *Adv. Cereal Sci. Technol.* **2**, 221–348.

Plattner, R. D., Spencer, G. F. and Kleiman, R. (1977). *J. Am. Oil Chem. Soc.* **54**, 511–515.

Plewa, M. J. and Weber, D. F. (1975). *Genetics* **81**, 277–286.

Poneleit, C. G. (1972). *Crop Sci.* **12**, 839–842.

Poneleit, C. G. and Bauman, L. F. (1970). *Crop Sci.* **10**, 338–341.

Priestley, D. A., Galinat, W. C. and Leopold, A. C. (1981). *Nature* **292**, 146–148.

Raghuveer, K. G. and Hammond, E. G. (1967). *J. Am. Oil Chem. Soc.* **44**, 239–243.

Ratković, S., Nikolić, D. and Fidler, D. (1978). *Maydica* **23**, 137–144.

Sahasrabudhe, M. R. and Farn, I. G. (1964). *J. Am. Oil Chem. Soc.* **41**, 264–267.

Sgarbieri, V. C., da Silva, W. J., Antunes, P. L. and Amaya-F., J. (1977). *J. Agric. Food Chem.* **25**, 1098–1101.

Shadley, J. D. and Weber, D. F. (1980) *Can. J. Genet. Cytol.* **22**, 11–19.

Sharma, B. N., Gopal, S., Paul, Y. and Bhatia, I. S. (1975). *J. Res. Punjab Agr. Univ.* **12**, 378–381.

Smouse, T. H. (1979). *J. Am. Oil Chem. Soc.* **56**, 747A–751A.

Sprague, G. F. (1971). *Proc. Corn. Sorghum Res. Conf.* **26**, 96–104.
Sun, D., Gregory, P. and Grogan, C. O. (1978). *J. Heredity* **69**, 341–342.
Tan, S. L. and Morrison, W. R. (1979). *J. Am. Oil Chem. Soc.* **56**, 531–535.
Timothy, D. H. and Goodman, M. M. (1979). In *The Plant Seed — Development, Preservation, and Germination* (I. Rubenstein, R. L. Phillips, C. E. Green and B. G. Gengenbach, eds), pp. 171–200. Academic Press, London and New York.
Trifunović, V., Ratković, S., Mišović, M., Kapor, S. and Dumanović, J. (1975). *Maydica* **20**, 175–183.
United States Department of Agriculture (1981). *Agricultural Statistics.* U.S. Government Printing Office, Washingdon, D.C.
United States Department of Agriculture, Foreign Agricultural Service (1982). In *Corn Annual, 1982*, p. 16. Corn Refiners Association, Inc., Washington, D.C.
United States Department of Agriculture (1982). *Crops Soils* **34**. 12–13.
van Nieuwenhuyzen, W. (1976). *J. Am. Oil Chem. Soc.* **53**, 425–427.
Vincent, L. E., Rawal, K. M. and McArthur, G. R. (1978). *Maydica* **23**, 75–87.
Vojnovich, C., Anderson, R. A. and Griffin, E. L. Jr. (1975). *Cereal Foods World* **20**, 333–335.
Waltking, A. F. and Zmachinski, H. (1977). *J. Am. Oil Chem. Soc.* **54**, 454–457.
Watson, S. A. (ed.) (1981). *Grain Quality Newsletter* **3**(3), 7–10.
Watson, S. A. and Freeman, J. E. (1975). *Proc. Corn. Sorghum Res. Conf.* **30**, 251–275.
Weber, E. J. (1973). In *Industrial Uses of Cereals* (Y. Pomeranz, ed.), pp. 161–206, American Association of Cereal Chemists, St. Paul. Minnesota.
Weber, E. J. (1978). *Cereal Chem.* **55**, 572–584.
Weber, E. J. (1979). *J. Am. Oil Chem. Soc.* **56**, 637–641.
Weber, E. J. (1980). In *The Resource Potential in Phytochemistry* (T. Swain and R. Kleiman, eds.), Vol. 14, pp. 97–137. Plenum Press, New York.
Weber, E. J. (1981). *J. Am. Oil Chem. Soc.* **58**, 898–901
Weber, E. J. (1983). *Biochem. Genetics* **21**, 1–13.
Welch, L. F. (1969). *Agron. J.* **61**, 890–891.
Widstrom, N. W. and Jellum, M. D. (1975). *Crop Sci.* **15**, 44–46.
Worthington, R. E. (1982). *J. Am. Oil Chem. Soc.* **59**, 317A.
Wu, G.-S., Stein, R. A. and Mead, J. F. (1979). *Lipids* **14**, 644–650.
Zuber, M. S. (1975). *Proc. Corn Sorghum Res. Conf.* **30**, 277–286.
Zuber, M. S. and Darrah, L. L. (1980). *Proc. Corn Sorghum Res. Conf.* **35**, 234–249.

18 Corn Oil Production, Processing and Use

T. L. MOUNTS and R. A. ANDERSON

Northern Regional Research Centre, Agricultural Research Service, U.S. Department of Agriculture, Peoria, Illinois, U.S.A.

I INTRODUCTION

Some historical figures on the annual disposition of corn (maize) in the U.S. are shown in Table 18.1 (Anon., 1978). About 80% of the corn crop was fed to hogs, cattle and poultry either on the farm where they were grown or at some other location. About 13% was exported and the remainder went for industrial uses and seed. Today, however, there has been a shift in corn disposition. Some 28% of the crop is exported and 63% of the crop is fed to animals. The remaining 9% is used for food, industrial and seed purposes. In 1977, it was predicted that the remaining corn (representative of the current 9%) would be about 13 million metric tons (about 516 million bushels). Of that, 9 million metric tons (about 350

"Lipids in Cereal Technology"
ISBN 0-12-079020-3

Table 18.1 Corn uses in the United States†

Uses	Before 1970 (%)	1977 (%)
Livestock feed	78.5	63
Wet-milling	4.8	5.9
Exports	12.7	28
Dry-milling	2.5	1.7
Alcohol, etc.	0.7	0.6
Seed	0.3	0.3
Breakfast food	0.4	0.4
Farm household	0.1	0.1

†Anon. (1978).

million bushels) would be used by the corn wet-milling industry to produce starch, oil, and animal feeds. Another 2.5 million metric tons (100 million bushels) would be used by the corn dry-milling industry. The products from dry-milling include grits, meal, flour, oil and animal feeds.

II GERM RECOVERY

A Wet-Milling Process

In the U.S., 22 corn wet-milling plants in 10 states convert the corn into sweeteners, starches, corn oil and feed ingredients. Five new plants have been constructed in the past 5 years. United States' corn comprises about 65% starch and 3–4% oil. Approximately 98% of the starch and 75% of the protein reside in the endosperm. The germ contains about 85% of the oil and 80% of the minerals. Therefore, mechanical separation of the various grain fractions does not in itself result in a complete separation of the individual chemical constituents. When the grain is wet-milled, one might expect to recover about 3% oil, 16% water, 26% feed and 55% starch from a bushel of corn.

The basic flowsheet of the wet-milling process has changed very little over the years (Fig. 18.1) (Anderson, 1970). After being steeped, the corn is ground and the various components are separated by screening, centrifuging, hydrocloning and drying operations. Many changes have been made in the equipment used in these operations. New and improved energy-conserving and higher capacity equipment has been developed.

The initial step in corn processing is steeping; the cleaned corn is soaked for about 40 h in large stainless-steel tanks (about 3,000-bushel capacity) filled with water containing a small amount of sulphur dioxide (0.1 to

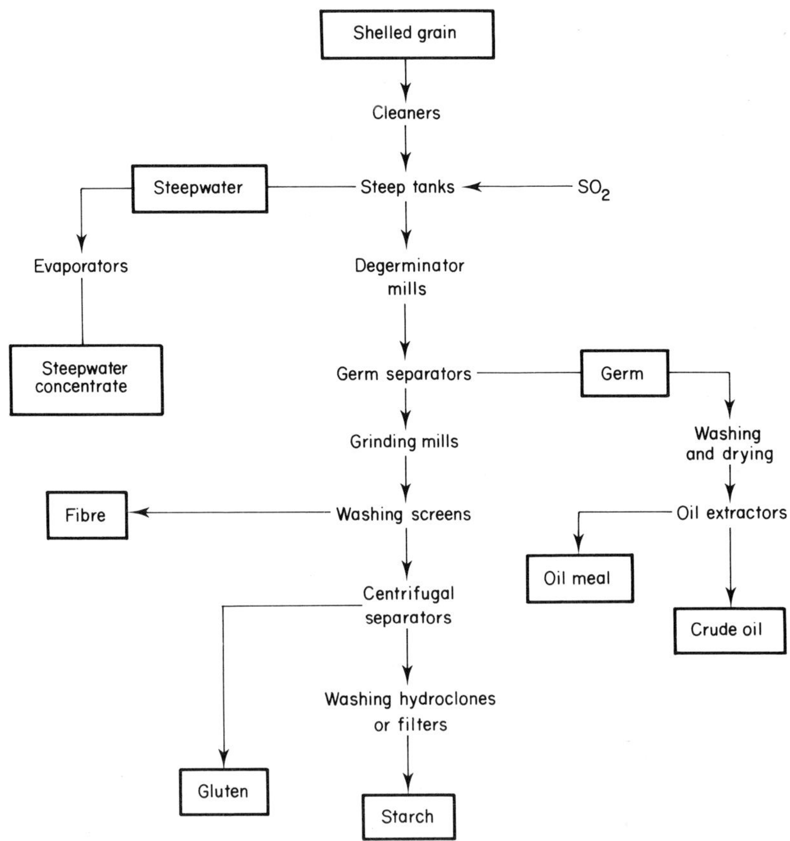

Fig. 18.1 Flowsheet of corn wet-milling process (Anderson, 1970).

0.2%), which is heated to 125 °F (52 °C). After the completion of the steeping cycle the corn is passed over a series of curved screens that separate the corn from the water. The steepwater is evaporated and the concentrate is dried and incorporated into livestock feed. The steepwater is also used by the fermentation industry as a nutrient in the growth and manufacture of antibiotics, vitamins, enzymes and other useful organic compounds.

The corn grain, softened in steeping, is run through degerminating mills that tear the grain apart. The ground pulpy material contains germ, hulls, starch and gluten. The germ is separated from other components by centrifugal action in hydroclones (Anon., 1976). The hydroclones have replaced the flotation method of separating the germ and give improved efficiency of separation, greater operating capacity, improved sanitary conditions and a cleaner germ fraction. The overflow from the hydroclone is the lighter weight germ, whereas the underflow is the heavier starch, gluten and hulls. The germ is washed and dewatered on Dutch States Mines (DSM) screens (Anon., 1957), and then dried in a rotary steam-tube dryer. The germ contains about 50% oil. The DSM screen consists of a stationary screen housing equipped with a concave wedge bar-type screen with provisions for introducing feed and withdrawing undersize and oversize products. It has no moving parts and gives efficient removal of residual starch and water from the germ.

The underflow from the hydroclones is reground and goes into the fibre wash system. Here a series of DSM screens has replaced the rotating or vibrating screens, called shakers or reels. This innovation has provided energy savings, a reduction in required floor space and increased operating capacity together with less equipment maintenance. An efficient separation of fibre from the starch-gluten slurry is obtained with this system. According to recent practice, dewatering of the fibre is accomplished in two stages. In the first stage a screening centrifuge is used; the use of centrifugal force acting on a perforated screen permits additional recovery of starch. In the second stage the dewatered fibre is transferred to a conveyor-type solid bowl centrifuge where the final dewatering is accomplished. At this point the fibre is mixed with steepwater, germ meal and gluten and then dried to form feed products.

The separation of gluten from starch is the key step in the corn wet-milling process. The economic success of the process demands consistent production of a high-protein gluten product with essentially no loss of starch. Much less time and space are required with this method than were needed formerly when long starch tables were used to carry out this separation. High-speed centrifuges, for example the Merco centrifugal separator (Anon., 1968), give efficient separation. The gluten is then concentrated in another centrifuge, filtered and dried in rotary or flash dryers. Gluten is a major component of mixed feeds.

The starch is washed free of gluten in a hydroclone washing system. This washing was previously done in secondary centrifuges; today it is done in most plants with equipment consisting of several hundred small hydroclone tubes in a partitioned housing (Anon., 1976). The tubes are identical in construction and operation but are much smaller than the hydroclones used for germ separation. After several washings and filterings, the starch is ready to be pumped to a finishing department where it is made into various dry starch products or converted into syrup or sugar.

B Dry-Milling Process

The size of corn dry-milling in the U.S. ranges from very small country mills, largely grinding whole corn to meal, to large modern mills using degerming equipment and a milling system that separates practically all of the hull, germ and tip cap from the corn endosperm fractions: grits, meal, and flour. The process used in the larger mills is usually referred to as the tempering-degerming system (TD), because moisture under controlled conditions is added to the grain before the dehulling-degerming step (Fig. 18.2). The Beall degerminator and corn dehuller is used for the initial break in 90% of the U.S. mills. The objectives of the TD system are: (1) to remove essentially all germ and hull and to leave the endosperm low in fat and fibre; (2) to recover a maximum amount of endosperm; and (3) to recover a high percentage of the germ as large, clean particles.

Basically, the TD system combines physical and mechanical treatments to release, separate and recover various components of the corn grain. The final products retain much of the physical identity of the grain component they are derived from. The amount of grits and meal produced depends upon the proportion of horny endosperm in the grain because each particle consists largely of horny endosperm in which recesses are filled with floury endosperm.

The basic processing steps in the TD system, after receipt of the corn, are:

(1) Dry cleaning and, if necessary, wet cleaning the corn.

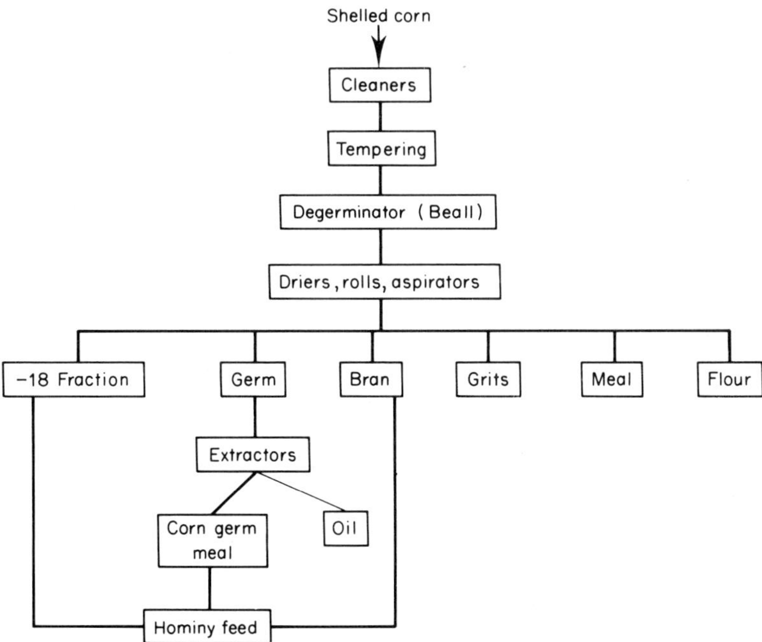

Fig. 18.2 Flowsheet of corn dry-milling process.

(2) Tempering the corn by controlled addition of moisture to toughen the germ and hull, to make them pliable and resilient and promote their release from the endosperm. A minimum addition of moisture to the endosperm is often desired.

(3) Releasing hull, germ and tip cap from the endosperm in a degermer.

(4) Drying and cooling the stock or product from the degermer in preparation for fractionation.

(5) Fractionating the degermer stock by multi-step milling through a series of roller mills, sifters, aspirators, gravity table separators and purifiers to separate and recover the various products. (Some mills do not use gravity table separators or purifiers.)

(6) Drying the products further when necessary.

(7) Processing the germ fraction for recovery of crude corn oil.

(8) Blending, packaging and shipping products.

A miller can sometimes omit or combine certain steps for various reasons.

A mill with appropriate equipment and conditions for the TD system can produce a wide variety of products including cereal flaking and puffing grits. The product spectrum can, and does, vary from day to day with the corn being milled and the product mix being sought. The corn dry miller has many options available since endosperm product ratios may be altered by omitting or combining certain steps as dictated by changes in the demand picture (Brekke, 1970). Other products can be produced from the primary ones such as fine meal (a combination of dusted meal and flour), 100% meal (i.e., fine meal and coarse meal combined) and brewers' grits. The brewers' grits are a blend of coarse and regular grits prepared to meet a customer's specifications; for example, a maximum of 0.75% fat and 14% moisture, and a granulation where at least 95% goes through a U.S. No. 12 sieve and not more than 5% through a U.S. No. 30 sieve.

The by-product, hominy feed, is a mixture of the hull fraction, residual meal after the germ has been processed for oil removal, the fines created during the degerming step, feed-quality material removed in cleaning the corn, tailings stream from the grit reduction system, and occasionally, any corn flour not otherwise sold. Hominy feed is guaranteed to contain, on an as-is moisture basis, a minimum of 5% fat and 10% protein and no more than 6% crude fibre.

The recovered germ contains from 20 to 25% oil.

III CORN OIL PROCESSING

A Corn Oil Extraction

The germ, as separated in both milling processes, is first dried and then subjected to high pressure in an auger-type machine (expeller) that breaks down the cell structure and extracts the oil (Fig. 18.3) (Anon., 1960). This expelling operation reduces the oil content of wet-milled germ from about

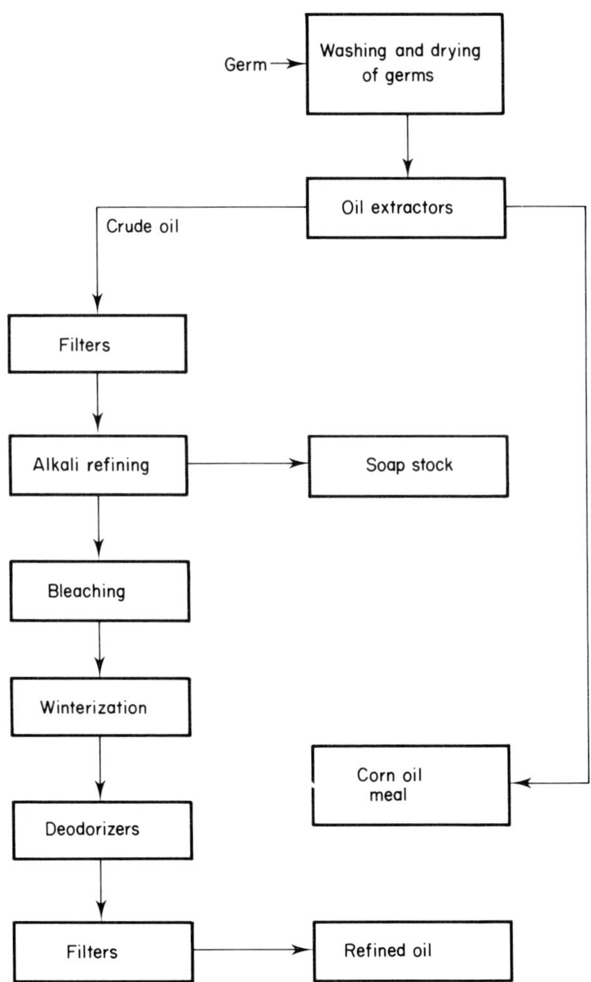

Fig. 18.3 Corn oil extraction and processing flowsheet (Anon., 1960).

Table 18.2 Composition of crude corn oil†

Component	Typical Value (%)
Triglycerides	95.6
Free fatty acids	1.7
Phospholipids	1.5
Phytosterols	1.2
Tocopherols	0.06
Waxes	0.05
Carotenoids	0.0008
Odor, Flavour	Trace

†Reiners (1978).

50% to approximately 15%. The residue, called "corn germ cake", is then heated and flaked by rolls and most of the remaining oil is extracted by a solvent (hexane). The residual solid material, containing about 1.5% oil, is ground into corn germ meal and used as an ingredient in livestock feed. The hexane is removed from the meal by vapour desolventization. The oil obtained by extraction is stripped of solvent by distillation, combined with that obtained by expelling, and filtered to remove suspended solid material.

The composition of the recovered crude corn oil is shown in Table 18.2 (Reiners, 1978).

The quality of the corn germ can affect the oil which is extracted from it. A study on oil recoveries and quality from blight-infected corn (*Helminthosporium maydis*) showed that the oil content of the germ decreased from 53% in sound corn to less than 30% in samples containing increasing percentages of damaged grain (5% to 86%) (Anderson *et al*., 1972). The extracted oils had increased free fatty acid (FFA) contents up to 12.3%, which increases the refining loss during processing.

B Corn Oil Refining

Refining can be said to start in the crude oil storage tank not only because oil-insoluble material is separated by gravity but because correct pumping and storage conditions produce improved quality, reduce losses and prevent fractionation that would affect final product consistency. Therefore, design features for crude oil storage tanks should include: lagging; closed tank with manholes at side and top and with breather pipe; intake pipe extending to the tank bottom; lacquered internal surface of either epoxy or polyurethane; temperature-controlled preferably with hot water heating, or alternatively by very low pressure steam (about 10 psig); water drainage; easy cleaning; and manifold outlet to reduce contamination.

Crude corn oil contains about 1.5% phospholipids, 75% of which can be precipitated by simple water or steam treatment of the oil. This process, referred to as degumming, partly removes phospholipids and other non-triglyceride material from the oil. After dehydration of the precipitated material, it is known as corn lecithin. Few processors degum corn oil; instead, the phospholipids are removed during alkali refining.

1 *Alkali Refining*

Alkali refining, or neutralization, of crude corn oil removes the FFA which represent 1.5–4%, of the oil. The vast majority of refiners in the U.S. are currently using the conventional caustic soda continuous refining system (Carr, 1976). A caustic soda strength of 18 °Be′ (12.68% NaOH) with an excess of 0.13% is recommended for corn oil. The caustic soda treatment is calculated according to the following equation

$$\text{wt. \% lye} = \frac{(\text{\% FFA crude} \times 0.142 + \text{\% excess NaOH}) \times 100}{\text{\% NaOH in lye}}$$

Crude oil is mixed continuously with a proportional stream of dilute caustic soda solution and heated to obtain a break in the emulsion. Corn oil should be heated rapidly to 82–88 °C immediately before it is pumped to the caustic-oil mixing step. The usual mixing system for corn oil provides a high-speed, in-line mixer that operates at 1750 rev min^{-1} to obtain intimate contact between the caustic soda and oil. A dwell mixer provides for full reaction with the FFA to form soapstock while hydrolysing phosphatides and removing unsaponifiable matter.

The resulting soap-in-oil suspension is fed to high-speed centrifuges for separation into low- and high-density phases. The low-density phase discharge is the refined oil containing traces of moisture and soap, whereas the high-density phase is primarily soap, free caustic soda, phosphatides, and small quantities of neutral oil.

Most centrifuges in the U.S. are of the pressure or hermetic type in which zone changes can be readily achieved by adjusting the back-pressure applied to the low-density phase discharge. Increasing the discharge back-pressure reduces the soap content in the oil phase but increases the amount of neutral oil lost in the soapstock. Reducing the back-pressure decreases the neutral oil loss in the soap phase but increases the quantity of soap in the refining oil to a level beyond the capacity of the subsequent water-washing step. Soapstock from the refining centrifuge is pumped to a soap tank prior to acidulation or sale.

Refined oil is reheated as necessary to 88 °C, and soft water at about

93 °C is proportioned into the refined oil at a rate of 10–20% by weight of the oil flow. This combination passes through another high-speed, in-line mixer to obtain intimate contact for maximum soap transfer from oil to the water phase. The soapy water–oil mixture is separated by centrifugation, and about 90% of the soap in the refined oil is removed by the single water-washing operation. Washed oil is passed through nozzles into the evacuated section of a continuous vacuum dryer to reduce the moisture content below 0.1%. The typical dryer operates at a vacuum of 70 cm (Hg). Dried oil is continuously cooled to about 49 °C before entering the refined oil storage tank.

2 *Bleaching*

Refined oil is subsequently bleached to remove soap, trace metals, sulphurous compounds and a part of the more stable pigment and pigment breakdown products that can result from raw material damage or oxidation (Young, 1978). Vacuum bleaching in semi-continuous systems is preferred. Bleaching earth, generally an activated clay, is mixed with oil at 1–2% by weight in a stainless-steel tank equipped with an agitator. Dark-coloured crude oils may require an activated clay concentration of up to 4% to achieve adequate colour removal. Residence time is a minimum of 20 min at 90–110 °C. The clay-soil mixture is then pumped to a filtration stage. The types of filters in use for this stage are the plate and frame, the vertical leaf tank or pressure filter, and the centrifugal self-cleaning filter. The oil obtained from bleaching-earth filtration should be clear and bright and as free of earth particles as possible.

3 *Winterization*

A dewaxing of the oil is usually performed (Neumunz, 1978). The bleached oil is passed through heat exchangers to a final cooling and chilling tank to lower the oil temperature to 10–12 °C. This cooling can be continuous with no retention time required. From this tank, the oil is pumped to a surge (or filter supply) tank to which filter aid is added, and the mixture is pumped at low pressure through a filter press or tank-type filter. Filtration takes place at a temperature of about 10 °C but not over 16.5 °C. The amount of wax removed can vary from 0.5–1.0% of the oil. The corn oil leaving the filter is wax-free and ready for deodorization. A properly dewaxed corn oil can normally stand a cold test (remains clear at 0 °C) (American Oil Chemists' Society, 1975) of 98 h or more. The amount of wax in corn oil is so small that often winterization is not practised when the oil is to be hydrogenated or used as a cooking oil.

4 *Deodorization*

Deodorization is the final step required in processing to produce a quality corn oil product with a low FFA content, light colour and stability against oxidation (Reiners and Gooding, 1970). The deodorization of a well-refined corn oil can be achieved by the vacuum (1–5 mm, Hg) steam process at 220–240 °C. Batch, semi-continuous and continuous deodorization systems are used in processing edible oils (Gavin, 1977). Deodorization is essentially a process of steam distillation, where odoriferous, flavoured substances are stripped from the relatively non-volatile oil. It is important that all metal surfaces that come into contact with the oil during deodorization be fabricated out of stainless steel, so that iron, an oxidation catalyst, will not contaminate oil. Generally, the oil is treated with a trace-metal sequestrant (citric acid) on the cooling side of deodorization (0.01% added at below 100 °C). The finished refined corn oil will meet the analytical specifications compiled in Table 18.3 (Anon., 1960). Although the introduction of new hybrids over the years has affected the fatty acid composition of the oil, Table 18.4 presents average data for corn oil. The

Table 18.3 Refined corn oil analytical data†

Property	Value	Recommended method (American Oil Chemists' Society, 1975)
Acidity (free fatty acid as oleic)	0.020–0.050	Ca 5a–40
Acid value	0.04–0.10	Ka 2–55
Colour (Lovibond)	20–35 yellow 2.5–5 red	Cc 13b–45
Flavour	Bland	
Cold test	Clear	Cc 11–53
Saponification value	189–191	Cd 3–25
Iodine value	125–128	Cd 1–25
Hehner value	93–96	‡
Titre test	64° to 68°F	Cc 12–41
Melting point	4° to 12°F	Cc 1–25
Smoke point	430° to 500°F	Cc 9a–48
Solidifying point	−4° to 14°F	‡
Flash point	575° to 640°F	Cc 9a–48
Fire point	590° to 700°F	Cc 9a–48
Specific gravity	0.918–0.925	Cc 10a–25
Pounds per gallon	7.672 at 70°F	‡

†Anon. (1960).
‡No official method, but suitable procedures are described in the oil and fat literature.

Table 18.4 Average fatty acid composition for refined corn oil†

Fatty acid	Composition (% fatty acid basis)
16 : 1	11.5
18 : 0	2.0
18 : 1	24.1
18 : 2	62.5
18 : 3	0.7
20 : 0	0.2

†Reiners and Gooding (1970).

high content of 18 : 2 fatty acid (linoleic acid) is claimed to be a dietary advantage (Beadle *et al*., 1965). The low content of 18 : 3 (linolenic acid) and high concentration of tocopherols give corn oil good oxidative stability (Reiners, 1978).

5 *Hydrogenation*

Liquid corn oil is used in the production of margarine by blending with partially hydrogenated corn, soyabean or cottonseed oils. It has also been proposed that production of margarine oils containing large amounts of liquid corn oil may be able to utilize a hardening fraction prepared by interesterification of hydrogenated palm kernel and palm oils.

Hydrogenation is a batch process, and the type of equipment used can vary from 5000 lb to 60 000 lb converters, from filter presses to a pressure leaf automated filter and from controls such as simple temperature/pressure recorders with manual controls to fully automated operation (Mounts, 1980). The process is started by evacuation of the converter with a vacuum jet. The feedstock of refined and bleached oil is pumped into the converter through a heat recovery section where it is heated by the previous batch of oil. During the loading, a portion of the oil is diverted to the catalyst mix tank. When the correct fill level is reached, the flow stops. The operator adds the desired amount of catalyst to the mix tank and the slurry is pumped to the converter.

Commercial hydrogenations use nickel catalysts and require between 0.02 to 0.10% nickel, based on the weight of the oil charge. The hydrogenation reaction is generally conducted at temperatures of 170–180 °C and hydrogen pressures of 5–45 psig. Two or more agitators provide stirring and intimate mixing of the hydrogen gas with the oil-catalyst charge. The progress of the hydrogenation is followed by taking a sample and checking the index of refraction, which is related to the iodine value of the oil.

When the hydrogenation is completed, the batch is pumped through the heat recovery system to heat the incoming feedstock. After being cooled to filtration temperature, the batch is recirculated through a previously precoated filter until it is clear and brilliant and then it is discharged to storage tanks for final deodorization.

III PRODUCTS AND MARKETS FOR CORN OIL

Production of corn oil in the U.S. has increased by 160% during the past 20 years, with the bulk of this growth occurring during the last 10 years (see Table 18.5) (Anon., 1979). Corn oil was not exported at all in 1959, but 57 million kg of the oil was exported during 1978. Domestic consumption has been relatively stable during the last 4 years.

The major food uses of corn oil are as a salad/cooking oil and as a component in margarines (see Table 18.6). Some corn oil is also used in the manufacture of salad dressings. The nutritional properties of corn oil are excellent; it is highly digestible and a good source of essential fatty acids. Most vegetable oils marketed in the U.S. are designed as dual-purpose salad/cooking oils so citric acid or isopropyl citrate is added to commercial corn oil to inactivate traces of metals that act as oxidation catalysts, and methyl silicone is added to reduce foaming and improve stability during frying.

A variety of margarine types are marketed in the U.S. About two decades ago, only a regular stick margarine was available, but today the consumer can purchase 10 different varieties of margarine and spreads (see

Table 18.5 U.S. supply and disappearance of corn oil 1969–1978 (million kg)†

Year	Production	Export	Domestic
1969	215	15	206
1970	220	20	202
1971	226	22	199
1972	237	20	223
1973	240	31	204
1974	211	38	181
1975	292	44	254
1976	303	42	264
1977	315	53	252
1978	333	57	284

†Anon. (1979).

Table 18.6 Corn oil use (million kg)†

Year	Salad/cooking oil	Margarine
1969	128	78
1970	118	84
1971	115	84
1972	135	88
1973	108	97
1974	96	85
1975	168	85
1976	165	99
1977	142	110

†Anon. (1978).

Table 18.7) (Massiello, 1978). The stick margarines comprise approximately 70% of the market. Liquid, partly-hydrogenated corn oil is used in the formulation of both high-polyunsaturated premium stick margarine and premium soft margarine, which constitute 13.0 and 6.6% of the total market, respectively. A diet imitation margarine is formulated with liquid, partly hydrogenated corn oil that contains only 40% fat, whereas the standard for margarine is 80% fat content.

Although the consumption of corn oil in the formulation of salad/cooking oil and margarines has increased in the last decade, the corn oil market share has changed little. The salad/cooking oil market is dominated

Table 18.7 Varieties of margarine and percentage share of market†

Types of margarine‡	1976 Sales (million kg)	% Share of market	
1. Regular stick margarine	200	22.0	70.2
2. Polyunsaturated stick margarine	301	33.2	
3. High-polyunsaturated stick margarine	118	13.0	
4. Whip stick margarine	18	2.0	
5. Regular soft margarine	109	12.0	21.6
6. Premium soft margarine	60	6.6	
7. Whipped soft margarine	27	3.0	
8. Diet imitation margarine	22	2.4	
9. Liquid margarine	12	1.3	
10. Vegetable oil spreads	41	4.5	
Total	907	100.0	

†Massiello (1978).
‡In 1955 only regular stick margarine was available. Sales: 544 million kg.

by soyabean oil (75%), whereas the corn oil market is relatively minor (7%). Corn oil (12%) is second to soyabean oil (82%) for use in margarines.

Although this paper has dealt primarily with corn oil, it must be remembered that the production of starch is the major objective of the wet-milling industry. The two by-products of this process are the assorted feed products and the edible oil. The production of corn oil, therefore, is very dependent on the demand for the starch component. However, the demand for corn oil in the U.S. exceeds the supply, so this by-product is a profitable item.

REFERENCES

American Oil Chemists' Society (1975). *Official and Tentative Methods*, 3rd Edn. (W. E. Link, ed.). American Oil Chemists' Society, Champaign, Illinois.

Anderson, R. A. (1970). In *Corn: Culture, Processing, Products*, Chapter 9 (G. E. Inglett, ed.). AVI Publishing Company, Inc., Westport, Connecticut.

Anderson, R. A., Ellis, J. J. and Griffin, E. L. Jr. (1972). *Cereal Sci. Today* **17**, 41–45.

Anon. (1957). Bulletin No. 2300, Dorr-Oliver Incorporated, Stamford, Connecticut.

Anon. (1960). *Corn Oil.* Corn Refiners' Association, Washington, D.C.

Anon. (1968). Bulletin No. FOOD-2, Dorr-Oliver Incorporated, Stamford, Connecticut.

Anon. (1976). Bulletin No. DC-1, Dorr-Oliver Incorporated, Stamford, Connecticut.

Anon. (1978). *Agricultural Statistics.* U.S. Department of Agriculture, Washington, D.C.

Anon. (February 1979). *Fats and Oils Situation.* U.S. Department of Agriculture, Washington, D.C.

Beadle, J. B., Just, D. E., Morgan, R. E. and Reiners, R. A. (1965). *J. Am. Oil Chem. Soc.* **42**, 90–95.

Brekke, O. L. (1970). In *Corn: Culture, Processing and Products*, Chapter 14 (G. E. Inglett, ed.). AVI Publishing Company, Inc., Westport, Connecticut.

Carr, R. A. (1976). *J. Am. Oil Chem. Soc.* **53**, 347–352.

Gavin, A. M. (1977). *J. Am. Oil Chem. Soc.* **54**, 528–532.

Massiello, F. J. (1978). *J. Am. Oil Chem. Soc.* **55**, 262–265.

Mounts, T. L. (1980). In *Handbook of Soy Oil Processing and Utilization* (D. R. Erickson, E. H. Pryde, O. L. Brekke, T. L. Mounts, and R. A. Falb, eds), pp. 131–144. American Oil Chemists' Society, Champaign, Illinois.

Neumunz, G. M. (1978). *J. Am. Oil Chem. Soc.* **55**, 396–398A.

Reiners, R. A. and Gooding, C. M. (1970). In *Corn: Culture, Processing, Products*, Chapter 13 (G. E. Inglett, ed.). AVI Publishing Co., Inc., Westport, Connecticut.

Reiners, R. A. (1978). Corn oil. In *Products of the Corn Refining Industry in Food* pp. 18–22, Seminar Proceedings. Corn Refiners' Association, Washington, D.C.

Young, V. (1978). *Chem. Ind.* **18**, 692–703.

19 Wheat Germ Oil

P. J. BARNES

The Lord Rank Research Centre, High Wycombe, Bucks, U.K.

I INTRODUCTION

Wheat germ oil is a speciality product manufactured in much smaller quantities than the common edible oils such as corn oil. This is due primarily to the relatively low yields of wheat germ and wheat germ oil. Typical yields of commerical wheat germ from flour milling are less than 1% of the grain weight, and the oil content is usually not more than 10% of the germ dry weight, in contrast to 30–50% of oil for maize (corn) germ. Although obtained in low yield, and therefore expensive, wheat germ oil finds a ready market in the health and cosmetic sectors. These applications are related to the high vitamin E activity and, more recently, to the content of octacosanol.

II PRODUCTION OF WHEAT GERM

The part of the wheat grain described as the *germ* consists of two major parts, the embryo axis and the scutellum (see Chapter 1), which, on aver-

"Lipids in Cereal Technology"
ISBN 0-12-079020-3

age, represent 1.2% and 1.5% of the grain dry weight respectively (Hinton, 1944). As described in Chapter 7, the scutellum is relatively friable and difficult to separate from the other milling fractions. In contrast, the embryo axis passes with coarse semolina (first and second break coarse middlings) to the reduction rolls where it is flaked and may be separated by sieving (see Chapter 7). Commercial wheat germ thus consists predominantly of embryo axis.

The commerical germ fraction is not pure wheat germ but also contains variable proportions of bran and endosperm. Oil is pressed out of the germ to some extent during flaking and is transferred to the flour. The lipid composition of commercial germ is thus different from that of the original embryo axis.

If commercial germ is to be used in foodstuffs, it must be stabilized against hydrolytic and oxidative rancidity. This may be partly achieved by drying the germ to 4% moisture content, cooking or defatting.

III PRODUCTION OF WHEAT GERM OIL

Oil is separated from wheat germ commercially either by pressure expelling or solvent extraction. Although expelling avoids the dangers of inflammable solvents, it recovers only half of the available oil and yields germ that requires further stabilization against rancidity. Solvent defatting is more efficient and the residual oil content of the extracted germ is less than 10 g kg^{-1}. Solvent defatted wheat germ is more valuable than the corresponding full-fat germ because of its stability, and therefore the production of wheat germ oil must always consider the quality and quantity of the defatted germ product.

Only germ having very low levels of bran contamination can be successfully pressed to yield oil; at higher bran contents the overall oil content of the germ is too low (bran contains much less oil than does germ). Solvent extraction is less influenced by the bran content because the yield of oil is greater, but the purity of germ used for extraction is governed by the specifications set for the oil and defatted germ products. High temperatures should be avoided during manufacture to minimize destruction of nutritionally important components. Most manufacturers fail to provide full details of the temperatures which the oil has been subjected to during processing, but one company claims to carry out solvent extractions at temperatures not exceeding 38 °C without any subsequent refining (Viobin Corp., U.S.A.). Hexane, 1,2-dichloroethane and ethanol are the preferred solvents for wheat germ oil extraction.

Most edible oils are refined to remove phospholipids, free fatty acids (FFA), pigments and volatile compounds. However, these treatments may not be necessary for wheat germ oil and may even be considered undesirable for a product that is marketed as "natural" and "unprocessed". Lecithin (phosphatidylcholine), colour and flavour are positive attributes in a product sold in the "health food" market. FFA have been removed from commerical wheat germ oil by alkali treatment but this also results in substantial losses of oil and tocols. An alternative approach is the use of molecular distillation to remove FFA and also to produce a tocol concentrate (Singh and Rice, 1979).

IV APPLICATIONS OF WHEAT GERM OIL

The market for wheat germ oil is based mainly on the high vitamin E content and, more recently, octacosanol. Wheat germ oil is the richest, readily available, natural source of α-tocopherol, the tocol with greatest vitamin E activity, and it is retailed in bottles or in gelatin capsules. It is also used in admixture with other materials such as lecithin and cod liver oil and as a component of shampoos and cosmetics. Wheat germ oil preparations are sold as diet supplements for farm animals, racehorses, pets and mink. Although synthetic α-tocopherol is readily available, the natural vitamin is often preferred and tocol concentates from wheat germ oil are used to fortify other preparations.

Vitamin E is regarded as a nutrient essential for human health but no clinical deficiency syndrome has been identified in adult man in the absence of other disease conditions or malnutrition. The Food and Nutrition Board of the United States National Research Council believes that "in as much as there is no clinical or biochemical evidence that vitamin E status is inadequate in normal individuals ingesting balanced diets in the United States, the vitamin E activity in average diets is considered satisfactory". (Committee on Dietary Allowances, 1980.) Similarly, in the U.K. the Department of Health and Social Security notes that vitamin E occurs in sufficient quantity in a large number of foods, and, in the light of present knowledge and the context of the U.K. diet, does not recommend a dietary allowance for this vitamin (Committee on Medical Aspects of Food Policy, 1979). Regardless of these pronouncements, wheat germ oil and α-tocopherol still find a ready market and frequent claims are made for the role of α-tocopherol, not only as a vitamin but as a treatment for certain diseases.

A further boost to the sales of wheat germ oil was given by reports indicating that the oil increased human physical fitness (Cureton, 1972). This effect is claimed to be due to the presence of long-chain *n*-alkanols (particularly octacosanol) in the oil and not by vitamin E (Levin *et al*., 1962).

V COMPOSITION AND PROPERTIES

A comprehensive review of the composition of wheat germ oil has been published recently (Barnes, 1982) and only a summary will be presented here.

A Yield of Oil

Typical values for the gross composition of commercial wheat germ are given in Table 19.1. The oil content is influenced by the extent of contamination with bran and endosperm, both of which have much lower oil contents than do the embryo axis, and by loss of oil during flaking. The most pure commercial germ usually yields about 100 g kg^{-1} (dry weight) of oil by extraction with hexane or light petroleum, whereas unpurified germ may yield only 60 g kg^{-1}. In contrast, the oil content of dissected embryo axis is approximately 150 g kg^{-1}.

Table 19.1 Gross composition of partly-dried commercial wheat germ†

Component	Content (g kg^{-1})
Protein	260
Starch	200
Sugars	160
Fat	100
Water	60
Ash	40
Crude fibre	30
Others	150

†Barnes, 1982.

B Physicochemical Properties

Wheat germ oil typically has a refractive index of from 1.4700 to 1.4800, specific gravity from 0.9000 to 0.9300, iodine value from 120 to 130 and saponification value from 184 to 185. The content of free fatty acids (FFA) is usually less than 60 g kg^{-1} but may be higher if the germ has been stored without stabilization. Solvent-extracted oil has a lower FFA content than oil produced by pressure-expelling, but the FFA content of the finished oil will depend on whether refining has been carried out. The content of non-saponifiable matter is greater than found in most other edible oils and ranges from 15 to 78 g kg^{-1}.

C Fatty Acid Composition

Wheat germ oil is rich in polyunsaturated fatty acids. The major fatty acid (FA) is 18 : 2 (linoleic acid), which accounts for about 60% of the total. Most of the saturated FA is represented by 16 : 0 (palmitic acid) and the content of 18 : 0 (stearic acid) is usually below 2%. The FA compositions shown in Table 19.2 were all determined in the same study and are tabulated in full to illustrate the range of variation in the commercial oils. The laboratory-extracted oils, derived from four diverse samples of commercial wheat germ, showed much less variation.

It is possible that some of the variation in the commercial oils is caused by adulteration with other edible oils, as discussed below. This is particularly the case for an oil listed in Table 19.2 with undetectable 18 : 3 (linolenic acid) and a relatively high proportion of 18 : 0.

The high proportion of polyunsaturated FA in wheat germ oil is an important factor in the "health food" market but the high content of 18 : 3 makes the oil susceptible to oxidative rancidity.

The major triglycerides of wheat germ oil are 1-palmito-2,3-dilinolein (29%), trilinolein (16%) and 1-palmito-2-linoleo-3-olein (12%) (Tamaki *et al*., 1971).

Table 19.2 Fatty acid composition of wheat germ oils†

	Fatty acid composition (%)				
Oil sample	16 : 0	18 : 0	18 : 1	18 : 2	18 : 3
Laboratory extracted	16.5	0.5	15.5	58.1	9.4
	17.4	0.9	12.3	58.0	11.4
	17.5	0.6	12.3	58.7	10.9
	17.5	0.5	13.8	59.3	8.8
Commercial	12.3	2.0	19.3	61.2	5.2
	13.7	1.5	21.8	57.9	5.1
	15.5	1.3	22.2	57.3	7.0
	21.0	1.0	18.8	52.2	3.7
	7.1	4.1	22.7	66.1	< 1.0

†Barnes and Taylor, 1980.

D Acyl Lipids

Most of the published data for wheat germ acyl lipid composition concerns lipids extracted with polar solvents rather than commercial pressure-expelled or hexane-extracted oils. Five commercial oils consisted almost completely of non-polar lipids with the total polar lipid content only 0.2% to 1.8% of the oil (Barnes and Taylor, 1980); by comparison, oils extracted with hexane in soxhlet contained from 3.6 to 10.1% of polar lipids. In dissected wheat germ the polar lipids consist mainly of phospholipids with only very low concentrations of glycolipids (Hargin and Morrison, 1980) but there is no comparative information for wheat germ oil.

Table 19.3 shows the percentage composition of non-polar acyl lipids in commercial and laboratory-extracted wheat germ oils. Triglycerides are the major component in all the oil samples, with variable proportions of FFA as a result of differences in the extent of hydrolytic rancidity, and possibly due to refining in the case of commercial oils.

Table 19.3 Composition of non-polar acyl lipids in wheat germ oil†

	Composition (% of total)	
Lipid class	Laboratory-extracted oils	Commercial oils
Steryl esters	5.1–5.8	1.9–5.7
Triglycerides	63.9–88.5	76.3–94.9
Free fatty acids	0.6–22.0	0.7–8.1
Diglycerides	1.8–6.9	2.6–10.9
Monoglycerides	0.3–1.1	0.1–0.7

†Barnes and Taylor, 1980.

E Non-saponifiable Lipids

The non-saponifiable fraction from wheat germ oil consists predominantly of 4-demethyl sterols, accompanied by much smaller quantities of methyl sterols, triterpenols, tocols, *n*-alkanols, carotenoids and hydrocarbons (Itoh *et al*., 1973; Lercker *et al*., 1977). Although quantitatively minor components of the non-saponifiable fraction, the tocols and *n*-alkanols are commercially the most important.

1 *Tocols*

As explained in Chapter 3, the α-tocopherol (α-T) and β-tocopherol (β-T) of the wheat grain are located in the germ; no other tocols are detectable in pure dissected wheat germ. Thus, the major tocols of wheat germ oil are α-T and β-T, but small amounts of α-tocotrienol (α-T-3) and β-tocotrienol (β-T-3) may also be found in the oil and are derived from bran and endosperm that contaminate the wheat germ (Table 19.4).

The γ- and δ-tocols are usually not detected in laboratory-extracted wheat germ oils (Russell Eggitt and Ward, 1955; Mason and Jones, 1958; Lercker *et al.*, 1977) or, in the case of γ-T, are only detectable as a very small proportion of the total tocols (Table 19.4). However, γ-T and δ-tocopherol (δ-T) are sometimes found in relatively high proportions in commercial wheat germ oils (Taylor and Barnes, 1981; Müller-Mulot, 1976; Abe *et al.*, 1975; Berndorfer-Kraszner, 1970), suggesting that the oils may be impure. This situation could arise in several ways. First, commercial wheat germ oils are sometimes fortified with tocopherol concentrates which may be derived from sources other than wheat germ. Second, the wheat germ used for extraction or expelling may contain fragments of other seeds; this should not occur to any significant extent in flaked germ from the reduction rolls of a flour mill, but would be more likely in germ separated from wheat grain before milling. Third, the wheat germ oil may have been adulterated with a less expensive edible oil that contains γ-T and δ-T, for example soyabean oil.

One of the commercial oils listed in Table 19.4 contained a very high concentration of α-T but no detectable β-T. This, together with the absence of 18 : 3 (linolenic acid), suggests that the oil is either not wheat germ oil or has been refined to the extent that all tocols have been removed or destroyed and then replaced by fortification with α-T. The very low contents of carotenoids found in some of the commercial oils are in accord with the use of refining (Barnes and Taylor, 1980).

Several oils listed in Table 19.4 contained α-tocopherol acetate, a commercial form of vitamin E commonly used in fortification and not detected in any significant quantity in plant lipid extracts. The acetate was not found in any of the laboratory-extracted wheat germ oils.

The results in Table 19.4 show the effect of bran content of the germ on the tocol composition of the extracted oil. The oils derived from unpurified wheat germ (oil content 60 g kg^{-1} dry weight) contained much more tocotrienols and less tocopherols compared with those from commercially purified germ (oil content 100 g kg^{-1}). It is notable that the rancid germ, although having a very high content of FFA and a strong rancid odour, yielded oil with a tocol content similar to that of the fresh germ.

Müller-Mulot *et al.* (1983) have detected α,β- and γ- tocomonoenols (dehydrotocopherols) in commercial wheat germ oils although it remains to be proven whether these compounds are derived from wheat.

2 *n-Alkanols*

There is no detailed information in the literature about the composition of *n*-alkanols (*n*-alcohols) in wheat germ oil. However, a mixture of

Table 19.4 Tocol composition of wheat germ oils†

Oil sample	Composition of tocols (% of total)							
	α-T-acetate	α-T	β-T	γ-T	δ-T	α-T-3	β-T-3	Total tocols ($g\ kg^{-1}$)
Laboratory-extracted from:								
Commercially-purified germ:	–	71	25	1	–	–	3	3.2
	–	69	25	2	–	–	3	3.1
Unpurified germ:	–	61	16	1	–	4	18	1.8
	–	60	21	2	–	3	14	2.6
Commercial oils:	1	60	34	–	–	–	5	3.5
	–	63	23	4	–	2	8	1.8
	–	66	26	8	–	–	–	2.8
	15	44	28	10	1	–	2	3.0
	17	41	23	14	1	–	4	2.0
	–	52	16	28	3	–	–	2.5
	–	46	18	32	5	–	–	2.8
	–	99	–	1	–	–	–	4.1

†Barnes and Taylor, 1980; Taylor and Barnes, 1981.

Table 19.5 Composition of sterols, 4-methyl sterols and triterpenols in commercial wheat germ oil†

Sterols	%‡	4-Methyl sterols	%‡	Triterpenols	%‡
Sitosterol	60	Gramisterol	25	24-Methyl cycloartanol	33
Campesterol	19	Citrostadienol	30	Cycloartenol	25
Δ^5-avenasterol	7	Obtusifoliol	14	β-amyrin	12
Δ^7-avenasterol	2	Cycloeucalenol	6	α-amyrin	7
Δ^7-stigmasterol	2	24-ethyl lophenol	5	Cyclobranol	2
Stigmasterol	4	Others	20	24-methylene-5α-lanost-8-en-3β-ol	7
Cholesterol	trace			Others	14
Brassicasterol	–§				
Others	5				

†Modified from Kornfeldt and Croon, 1981.
‡Percentage of each fraction.
§Not detected.

octacosan-1-ol and triacontan-1-ol is claimed to have been separated from a wheat germ oil non-saponifiable fraction (Levin *et al*., 1962). Data provided by a manufacturer indicates that unrefined, unadulterated, solvent-extracted wheat germ oil contains 80 mg kg^{-1} of octacosan-1-ol (Viobin Corporation Inc., Monticello, Illinois, USA, 1982), and that this concentration is more than eight times greater than found in a sample of partly-refined oil and a sample of pressure-expelled oil. With the current interest in the physiological effects of octacosanol, it can be expected that future studies of wheat germ oil composition will include analysis of the alkanols.

3 *Sterols, 4-Methyl Sterols and Triterpenols*

Sitosterol is the major sterol in wheat germ oil (60–70% of total sterols) accompanied by campesterol (20–30% of total). The compositions of the sterol, 4-methyl sterol and triterpenol fractions are shown in Table 19.5. These results were obtained using a capillary column gas chromatograph coupled to a mass spectrometer (Kornfeldt and Croon, 1981) and confirm the results of earlier studies (Itoh *et al*., 1973a, 1973b; Lercker *et al*., 1977). The percentage compositions of these fractions are similar to the corresponding fractions of many of the other seed oils examined, but wheat germ and maize germ oil characteristically have higher proportions of gramisterol. It is possible that the two oils could be distinguished from one another by means of the much higher proportion of β-amyrin in the wheat germ oil (Kornfeldt and Croon, 1981).

4 *Hydrocarbons*

There have been two detailed analyses of the hydrocarbons in wheat germ oil. In one study, 50% of the hydrocarbon fraction consisted of squalene and the remainder comprised alkanes with n-C_{29} as the major component (Kuksis, 1964). In the other study, squalene was not detected and the carbon number distribution of the alkanes was more widely spread with a maximum at n-C_{24} (Lercker *et al*., 1977).

The hydrocarbons are minor components of wheat germ oil and the composition could be considerably influenced by contamination with trace amounts of mineral oil during processing of the germ and germ oil.

5 *Pigments*

In 1933, Bowden and Moore crystallized the pigment responsible for the colour of wheat germ oil and obtained an absorption spectrum typical of a xanthophyll. The pigment was present in the oil at a concentration of 60 mg kg^{-1}. It was later shown that the xanthophylls lutein and cryptoxanthin were present in wheat germ oil (Drummond *et al.*, 1935). No further information on the carotenoid composition of the oil is available, but it is known that the carotenoids of wheat germ consist of 71 to 88% xanthophylls, 2 to 17% xanthophyll esters and 10 to 12% carotene (Chen and Geddes, 1945).

In a recent investigation of the total carotenoid content of wheat germ oils by measurement of absorbance value at 440 nm, oils extracted in the laboratory all had much higher carotenoid content than did the samples of commercial oils (Barnes and Taylor, 1980). The low carotenoid contents of the commercial oils may result from oxidation during pressure-expelling or from refining treatments carried out on the oils.

In addition to carotenoids, flavanoid glycosides also contribute to the yellow colour of wheat germ (King, 1962) but there is no published evidence for the presence of these compounds in wheat germ oil.

ACKNOWLEDGEMENT

I wish to thank the President of the Viobin Corporation, Monticello, Illinois, U.S. for providing information on wheat germ oil.

REFERENCES

Abe, K., Yuguchi, Y. and Katsui, G. (1975). *J. Nutr. Sci. Vitaminol.* **21**, 183–188.

Barnes, P. J. (1982). *Fette. Seifen. Anst.* **84**, 256–269.

Barnes, P. J. and Taylor, P. W. (1980). *J. Sci. Food Agric.* **31**, 997–1006.

Berndorfer-Kraszner, É. (1970). *Elelmiszervizsgalati Kozlem* **16**, 193–202.

Bowden, F. P. and Moore, T. (1933). *Nature* **131**, 204–205.

Chen, K. T. and Geddes, W. F. (1945). M.Sc. Thesis, University of Minnesota, St. Paul, Minnesota, U.S.A.

Committee on Dietary Allowances (1980). In *Recommended Dietary Allowances*, 9th edn, p. 66. National Academy of Sciences, Washington, D.C.

Committee on Medical Aspects of Food Policy (1979). In *Recommended Daily Amounts of Food Energy and Nutrients for Groups of People in the United Kingdom.* Department of Health and Social Security, H.M.S.O., London, U.K.

Cureton, T. K. (1972). *The Physiological Effects of Wheat Germ Oil on Humans in Exercise*. Charles C. Thomas, Springfield, Illinois.

Drummond, J. C., Singer, E. and MacWalter, R. J. (1935). *Biochem. J.* **29**, 456–471.

Hargin, K. D. and Morrison, W. R. (1980). *J. Sci. Food Agric.* **31**, 877–888.

Hinton, J. J. C. (1944). *Biochem. J.* **38**, 214–217.

Itoh, T., Tamura, T. and Matsumoto, T. (1973a). *J. Amer. Oil Chem. Soc.* **50**, 122–125.

Itoh, T., Tamura, T. and Matsumoto, T. (1973b). *J. Amer. Oil Chem. Soc.* **50**, 300–303.

King, H. G. C. (1962). *J. Food Sci.* **27**, 446–454.

Kornfeldt, A. and Croon, L.-B. (1981). *Lipids,* **16**, 306–314.

Kukis, A. (1964). *Biochemistry* **3**, 1086–1093.

Lercker, G., Capella, P., Conte, L. S. and Folli, B. (1977). *Riv. Ital. Sostanze Grasse* **54**, 177–182.

Levin, E., Collins, V. K., Varner, D. S., Moser, J. D. and Wolf, G. (1962). U.S. Patent No. 3,031,376.

Mason, E. L. and Jones, W. L. (1958). *J. Sci. Food Agric.* **9**, 525–527.

Müller-Mulot, W. (1976). *J. Amer. Oil Chem. Soc.* **53**, 732–736.

Müller-Mulot, W., Rohrer, G., Oesterhelt, G., Schmidt, K., Alleman, L. and Maurer, R. (1983). *Fette. Seifen. Anst.* **82**, 66–72.

Russell-Eggitt, P. W. and Ward, L. D. (1955). *J. Sci. Food Agric.* **6**, 329–336.

Singh, L. and Rice, W. K. (1981). U.S. Patent No. 4,298,622.

Tamaki, Y., Loschiavo, S. R. and McGinnis, A. J. (1971). *J. Agr. Food Chem.* **19**, 285–288.

Taylor, P. W. and Barnes, P. J. (1981). *Chem. Ind.* **20**, 722–726.

Appendix 1 Extraction and Analysis of Cereal Lipids

P. J. BARNES
The Lord Rank Research Centre, High Wycombe, Bucks, U.K.

Three major precautions which must be observed are the inactivation of destructive enzymes, the protection of cereal lipids from autoxidation and the use of pure solvents.

If enzymes are not inactivated, artefacts will be formed during extraction and the composition of the lipid extract will not be representative of the lipid composition of the sample. Steeping grains prior to dissection will also permit enzyme-catalysed tranformation of lipids, and it is preferable to soften grains for dissection by immersing them in boiling water.

Cereal lipids characteristically contain large proportions of polyunsaturated fatty acids together with carotenoids and tocols, all of which are highly susceptible to autoxidation. The lipids are only slowly oxidized while the grain tissues are intact, but once the tissues have been disrupted, or the lipids extracted, oxidation is much more rapid. Oxidation can be minimized by keeping liquids in solution and under a N_2 atmosphere, and by working at low temperatures if possible.

Only pure solvents and reagents should be used. Solvents, including water, should be distilled in glass prior to use and blank analyses should be carried out to confirm the absence of contamination.

All samples should be in a finely divided state (preferably ≤ 75 μm) for extraction if maximum lipid yield and a representative composition are to be obtained. This may be achieved by fine grinding in a laboratory mill or by homogenizing in the boiling extraction solvent.

For further details of the quantitative analysis of cereal lipids and of methods for preventing the formation of artefacts, refer to the publications

"Lipids in Cereal Technology"
ISBN 0-12-079020-3

by Morrison *et al*. (1975; 1980); Hargin and Morrison (1980); Morrison (1982) and Colborne and Laidman (1975). Details of the general aspects of lipid analysis will be found in the books by Christie (1982) and Marinetti (1967).

A distinction is now made between non-starch lipids, starch granule surface lipids and starch granule internal lipids. For efficient extraction of the internal lipids, it is essential that the solvent mixture contains water and that extraction is carried out at high temperature. Appropriate methods are discussed by Morrison (1981) and Morrison *et al*. (1980). It is important to optimize the alcohol : starch and water : starch ratios and to take into account the type of starch being extracted (W. R. Morrison, personal communication).

REFERENCES

Christie, W. W. (1982). *Lipid Analysis*. 2nd ed. Pergamon Press, Oxford.
Colborne, A. J. and Laidman, D. L. (1975). *Phytochemistry* **14**, 2639–2645.
Hargin, K. D. and Morrison, W. R. (1980). *J. Sci. Food Agric.* **31**, 877–888.
Marinetti, G. V. (1967). *Lipid Chromatographic Analysis*. Dekker, New York.
Morrison, W. R. (1981). *Stärke* **33**, 408–410.
Morrison, W. R. (1982). *J. Am. Oil Chem. Soc.* **59**, 102.
Morrison, W. R., Mann, D. L., Soon, W. and Coventry, A. M. (1975). *J. Sci. Food Agric.* **26**, 507–521.
Morrison, W. R., Tan, S. L. and Hargin, K. D. (1980). *J. Sci. Food Agric.* **31**, 329–340.

Appendix 2 Assays for Lipid-Degrading Enzymes

T. GALLIARD
The Lord Rank Research Centre, High Wycombe, Bucks, U.K.

I TRIGLYCERIDE (TG) LIPASE

A General Considerations

TG lipases are distinguished from most other hydrolytic enzymes by two factors discussed in Chapter 6: (a) lipases do not act on dissolved substrates but at lipid-water interfaces; (b) lipases operate at low moisture levels and, in some cases, are inhibited by excess water. A considerable amount of literature on "lipases" in cereals is based on enzyme assays that have not distinguished TG lipase activity from "esterases" that occur widely in plants, including cereal grains.

Assays using water-soluble substrates (e.g. triacetin) or dispersions of acyl esters at concentrations below the critical micella concentration, will measure esterase rather than TG lipase activity. Thus, enzyme assays of this type can give misleading information about the potential for hydrolysis of natural, long-chain TG in materials. Many types of artificial substrates have been used to measure "lipase" activity, taking advantage of coloured or fluorescent groups in the substrates used: esters of p-nitrophenol, β-naphthol, indoles, fluorescein, etc. The use of such substrates for the assay of lipolytic enzymes is valid only if it has been proved that the enzyme under investigation does hydrolyse the substrates and that the relative activities towards the artificial substrate and the naturally occurring lipids, is known. Unfortunately, this is not true for most lipolytic enzymes in cereals because the enzymes have not been fully characterized.

"Lipids in Cereal Technology"
ISBN 0-12-079020-3

Two different approaches have been used in the measurement of TG lipases in cereal products. One which is more suitable for materials with relatively high lipase activity uses stabilized emulsions of TG; the other, which has been used for the measurement of low-lipase activity, employs non-aqueous systems.

B TG Lipase Assays

1 *TG Emulsion Methods*

Methods based on the one originally developed for pancreatic lipase (Desnouelle *et al.*, 1955) have been used for a range of cereal grains and products. In general, methods of this type use aqueous emulsions of TG (e.g. triolein, olive oil, maize oil, etc.) stabilized with gum arabic, gum acacia or Triton X-100. The stabilized emulsion is incubated with a ground sample, or aqueous extract, of the material under investigation. The FFA, released by hydrolysis, are measured by titration (pH stat) or by colorimetric measurements of derivatives.

This TG emulsion method has been used successfully for materials with appreciable lipase activity including wheat bran (Caillat and Drapron, 1970), wheat aleurone grains (Jelsema *et al.*, 1977) germinating wheat grains (Tavener and Laidman, 1972), barley (MacLeod and White, 1962), rice bran (Aizono *et al.*, 1976) and rice germ (Aoyogi *et al.*, 1979). Recently, Matlashewski *et al.* (1982) have described a sensitive assay using ^{14}C-labelled triolein for lipase in oats and other cereals.

2 *Non-aqueous Methods*

Since the rate of lipase-catalysed reactions depends more on the availability to the enzyme of liquid oil than upon free water content, an increased concentration of liquid-phase TG can accelerate the rate of FFA production in lipase-containing materials. This forms the basis of non-aqueous assays in which a flour or ground sample of the material of interest is blended with an oil, the mixture allowed to incubate and the resulting FFA measured by titration, colorimetric methods or gas chromatogaphy.

Methods based on this principle have been used to measure lipase activities in wheat biscuit flours (Halton *et al.*, 1959) and bread flours (Bell *et al.*, 1979). Drapron and Sclafani (1969) and Drapron (1983) have used a more controlled modification of the non-aqueous technique on a range of cereal grains and products. Their technique involves incubating a mixture of ground, defatted sample with olive oil at a fixed a_w = 0.8. The FFA are subsequently determined by a combination of thin-layer chromatography and gas-liquid chromatography. A method that uses endogenous lipid as substrate and is applicable to single grains of cereals, has been described recently by Sahasrabudhe *et al.* (1982); this method will detect lipolytic activity, other than TG lipase, if the endogenous polar lipids are hydrolysed under the conditions used (30% moisture).

II OXIDATION REACTIONS

A Lipoxygenase (LOX)

Several different methods have been used for LOX determinations and these have been reviewed recently (Grossman and Zakut, 1979; Galliard and Chan, 1980; Nicolas and Drapron, 1981). Essentially, the various methods are based on four characteristics of the LOX reaction, but all use linoleic acid as substrate:

$$\text{linoleic acid} + O_2 \xrightarrow{\text{LOX}} \text{linoleic acid hydroperoxide (u.v. absorbing; } \lambda_{max} \text{ 234 nm)}$$

$$\downarrow \text{pigment bleaching}$$

Reaction rates can be measured by (a) O_2 consumption using an oxygen electrode cell; (b) u.v. chromophore formation at 234 nm; (c) colorimetric determination of the hydroperoxide product by reaction with iodine or thiocyanate; (d) loss of pigment colour by the co-oxidation reaction.

The oxygen electrode assay is the method of choice for LOX measurement in crude extracts including extracts of cereal grains and products. Appropriate controls are necessary with unfamiliar materials to eliminate contributions to O_2-uptake from reactions other than LOX (e.g. α-oxidation of fatty acids, autoxidation, oxidation of non-lipid materials in crude extracts). In my experience, such interfering reactions in cereal grain or soyabean extracts are uncommon. Experimental details of the method used in my laboratory for LOX assays on cereal extracts are given in a recent review (Galliard, 1983).

Spectrophotometric analysis of the conjugated diene hydroperoxide product is a widely used method for LOX assay. This method is suitable for pure LOX enzyme preparations but is not valid if, as in crude extracts of cereals, the hydroperoxide is further converted to products lacking the 234 nm chromophore. The same limitation applies to methods based on colorimetric reactions of hydroperoxides.

Bleaching β-carotene emulsions or gels or of crocin (a water-soluble carotene glycoside) is a useful qualitative assay for those LOX isoenzymes that have high co-oxidation activity (see p. 128). However, the reaction lacks stoichiometry and is not recommended for LOX asssays. Conversely, when the pigment-bleaching capacity of a material is of interest, the method of Ben Aziz *et al*. (1973) appears to be suitable; the assay involves spectrophotometric analysis of an emulsion containing β-carotene, linoleic acid and enzyme extract.

B Other Oxidation Reactions

Details of methods for monitoring secondary reactions of lipid hydroperoxides (isomerization, cyclization, chain cleavage, etc.) may be obtained from references quoted in relevant sections of Chapter 6. The numerous methods for measuring autoxidation, and off-flavour formation are reviewed in recent publications by Simic and Karel (1980), Chan (1983) and Allen and Hamilton (1983).

REFERENCES

Aizono, Y., Funatsu, M., Fujiki, Y. and Watanake, M. (1976). *Agric. Biol. Chem.* **40**, 317–324.

Allen, J. C. and Hamilton, R. J. (1983). *Rancidity in Foods*, Applied Science Publishers, London (in press).

Aoyaki, Y., Yamashita, H., Matsumoto, S. and Obara, T. (1979). *Agric. Biol. Chem.* **43**, 1771–1772.

Bell, B. M., Chamberlain, N., Collins, T. H., Daniels, D. G. H. and Fisher, N. (1979). *J. Sci. Fd. Agric.* **30**, 1111–1122.

Ben Aziz, A., Grossman, S., Ascarelli, I. and Budowski, P. (1971). *Phytochemistry*, **10**, 1445–1452.

Caillat, J. M. and Drapron, R. (1970). *Bull. Soc. Chim. Biol.* **52**, 59–73.

Chan, H. W.-S. (1983). *Autoxidation of Unsaturated Lipids.* Academic Press, London and New York (in press).

Desnuelle, P., Constantin, M. J. and Baldy, J. (1955). *Bull. Soc. Chim. Biol.* **37**, 285– .

Drapron, R. (1983). Proc. 7th World Cereals and Bread Congress (in press).

Drapron, R. and Sclafani, L. (1969). *Ann. Technol. Agric.* **18**, 5–16.

Galliard, T. (1983). In *Rancidity in Foods* (J. C. Allen and R. J. Hamilton, eds), Applied Science Publishers, London (in press).

Galliard, T. and Chan, H. W.-S. (1980). In *The Biochemistry of Plants* Vol. 4. Lipids: Structure and Function (P. K. Stumpf, ed.), pp. 131–161, Academic Press, London and New York.

Grossman, S. and Zakut, R. (1979). In *Methods of Biochemical Analysis* (D. Glick, ed.), Vol. 25, pp. 303–329. J. Wiley and Sons, New York.

Halton, P., Knight, R. A., Martin, H. F. and Ottaway, F. J. H. (1959). *J. Sci. Fd. Agric.* **10**, 401–403.
Jelsema, C. L., Morre, D. J., Ruddat, M. and Turner, C. (1977). *Bot. Gaz.* **138**, 138–149.
MacLeod, A. M. and White, H. B. (1962). *J. Inst. Brew.* **68**, 487–495.
Matlashewski, G. J., Urquhart, A. A., Sahasrabudhe, M. R. and Altosaar, I. (1982). *Cereal Chem.* **59**, 418–422.
Nicolas, J. and Drapron, R. (1981). *Sciences des Aliment.* **1**, 91–168.
Sahasrabudhe, M. R. (1982). *J. Amer. Oil Chem. Soc.* **59**, 354–355.
Simic, M. G. and Karel, M. eds. (1980). *Autoxidation in Food and Biological Systems*, Plenum Publishing Corp., New York.
Tavener, R. J. A. and Laidman, D. L. (1972). *Phytochemistry*, **11**, 981–987.

Appendix 3 Extraction and Reconstitution of Cereal Lipids for Baking Studies

F. MacRITCHIE
CSIRO, Wheat Research Unit, North Ryde, New South Wales, Australia

I INTRODUCTION

Two main approaches to elucidate the role of lipids in baking may be distinguished. In the first, the baking performance of a large number of flours is evaluated, corresponding analytical data for lipid content and composition are compiled and correlations are sought between the two sets of data. The second approach – a more direct one – is to separate the lipid and evaluate its contribution to baking by either varying its amount in a given flour or interchanging it between flours of varying baking potential. This procedure may then be extended to particular fractions of the lipid. For this type of study to yield reliable conclusions, it is imperative to adopt experimental procedures that do not alter the functional properties of any of the flour components, including the lipid itself.

II LIPID EXTRACTION

The choice of extraction solvent is governed not only by efficiency of lipid extraction but also by effects on the functional properties of the flour. Non-polar solvents e.g. light petroleum, have little or no effect on the functional properties of the flour but they do not efficiently extract the

"Lipids in Cereal Technology"
ISBN 0-12-079020-3

non-starch lipids. Conversely, alcohol-water mixtures extract virtually all of the non-starch lipids but cause large increases in the peak dough development times (Finney *et al.*, 1976; MacRitchie and Gras, 1973). In addition, prolonged extraction with alcohols will result in removal of a proportion of the starch lipids. Chloroform appears to extract practically all the non-starch lipids and has no significant effect on the functional properties of the flour (MacRitchie and Gras, 1973).

In baking studies, batch extractions with chloroform at room temperature are effective and convenient, the extracted flour is separated from the slurry by filtration. The solvent should be evaporated at low temperature on a rotary evaporator until the lipid remains in a small volume of solvent. During storage, the lipid extracts should be in solution, never dry, and kept under a N_2 atmosphere to avoid oxidation. It is preferable (if it is convenient), to extract lipids on the same day as they are used in the baking experiments.

III FRACTIONATION

Both column and batch methods have been used to prepare specific flour lipid fractions for baking evaluation tests. Both methods are based on the successive elution from adsorbates such as silica gel by liquids of increasing solvent power. Daftary *et al.* (1968) fractionated flour lipid into polar and non-polar fractions by silicic acid column chromatography followed by sub-fractionation using DEAE-cellulose columns. This allows better resolution than batch methods. However, batch methods have the advantage of being more easily scaled up to give the quantities of material required for use in baking tests. Batch methods minimize exposure of the lipid to adverse conditions because of their rapidity, and they are probably most suitable for preliminary evaluation of crude fractions. Ponte and De Stefanis (1969) introduced a simple batch fractionation scheme for separating wheat flour lipids into two major fractions, the polar and non-polar lipids. Further separation of the lipid extract into five fractions can be achieved by elution from silica gel using a method devised by De Stefanis and Ponte (1969). Preparations from this fractionation scheme have been evaluated in baking tests (MacRitchie, 1977).

IV RECONSTITUTION

Although the original state of dispersion of lipids in flour cannot be achieved by reconstitution, the important criterion is that the functional properties (i.e. dough-mixing characteristics and baked loaf volume and texture) be unchanged by the procedures.

Lipid fractions may be blended with extracted flour in a Stein mill (Chung *et al*., 1979) or added in solution to the flour (MacRitchie, 1976). In the latter case, the flour is slurried with the lipid solution and the solvent is then allowed to evaporate; this ensures a uniform dispersion of lipid but evaporation must be rapid if oxidation is to be avoided.

The restoration of functional properties of a flour reconstituted with its full complement of extracted lipid may be easily tested by comparison with the original untreated flour. An additional check was used by MacRitchie and Gras (1973) to confirm that the extracted lipid had not been altered and that the unusual loaf volume-lipid content relationship (see Chapter 8) did not result from artefacts. In this method, the lipid content of the flour was varied by addition of extracted lipid to defatted flour, and by admixing different ratios of the defatted and the untreated flour. Coincidence of the loaf volume-lipid curves then showed that the relationship was genuine.

REFERENCES

Chung, O. K., Pomeranz, Y., Hwang, E. C. and Dikeman, E. (1979). *Cereal Chem.* **56**, 220–226.
Daftery, R. D., Pomeranz, Y., Shogren, M. and Finney, K.F. (1968). *Food Technol.* **22**, 79–82.
De Stefanis, V. A. and Ponte, J. G. (1969). *Biochem. Biophys. Acta*. **176**, 198–201.
Finney, K. F., Pomeranz, Y. and Hoseney, C. (1976). *Cereal Chem.* **53**, 383–388.
MacRitchie, F. (1976). *Cereal Chem.* **53**, 318–326.
MacRitchie, F. (1977). *J. Sci. Food Agric.* **28**, 53–58.
MacRitchie, F. and Gras, P. W. (1973). *Cereal Chem.* **50**, 292–302.
Ponte, J. G. and De Stefanis, V.A. (1969). *Cereal Chem.* **46**, 325–329.

Index

B

C

D

M

N

O

Q

R

S

T

U

V

W

X

Z

FOOD SCIENCE AND TECHNOLOGY

A SERIES OF MONOGRAPHS

Maynard A. Amerine, Rose Marie Pangborn, and Edward B. Roessler, Principles of Sensory Evaluation of Food, 1965.

S. M. Herschdoerfer, Quality Control in the Food Industry. Volume I—1967. Volume II—1968. Volume III—1972.

Hans Reimann, Food-Borne Infections and Intoxications. 1969.

Irvin E. Leiner, Toxic Constituents of Plant Foodstuffs. 1959.

Martin Glicksman, Gum Technology in the Food Industry. 1970.

L. A. Goldblatt, Aflatoxin. 1970.

Maynard A. Joslyn, Methods in Food Analysis, second edition. 1970.

A. C. Hulme (ed.), The Biochemistry of Fruits and Their Products. Volume 1—1970. Volume 2—1971.

G. Ohloff and A. F. Thomas, Gustation and Olfaction. 1971.

George F. Stewart and Maynard A. Amerine, Introduction to Food Science and Technology. 1973.

C. R. Stumbo, Thermobacteriology in Food Processing, second edition. 1973.

Irvin E. Liener (ed.), Toxic Constituents of Animal Foodstuffs. 1974.

Aaron M. Altschul (ed.), New Protein Foods: Volume 1, Technology, Part A—1974. Volume 2, Technology, Part B—1976. Volume 3, Animal Protein Supplies, Part A —1978. Volume 4, Animal Protein Supplies, Part B—1981.

S. A. Goldblith, L. Rey, and W. W. Rothmayr, Freeze Drying and Advanced Food Technology. 1975.

R. B. Duckworth (ed.), Water Relations of Food. 1975.

Gerald Reed (ed.), Enzymes in Food Processing, second edition. 1975.

A. G. Ward and A. Courts (eds.), The Science and Technology of Gelatin. 1976.

John A. Troller and J. H. B. Christian, Water Activity and Food. 1978.

A. E. Bender, Food Processing and Nutrition. 1978.

D. R. Osborne and P. Voogt, The Analysis of Nutrients in Foods. 1978.

Marcel Loncin and R. L. Merson, Food Engineering: Principles and Selected Applications. 1979.

Hans Riemann and Frank L. Bryan (eds.), Food-Borne Infections and Intoxications, second edition. 1979.

N. A. Michael Eskin, Plant Pigments, Flavors and Textures: The Chemistry and Biochemistry of Selected Compounds. 1979.

J. G. Vaughan (ed.), Food Microscopy. 1979.

J. R. A. Pollock (ed.), Brewing Science, Volume 1—1979. Volume 2—1980.

Irvin E. Liener (ed.), Toxic Constituents of Plant Foodstuffs, second edition. 1980.

J. Christopher Bauernfeind (ed.), Carotenoids as Colorants and Vitamin A Precursors: Technological and Nutritional Applications. 1981.

Pericles Markakis (ed.), Anthocyanins as Food Colors. 1982.

Vernal S. Packard, Human Milk and Infant Formula. 1982.

George F. Stewart and Maynard A. Amerine, Introduction to Food Science and Technology, Second Edition. 1982.

Colin Dennis, Post-Harvest Pathology of Fruits and Vegetables.